Development and Planning Law

The development control and planning law system of the United Kingdom is one of the most comprehensive and detailed in the world. Development control is one of the most significant matters concerning anyone involved in the development of land, and an understanding of the legislation and enforcement of these powers is essential to the success of any development project.

This book is the fourth edition of a highly regarded work widely used by students and practitioners of real estate management, development, surveying, valuation, planning and law. Written by two experienced experts on law and the UK planning system, *Development and Planning Law* is essential reading for anyone involved in building and construction, surveying, planning and development, and who needs to know the law as it relates to their everyday professional practice.

It has been extensively updated to reflect the most recent legal developments, including the 2011 Localism Act.

Barry Denyer-Green, LLM, PhD, FRICS, Barrister of the Middle Temple, is the author of *Compulsory Purchase and Compensation* and the co-author of *Law of Commons and Town and Village Greens*.

Navjit Ubhi, LLB (Hons), Barrister of the Middle Temple, is the co-author of *Law of Commons and Town and Village Greens*.

Development and Planning Law

Fourth edition

**Barry Denyer-Green
and Navjit Ubhi**

LONDON AND NEW YORK

First edition published 1982
by Estates Gazette

Fourth edition published 2013
by Routledge
2 Park Square, Milton Park, Abingdon, Oxon OX14 4RN

Simultaneously published in the USA and Canada
by Routledge
711 Third Avenue, New York, NY 10017

Routledge and EG Books are imprints of the Taylor & Francis Group, an informa business

British Library Cataloguing in Publication Data
A catalogue record for this book is available from the British Library

Library of Congress Cataloging in Publication Data
Denyer-Green, Barry.
Development and planning law / Barry Denyer-Green, Navjit Ubhi. – 4th ed.
    p. cm.
Includes bibliographical references and index.
1. City planning and redevelopment law–Great Britain. 2. Regional planning–Law and
legislation–Great Britain. 3. Land use–Law and legislation–Great Britain.
I. Ubhi, Navjit. II. Title.
KD1125.D46 2012
346.4104'5–dc23   2012015822

ISBN: 978-0-415-53828-2 (hbk)
ISBN: 978-0-7282-0526-0 (pbk)

Typeset in Times New Roman
by Cenveo Publisher Services

MIX
Paper from
responsible sources
FSC
www.fsc.org   FSC® C004839

Printed and bound in Great Britain by
TJ International Ltd, Padstow, Cornwall

Contents

12 Trees, minerals and caravans

13 Environmental controls

PART IV
Positive planning and enforcement

14 Revocation, modification and discontinuance orders

x   *Contents*

Table of cases

Table of statutes

Table of statutory instruments

Table of circulars

1 Introduction

This book is divided into seven principal parts to deal with the wide range of topics that fall within the scope of its title. Inevitably, development control is one of the most significant matters that concerns any person involved in the development of land. Before 1925, the biggest impediment to the full economic realisation of land was the complicated forms of land ownership; today, attention is addressed to development controls, under the planning Acts.

Nowhere outside the countries of the communist world will one find a form of statutory development control as comprehensive and detailed as that of the United Kingdom. The doctrine of supremacy of Parliament means that the Secretary of State for Communities and Local Government, the minister responsible for development control policy and the initiation of legislation, enjoys wider powers in this field than is customary in other countries. The minister has acquired these powers, including the power to make orders, rules and regulations, by way of delegated legislation, through a gradual process over the last 60 odd years since controls commenced in 1948. This process has been almost imperceptible to the casual observer; the young student of today is conditioned to accept that the State, through its various organs and agencies, exercises wide powers; others are more deeply concerned by this trend.

It is possible that this trend might reverse following the Localism Act 2011, one purpose of which is to shift control to local communities. The Act contains a radical set of provisions for neighbourhood planning where local people will have more input. Until 2012, the policy guidance to local authorities and other decision-makers was extensive and found in the very many planning policy guides or statements published by central government. The National Planning Policy Framework reduces the bulk of these guides and statements to just one document. This document sets out a number of priorities, and contains a presumption in favour of sustainable development.

The growth of ministerial powers has not gone unhindered by the courts; although it is noticeable that where the minister loses in the courts, he later seeks legislation to avoid further judicial interference. The courts have been busy considering development control decisions where it is alleged that a decision is *ultra vires* the planning Acts, or there has been some procedural irregularity. The consequence of this judicial activity is a fast-growing body of case law, as the reader will in due course discover. Although most of this case law is concerned with controlling administrative discretion (within statutory powers) on the part of a local planning authority or the minister, it is invariably accepted by all concerned with development control in the same way as the precedent of previous cases is accepted in a primarily common law topic such as the law of torts. New concepts have evolved in development control cases, such as the concepts of abandonment, the planning unit or the planning history, which amount to purposive statutory interpretation; they are necessary to make complicated legislation work. All these matters concerned with development control are dealt with in Part I.

At the conclusion of Part I, the reader will have some idea of the scope of development control. Part II deals with the decision-making process: the authorities, the development plans in their present form and the procedure in relation to planning applications. Development plans have moved through at least three generations, from the rather simple prescriptive land use plans introduced in 1947, to the structure and local development plans introduced in 1968, to the local development framework provided for under the Planning and Compulsory Purchase Act 2004. The radical powers available under the Planning Act 2008, in respect of nationally significant infrastructure projects, are also dealt with in Part II. Some of the special controls that affect the development of land are found in Part III.

Positive planning and the enforcement of development controls are considered in Part IV. In this area, the initiative lies with the local planning authorities to take action in respect of activities that are not in conformity with development plans and policies. Additionally, development is carried out in some cases directly or indirectly by public authorities.

Part V deals with the rights and remedies of developers and landowners. Some people regard the development control process as involving the making of political decisions about development, a view that is tenable in the sense that the primary decisions are those of local authorities, and these authorities are part of the political democratic institutions in this country. Other people regard the process as purely administrative decision-making; a technical and objective exercise within powers conferred by Parliament. The weakness in this view lies in the nature of the plan and policy-making part of the process where local planning authorities, and where appropriate, the minister, make policy choices. In fact, the truth probably lies somewhere between these two views. As the principal right of the developer or landowner against development control decisions is an appeal to the Secretary of State, his determination of an appeal is therefore partly political and partly objective administrative decision-making. This has not always been recognised by the courts when exercising their powers of judicial supervision. One can foresee that this dichotomy will become increasingly the subject of debate in the future.

Any developer will know that in putting together a development scheme, development control is only one of the matters to be resolved. He may require connections to public sewers or water supplies, or to the other services: and that the inadequacy of some of these facilities may influence the result of an application for planning permission. The developer may build the sewers and roads, and he will want to be satisfied that these can be taken over by the appropriate authorities on the completion of his scheme. The provision, requisition or adoption of these services is set out in Part VI together with the relevant aspects of the law of highways of interest to developers.

Part VII of this book includes only one chapter. Planning the use and development of land causes changes in the underlying land values; this phenomenon has a number of practical and political consequences. This chapter provides an outline of the problem and the way it has been handled.

Inevitably, there may be legislative changes that affect some part of this book after it is published. This is a problem that all authors writing on a fast-moving subject of a statutory nature have to face. Hopefully, the problem has been minimised by concentrating on general principles that seem to possess a certain durability; it cannot be entirely solved so long as Parliament takes the active interest that it does in development and planning law. This is stated as on 1 November 2009.

The main statute concerned with development control and planning is at present the Town and Country Planning Act 1990 (as amended), and, unless otherwise indicated, most references are to this Act. A number of specialised areas of development control, such as the protection of historic buildings, have separate and appropriate legislation. However, important changes and additional provisions were introduced in the Planning and Compensation Act 1991, the Planning and Compulsory Purchase Act 2004 and the Planning Act 2008.

Part I

The scope of development control

2 Development control and the carrying out of operations

2.1. Introduction

One of the most important provisions of the Town and Country Planning Act 1990 (the 1990 Act) is section 57(1); it states 'planning permission is required for the carrying out of any development of land'. The word 'development' is the key to the scope of development control; it covers two activities: the carrying out of certain operations (considered in this chapter) and the making of a material change of use (see Chapter 3). There is an elaborate enforcement procedure where there has been a breach of development control (see Chapter 15). Planning permission for a use of land will not authorise the erection of buildings, as this entails the carrying out of operations (see *Sunbury-on-Thames Urban District Council* v *Mann* [1958] and *Wivenhoe Port Ltd* v *Colchester Borough Council* [1985]). Lord Denning MR explained the distinction between uses and operations in *Camrose* v *Basingstoke Corporation* [1966].

Section 55(1) of the 1990 Act defines development as to include: 'the carrying out of building, engineering, mining or other operations in, on, over or under land …'. However, in any particular case, the question of what amounts to development is a question of fact, and is decided initially by the local planning authority. They may be asked to determine whether any particular activity is development upon an application for that purpose under sections 191 and 192 of the Act. They may also make a preliminary determination because they have decided to take enforcement proceedings against an activity that has no planning permission. It is only if the local planning authority, or the Secretary of State upon appeal, makes a decision using the wrong criteria, or reaches an unreasonable decision in relation to the facts, that a developer will have a case for consideration in the High Court (*Pyx Granite Co* v *Ministry of Housing and Local Government* [1958]). Legal decisions are helpful as to the meaning of development insofar as they indicate the right criteria or the bounds of unreasonableness; they are not conclusive that any particular activity will or will not be development in every case.

2.2. Operations excluded from development control

Because the definition of the operations that constitute development is so wide, the 1990 Act excludes certain activities that would otherwise need planning

permission: 'the carrying out for the maintenance, improvement or other alteration of any building, of works which ... affect only the interior of the building
or do not materially affect the external appearance of the building ...' (section
55(2)(*a*)), subject to a limitation discussed in the following text; certain maintenance work carried out by a local highway authority within the boundaries of a
road (section 55(2)(*b*)); and, work of inspection, repair or renewal of sewers,
mains, pipes, cables or other apparatus by a local authority or statutory undertaker (section 55(2)(*c*)). Section 55(2A) provides that a development order may
specify limitations on floor areas in any of operations within section 55(2)(*a*),
mentioned above. The Town and Country Planning (Development Management
Procedure) Order 2010 imposes a limit on 200 square metres of additional floor
space.

It can be seen that internal building work, of the type described earlier, is
outside development control. However, such work may require consent under the
building regulations, or, if the building is listed, listed building consent (see
Chapter 10). Internal building work may be accompanied by a change of use of
the building; the change of use may be within development control (see Chapter 3).
If there has been an unauthorised change of use, and internal works have been
carried out, the local planning authority may be entitled to require that the internal
works be removed, even if they were not originally within development control:
see *Somak Travel Ltd* v *Secretary of State for the Environment* [1987].

Work that does not materially alter the external appearance of a building is not
always easy to determine. The first question to ask is whether such work is maintenance, improvement or other alteration. In *Street* v *Essex County Council*
[1965], it was held that the rebuilding of a cottage from damp-proof course up, to
comply with housing fitness standards, was not work of maintenance. And, in
Larkin v *Basildon District Council* [1980], where the building owner first pulled
down and rebuilt two walls of a dwellinghouse and then subsequently did the
same to a further two walls, the Divisional Court upheld the Secretary of State's
decision that this was not work of improvement. Lord Parker CJ, in the *Street*
case, said, 'whether works could fairly be said to amount to maintenance or were
properly called reconstruction, must be a matter of fact and degree'. The same
point was made by the Court of Appeal in *Hewlett* v *Secretary of State for the
Environment* [1985], where Sir John Donaldson MR said in relation to a series of
alterations to a building that the Secretary of State was entitled to look at the
totality of the work, and whether this resulted in a new building; it was not a
question of law, but of fact and degree. External painting is an operation normally
undertaken by a builder, but whether it is development depends on whether it
materially affects the external appearance of a building: see *Maidenhead Royal
Borough Council* v *Secretary of State for the Environment* [1988].

In *Burroughs Day* v *Bristol City Council* [1996], the installation of a lift shaft
and consequential alterations to the roof were held not to be development.
In *Church Commissioners for England* v *Secretary of State for the Environment*
[1995], it was decided that a large shopping centre was 'a building' so that
internal works not affecting the external appearance was not development.

Unsatisfactory as it may seem to the developer, questions of fact and degree are first decided by the local planning authority (or the Secretary of State upon an appeal). Only if the decision comes within a very narrow category of decisions can the courts consider a review of the issue (more fully explained in Chapter 20).

2.3. Building operations

To developers, building operations will be the most important of the operations that constitute development. Building operations are defined as including 'demolition of buildings, rebuilding, structural alterations of or additions to buildings, and other operations normally undertaken by a person carrying on business as a builder ...' (section 55(1A) of the 1990 Act). This wide definition is made wider, because a building includes 'any structure or erection and any part of a building, but does not include plant or machinery comprised in a building ...' (section 336(1)).

If what is erected can be called a building, then the Court would need to be persuaded that the erection of it was not a building operation: see Bridge J in *Barvis Ltd* v *Secretary of State for the Environment* [1971]. Thus, in *R (Westminster City Council)* v *Secretary of State for the Environment, Transport and Regions* [2002], the erection of a kiosk 2.14 m square and 2.7 m high was a building operation, because the resulting structure was a building. A building is something normally constructed on site, causes a physical change of some permanence and may have a degree of physical attachment, although this last point is not decisive: see *Skerrits of Nottingham Ltd* v *Secretary of State for the Environment, Transport and the Regions (No 2)* [2000] CA, where the erection of a marquee for an eight-month period was a building operation. However, the erection of a mobile shed was not a 'structure or an erection' in *Tewkesbury Borough Council* v *Keeley* [2004]. Even the erection of a model village may be a building operation: see *Buckinghamshire County Council* v *Callingham* [1952]. Plant or machinery erected outside a building may be regarded as structures or erections so as to involve building operations, as in *Barvis Ltd* v *Secretary of State for the Environment* [1971] DC (an 89-foot high crane, running on rails 120 feet long). In *Cheshire County Council* v *Woodward* [1962] DC, Lord Parker CJ had indicated tests, additional to the language of the statute, to decide the meaning of 'operations': he said that the question was whether the physical character of the land has been changed by operations in or on it; and the problem of determining the sort of operations likely to change the physical characteristics of land is analogous to the problem of deciding whether a fixture passes with the freehold – although one must undoubtedly look at the [1990] Act, the degree of permanency is extremely relevant in determining whether an operation constitutes development.

In *Parkes* v *Secretary of State for the Environment* [1979], Lord Denning MR applied similar tests in deciding that the storage of scrap was not an operation. He suggested that to be an operation there must be an activity which results in some physical alteration to the land, and which has some degree of permanence to the land itself.

The stationing of reasonably mobile caravans was held in *Guildford Rural District Council* v *Penny* [1959] not to be an operation, although the stationing of the more permanent form of 'mobile' home would seem to come within Lord Parker's test (see Chapter 12 for additional controls over caravans).

The installation of external floodlights was held not to be development in *Kensington and Chelsea Royal Borough Council* v *CG Hotels* [1980]. The work was too minor to be called an operation, and the real environmental consequence was due to the use of electricity in floodlighting. Because external lighting creates environmental problems (that at petrol-filling stations is a good example), levels of luminosity may be controlled by planning conditions imposed with a planning permission needed for other building work.

In *R (Hall Hunter Partnership)* v *First Secretary of State* [2006], it was held that the erection of many hectares of large, linked, walk-in polytunnels for growing soft fruit on a farm was a building operation. In *Tewkesbury Borough Council* v *Keeley* [2004], it was held that certain sheds on wheels did not constitute buildings.

For ministerial decisions on what constitutes a building operation, reference should be made to the Bulletins of Selected Appeal Decisions. It would appear that these decisions invariably involve Lord Parker's tests that an operation involves a change in the physical characteristics of the land and a degree of permanence.

A number of uses and changes of use are excluded from the meaning of development by section 55 (these are considered in Chapter 3).

Some operations, which are 'building operations' for the purposes of the Act, can be carried out without an express grant of planning permission if they fall within the classes of 'permitted development' (see Chapter 4).

2.4. Demolition

This activity has presented special problems in the past. The developer is often tempted to demolish a building occupying a suitable site for redevelopment in order to force the hand of the local planning authority, or to avoid a building being listed where it may have a special interest. The local planning authority may not wish to see a building demolished that appears to the authority to have further use or which makes a contribution to the townscape or landscape.

If a building is listed because of its special interest, listed building consent is then needed for its demolition, and it is a criminal offence to proceed otherwise. In conservation areas, with some exceptions, consent is also required for the demolition of buildings: see section 74 of the Planning (Listed Buildings and Conservation Areas) Act 1990 (Listed Buildings Act 1990). The exceptions include listed buildings (already protected), ecclesiastical buildings used for ecclesiastical purposes, buildings which are ancient monuments (already protected) and a description of buildings specified in a direction made by the Secretary of State (section 75(2) of the Listed Buildings Act 1990). It is otherwise an offence to demolish a building in a conservation area without the appropriate consent (sections 9 and 74 of the Listed Buildings Act – see more generally Chapter 10).

In *Cambridge City Council* v *Secretary of State for the Environment* [1992], the Court of Appeal held that works for the demolition of a building might, in appropriate cases, constitute 'development'. At about the same time as the *Cambridge* case, the government introduced provisions in the Planning and Compensation Act 1991 to ensure that demolition required planning permission in certain specified circumstances. The definition of 'building operations' in section 55 of the 1990 Act now includes the 'demolition of buildings'. However, section 55(2)(*g*) provides that the Secretary of State may specify in a direction descriptions of buildings the demolition of which does not amount to development. The Town and Country Planning (Demolition – Description of Buildings) Direction 1995 provides that the demolition of the following description of buildings is not 'development'; planning permission is not therefore required for such work:

1.   Any listed building defined as such by the Planning (Listed Buildings and Conservation Areas) Act 1990 – for which there are separate controls (see Chapter 10).
2.   Any building in a conservation area – for which there are separate controls (see Chapter 10).
3.   Any building which is a scheduled monument under the Ancient Monuments and Archaeological Areas Act 1979 – for which there are separate controls.
4.   Any building other than a dwellinghouse or a building adjoining a dwellinghouse.
5.   Any building the externally measured cubic content of which does not exceed 50 cubic metres.
6.   The whole or any part of any gate, fence, wall or other means of enclosure.

The effect of the direction is that the control of the demolition of buildings is restricted to dwellinghouses and buildings adjoining dwellinghouses. However, the Town and Country Planning (General Permitted Development) Order 1995 (Schedule 2, Part 31) grants planning permission for any building operations consisting of the demolition of a building. Because of the directions considered in the preceding text, only dwellinghouses and adjoining buildings are in practice covered by this permitted development right. Except in cases of urgency, or other excluded buildings, where demolition of a dwellinghouse is proposed, the local planning authority must first be notified of a proposal to demolish (see Appendix B).

2.5. Engineering and mining operations

Engineering operations are defined to include 'the formation or laying out of means of access to highways', otherwise engineering must be given its ordinary meaning, and it will be for the local planning authority (or the Secretary of State on appeal) to decide as a question of fact and degree whether an activity is an engineering operation. In *Fayrewood Fish Farms* v *Secretary of State for the Environment* [1984], the judge said that an activity would be regarded as an 'engineering operation' if the ordinary layman would so regard it, and not because the activity was supervised by a qualified engineer. Under section 55(4A) of the 1990 Act, the placing of a fish tank in inland waters will be an engineering operation if not otherwise development.

In *Coleshill & District Investment Co Ltd* v *Minister of Housing and Local Government* [1969], the House of Lords upheld the minister's decision that the *demolition* of an embankment surrounding an explosive store was an 'engineering operation'. The *construction* of an earth embankment was also held to be an 'engineering operation' in *Ewen Developments Ltd* v *Secretary of State for the Environment* [1980].

Work carried out to alter the track of a greyhound racing stadium for stock-car racing was held by the minister to be 'engineering operations' in an appeal case (*Re Salford City Borough Council* [1972] - a Ministerial decision).

Tipping of refuse or waste material would seem to be an operation: it satisfies both Lord Parker's tests of a change in the physical characteristics of the land and is permanent. In *Ratcliffe* v *Secretary of the Environment* [1975], the minister concluded that a particular mode of tipping waste materials amounted to an 'engineering operation'. Paradoxically, section 55(3) of the 1990 Act provides that the deposit of refuse or waste materials on land involves a material change of use if either the superficial area of the deposit is extended or the height of the deposit is extended and exceeds the adjoining ground level. This problem is further considered in Chapter 3. In many cases, there will be an overlap between building and engineering operations, as the definition of a building includes a structure or erection.

Although mining operations are within the scope of development control, there are now special provisions that make it more convenient to deal with mining in a later chapter (Chapter 12).

2.6. Other operations

The meaning of 'other operations' in section 55(1)(*d*) is usually interpreted under a rule of statutory interpretation as not adding any further operations to those of building, engineering or mining operations already specified: the *ejusdem generis* rule. The rule is that where particular words are followed by general words, the general words are limited to the same kind or nature as the particular words. Although this was doubted by the House of Lords in *Coleshill & District Investment Co Ltd* v *Minister of Housing and Local Government* [1969] as 'building, engineering', and 'mining' operations did not share a common genus; the maxim *noscitur a sociis* probably applied, and an activity must be construed by reference to these operations to be 'other operations'. This maxim means that the meaning of the words can be gathered from the context.

However, in a ministerial decision ([1985] JPL 129), where the activity in issue was the installation of a protective grill over a shop window, the minister decided the work was too specialised to be a building operation, and he considered it was an 'other operation' and needed planning permission.

One of the issues in *Bedfordshire County Council* v *Central Electricity Generating Board* [1985] was whether Nirex, the Nuclear Industry Radioactive Waste Executive, needed planning permission to carry out site investigations by means of boreholes; the work would only have taken a few days. The judge considered whether such work could be an 'other operation' but seemed to reject the idea, because the work was so trivial: it did not change the physical appearance of the land, nor was there any degree of permanence.

3 Development and change of use

3.1. Introduction

Planning permission is required for development, and development includes 'the making of any material change in the use of any buildings or other land' (section 55(1) of the Town and Country Planning Act 1990).

This chapter is largely concerned with the meaning of these words in order to understand the scope of development control over a material change of use. However, some uses are specifically excluded from development control; or are uses that do not need planning permission, because they existed in 1948; or are uses that are now immune from enforcement action.

The Town and Country Planning (Use Classes Order) 1987 (UCO) is separately considered: it does not determine whether certain changes of use are development requiring planning permission; its purpose is to exclude from development control certain changes of use within specified classes (see Appendix A). The Town and Country Planning (General Permitted Development) Order 1995 is considered in Chapter 4. Its purpose is not to determine whether any activities are or are not material changes of use; its purpose is to deem the grant of planning permission for, inter alia, certain changes of use.

3.2. Uses excluded from development control

Section 55(2) excludes certain uses from the meaning of development. The effect of this is that in either of the circumstances described in the following text, planning permission is not required for these uses of land or buildings.

3.2.1. Residential use within the curtilage of a dwellinghouse

> The use of any buildings or other land within the curtilage of a dwellinghouse for any purpose incidental to the enjoyment of a dwellinghouse as such (section 55(2)(d)).

Although the 'curtilage' was said to mean a small area in which a building is situated in *Dyer* v *Dorset County Council* [1989], the Court of Appeal in *Skerritts of Nottingham Ltd* v *Secretary of State for the Environment, Transport and the Regions (No 1)* [2000] said that the notion of 'smallness' was not relevant. In *McAlpine* v *Secretary of State for the Environment* [1995], it was held that the inspector was entitled to make a current and visual appraisal, in reaching his conclusion as to the extent of the 'curtilage'. Although curtilage can equate with ownership, ownership is only a factor, but is not a determinative factor of the extent of a building's curtilage; see *Lowe* v *First Secretary of State* [2003]. In its review of the meaning of 'curtilage' in the *Skerritts* case, the Court of Appeal did not refer to the test considered in the earlier decision of the House of Lords in *Sinclair Lockart's Trustees* v *Central Land Board* [1950], where it was said that the ground which is used for the comfortable enjoyment of a house or other building, and which serves the purposes of the house or the building in some necessary or reasonably useful way, is the curtilage. The true test in *Sinclair Lockart's Trustees* v *Central Land Board* is that there is no requirement that a subsidiary building like an annex must serve the house or building in some necessary or useful way, it is enough that such a building serves the purposes of the principal house or building in some necessary or reasonably useful way: see *Wheeler* v *First Secretary of State* [2003].

Although the construction or alteration of buildings may be development, the use of buildings within the curtilage for the enjoyment of the dwellinghouse is not. Therefore, a loft room above a garage could be used as an extra bedroom if no building work was involved that materially affected the external appearance of the building. A vehicle could be parked on land within the curtilage if it was there for the enjoyment of the occupiers of the dwellinghouse as such; but this exclusion would not apply if the vehicles were there for business purposes – it would then become necessary to determine the substantive point as to whether a material change of use had occurred (see the following text). In *Croydon Borough Council* v *Gladden* [1994], a large replica Spitfire aircraft in the garden of a house was not a reasonable use of the curtilage.

A hut in *Prosser* v *Sharp* [1985] was not regarded as a dwellinghouse so that a caravan parked nearby could not be regarded as 'incidental to the enjoyment of the dwellinghouse as such'. In *Secretary of State for the Environment, Transport and the Regions* v *Thurrock BC* [2002], the Court of Appeal had doubts as to whether an aircraft hanger was incidental to the enjoyment of a dwelling.

3.2.2. Agriculture and forestry use

The use of any land for the purposes of agriculture or forestry (including afforestation) and the use for any of those purposes of any building occupied together with land so used (section 55(2)(e)).

Any land may be used for agriculture, or for forestry; or agricultural land may be afforested and forestry land (subject to any tree preservation orders and the need for a felling licence) may be converted into agricultural land, all without planning permission, as such uses are not development. Agriculture is widely defined and includes horticulture, fruit growing, seed growing, dairy farming, the breeding and keeping of livestock (including any creature kept for the production of food, wool, skins or fur, or for the purpose of its use in the farming of land), the use of land as grazing land, meadow land, osier land, market gardens and nursery grounds, and the use of land for woodlands where that use is ancillary to farming of land for other agricultural purposes (section 336(1) of the 1990 Act). In *Crowborough Parish Council* v *Secretary of State for the Environment* [1981], the conversion of agricultural land to allotments was held not to involve development as the use of land for allotments was within the definition of agriculture. In *North Warwickshire Borough Council* v *Secretary of State for the Environment* [1984], the word 'land' was said to include buildings, so that buildings can be used for agriculture, such as the breeding of animals for their fur.

The subdivision of agricultural land and its use for leisure use is not an agricultural use and will require planning permission: see *Pittman et al* v *Secretary of State for the Environment* [1988].

The use of land or buildings as a stud farm for horses was considered in *Belmont Farm Ltd* v *Minister of Housing and Local Government* [1962] to be outside the meaning of agriculture as the keeping and breeding of livestock within that definition was restricted to creatures kept for the production of food, wool, skins or fur, or for the purposes of their use in the farming of land. However, in *Sykes* v *Secretary of State for the Environment* [1981] where land was used for grazing horses, the Secretary of State's decision, that this use was within the definition of agriculture, was upheld by the court. Donaldson LJ stated that the question in such cases was whether the predominant use of land kept for horses was for grazing – which is not development – or some other purpose which might involve development. Thus, in *Fox* v *Secretary of State* [2003], it was held that the grazing of horses, whether the working, the race or the recreational horse, was agriculture, even if the use of land for keeping anything other than working horses was not an agricultural use.

In *Wealden District Council* v *Secretary of State for the Environment* [1988], a caravan on agricultural land used as a store for food did not involve a material change of use, but remained an agricultural use. The sale of produce on agricultural land will be ancillary to the principal agricultural use unless the produce is not substantially produced on the land: see *Wood* v *Secretary of State for the Environment* [1973].

In *Farleyer Estate* v *Secretary of State for Scotland* [1992], it was held that the stock piling of extracted timber was a 'forestry' use of land. In *Millington* v *Secretary of State for the Environment* [1999], the making of wine was seen to be a perfectly normal activity for a farmer engaged in growing wine grapes.

3.3. Use Classes Order

> ... *in the case of buildings or other land which are used for a purpose of any class specified in an order ... the use of the buildings or other land or, subject to the provisions of the order, of any part of the buildings or any other land, for any purpose of the same class [does not involve development]* (section 55(2)(f)).

The Town and Country Planning (Use Classes) Order 1987 came into effect on 1 June 1987 (see Appendix A). The UCO classifies a number of uses of buildings and other land, and by section 55 of the Act, and article 3 of the order, where a building or other land is used for a purpose within one of the classes, a change of use to another purpose within the same class is not development. In *R (Harbige)* v *Secretary of State for Communities and Local Government* [2012], it was held that the change of use from an existing use to a new use, permitted by section 55(2)(f) of the 1990 Act, does not require the existing use to be necessarily lawful. It does not follow that a change from one class to another will necessarily amount to development that depends on general principles and whether there is a material change of use: *Rann* v *Secretary of State for the Environment* [1980]. Certain changes between classes are granted deemed planning permission under the Town and Country Planning (General Permitted Development) Order 1995 (see Chapter 4 and Appendix B).

The principal classes are reproduced in full in Appendix A, but can be summarised thus:

- Class A1 Shops – use for a number of specified retail purposes where the sale, display or services is to visiting members of the public. 'Shop' means uses taking place in a building: see *Cawley* v *Secretary of State* [1990]. However, Class A1 consists of a number of activities that would not normally be regarded as shops, e.g., a post office, travel agency and use as a funeral directors, use for the sale of sandwiches or other cold food for consumption off the premises, and as an Internet cafe.
- Class A2 Financial, Professional Services and betting services – this class applies where the services are provided principally to visiting members of the public. In appropriate circumstances, a solicitor's office providing services to visiting members of the public will fall within this class: see *Kalra* v *Secretary of State* [1996].
- Class A3 Food and Drink – this applies where food or drink is consumed on the premises.
- Class A4 Use as a public house, wine-bar or other drinking establishment.
- Class A5 Use for the sale of hot food for consumption off the premises.
- Class B1 Business – this includes offices, research and development, and any industrial process being uses which can be carried on in any residential area.
- Class B2 General Industrial.
- Classes B3–B7 Special Industrial Groups A–E.
- Class B8 Storage or Distribution.
- Class C1 Hotels, boarding or guest houses, where in each case no significant care is provided.
- Class C2 Residential accommodation and care to people in need of care (other than C3 dwellinghouses); hospital or nursing home; residential school, college or training centre.
- Class C2A Secure residential accommodation.
- Class C3 Dwellinghouses – use by a single person or by people living together as a family or by not more than six residents living together as a

single household (including a household where care is provided for residents). In *R (Hossack)* v *Kettering BC* [2002], three adjoining terrace properties had been used to accommodate homeless people; the Court of Appeal concluded that the expression 'living together as a single household' did not necessarily exclude residents, who were not related, having a common need of accommodation, support and resettlement; their precise relationship was clearly a material consideration, and if there were no more than six residents living together, they probably do so as a single household. In *Moore* v *Secretary of State for Communities and Local Government* [2012], the renting out of a house for self-catering holidays was not a change of use permitted within Class C3.

A planning condition or planning agreement can override the right to change uses within this class: see *R* v *Tunbridge Wells Borough Council, ex parte Blue Boys* [1990], see also the following text on this point.

- Class D1 Non-residential Institutions (e.g., medical centre, day nursery, museum, public worship and law court).
- Class D2 Assembly and Leisure (e.g., cinema, concert, bingo or dance hall, swimming pool and indoor or outdoor sports and recreation).

Article 3(6) provides that none of the specified classes include use as:

- A theatre;
- Amusement arcade or centre, or a funfair;
- Launderette;
- Petrol-filling station;
- Motor vehicle sales;
- Taxi or hire business;
- Scrap yard or yard for storage or distribution of minerals or breaker's yard;
- For work under the Alkali, etc., Works Regulation Act 1906;
- A hostel;
- A waste disposal installation for incineration or certain landfill waste;
- A retail warehouse club for the sale or display for sale of goods to members of that club;
- A night club;
- A casino.

Article 3(3) of the UCO provides that where a use is ordinarily incidental to any use that falls within one of the classes, it is not excluded from the use to which it is incidental, merely because it is specified as a separate class. It is the primary use of a building or other land that identifies the class: *Vickers Armstrong* v *Central Land Board* [1957]. Moreover, an incidental use cannot be turned into a primary use in its own right: *Percy Trentham Ltd* v *Gloucestershire County Council* [1966]. However, the business class B1, which covers offices, research and light industrial, means that an incidental office use may now have independent use rights.

Whether a particular activity falls within a class is determined as a matter of fact and degree. In *Tessier* v *Secretary of State for the Environment* [1975], it was accepted that the Secretary of State was entitled to decide that a sculptor's use of a building was a use s*ui generis* and was not an industrial process falling within the general industrial class. Following amendments introduced by the Housing and Planning Act 1986, where premises are subdivided, and the new units continue to be used for purposes within the use class of the original use, then this will not involve development even if there is an intensification of uses that might on general principles involve a material change of use. This does not authorise any building operations that would otherwise amount to development. Dwellinghouses are excluded from the effect of the 1986 amendment; Article 4 of the UCO provides that whenever a dwellinghouse is subdivided, e.g., by the sale of a 'granny annex' lodge, nothing in the order has the effect of excluding the subdivision from the meaning of development. The Court of Appeal's decision in *Wakelin* v *Secretary of State for the Environment* [1978], that the inspector was entitled to decide that the subdivision of a dwellinghouse and a lodge was a material change of use requiring planning permission, is therefore still relevant.

Although a change of use within a class is not development, and does not need planning permission, it was held in *City of London Corporation* v *Secretary of State for the Environment* [1972] that it was lawful to impose a planning condition (see Chapter 9) restricting a change of use within the same use class.

3.4. Three doubtful cases: the division of dwellinghouses, tipping and advertisements

Section 55(3) of the 1990 Act declares that, for the avoidance of doubt, two matters are to be considered as involving a material change of use:

(1) The use as two or more separate dwellinghouses of any building previously used as a single dwelling involves a material change in the use of the building and of each part of it which is so used (section 55(3)(a)).

The conversion of a house into two flats is therefore a material change of use, and requires planning permission. A building which is in multiple-paying occupation, with the tenants and lodgers living separately, although perhaps sharing a communal kitchen, a bathroom and lavatory, is not necessarily a building used as two or more separate dwellinghouses: *Ealing Borough Council* v *Ryan* [1965]. The borough council need not have relied on section 55(3) as a change of use to multiple-paying occupation is likely to involve a material change of use: *Birmingham Corporation* v *Minister of Housing and Local Government and Habib Ullah* [1964] (see also p. 22).

(2) The deposit of refuse or waste materials on land, even if the land is comprised in a site already so used, if either the superficial area of the

deposit is extended, or the height of the deposit is extended and exceeds the
adjoining ground level (section 55(3)(b)).

This provision may leave in doubt whether tipping that is outside its scope involves a material change of use or is an operation. The Court of Appeal decided that tipping was a material change of use in the case of *Bilboe* v *Secretary of State for the Environment* [1980]. Paradoxically, the opposite activity, excavation work, is an operation: *Thomas David (Porthcawl) Ltd* v *Penybont Rural District Council* [1972]. The importance of this distinction has already been noted (p. 12).

Section 55(5) provides that:

Without prejudice to the control of advertisement regulations, the use for
display of advertisements of any external part of a building which is not
normally used for that purpose shall be treated as involving a material
change of use of that part of the building.

However, where under the Town and Country Planning (Control of Advertisements) Regulations 2007, consent is given for an advertisement, there is deemed planning permission for any development, such as a material change of use: see section 222 of the 1990 Act. However, if no such consent is granted, the use of a part of building will be a breach of the Advertisement Control Regulations as well as being development (material change of use) without planning permission, and liable to planning enforcement.

3.5. The meaning of material change of use

So far, we have considered certain uses which are specifically excluded from the meaning of material change of use, and two matters which are stated to involve such a change. Now we must consider the principles for deciding whether there is a material change of use in all other cases.

Although material change of use relates to land and buildings, it will include the use of a moveable and floating helicopter-landing vessel. This is because the land will be used for supporting the weight of water to support the helicopter traffic: see *Thames Heliport plc* v *Tower Hamlets London Borough Council* [1997].

3.5.1. *Question of fact and degree for the planning authority*

It must be appreciated that a change of use is only a material change of use, and therefore within development control, when the local planning authority or the Secretary of State so decide. Such a decision will be made when a developer seeks a determination under sections 191 and 192 of the 1990 Act (certificates of lawful use, existing and proposed uses) as to whether a proposal involves development, or when an enforcement notice is served because of an alleged breach of planning control. Developers apply for planning permission to avoid enforcement action and to establish lawful use rights in the interest of the marketability of property.

The role of the law is therefore limited to establishing the matters that planning authorities should or should not take into account in deciding the question on the facts they have before them. Provided they keep within these legal boundaries, whether any particular change of use is development is therefore a factual rather than legal question. In *Blackpool Borough Council* v *Secretary of State for the Environment* [1980], the inspector, in her report to the Secretary of State, stated she was making a practical judgment of the essential nature of an activity and was not laying down any principle of law when deciding that the use of a house for holiday letting was not a material change of use. The Divisional Court agreed that she was entitled to reach this conclusion as a question of fact and degree.

However, it is important to distinguish between the question as to whether a change from use A to use B is a material change of use from the question as to whether in planning law use A and use B are different uses. Thus, in *Crawley BC* v *Hickmet Ltd* [1998] JPL 210, the Court of Appeal decided that there was a clear distinction between use of land for storage purposes and the use of land for car parking: the difference was primarily one of law and not exclusively of fact.

The court will not make declarations as to the identification of a planning unit so as to pre-empt the decision of the local planning authority: see *Thames Heliport plc* v *Tower Hamlets London Borough Council* [1997].

Early ministerial guidance as to the intended meaning of material change of use is found in Circular 67 of 1949; although now withdrawn, it is still a useful starting point to the problem:

> … *in considering whether a change is a material change, comparison with the previous use of the land or building in question is the governing factor and the effect of the proposal on a surrounding neighbourhood is not relevant to the issue.*

There are two points here: the effect on the neighbourhood is irrelevant in first deciding whether a change is a material change of use; but if that is answered affirmatively, the effect on the neighbourhood may well be a relevant consideration in deciding whether to permit a change of use. The Circular continues:

> *The effect of the [word material] is to make clear that a proposed change of use constitutes development only if the new use is substantially different from the old. A change in kind will always be material – e.g. from house to shop. A change in the degree of an existing use may be 'material' but only if it is very marked.*

This extract identifies two ideas that now need considering: that material change of use occurs where a new use commences which is substantially different from the old; or, alternatively, it may occur where there is an intensification of an existing use.

3.5.2. The new use must be substantially different from the old

In *Guildford Rural District Council* v *Penny* [1959], the Court of Appeal decided that a material change of use depended on whether there was a material change in the character of the use of the property in issue. *East Barnet Urban District Council* v *British Transport Commission* [1962] was one of the first cases on this problem. Certain land had been used for the storage and distribution of coal; it was then let to Vauxhall Motors Ltd for the storage of boxed motor vehicles. In dealing with the word 'material', Lord Parker CJ said it must refer to material as material for planning purposes, and he added:

> *What really is to be considered is the character of the use of the land, not the particular purpose of a particular occupier ... the mere fact that the commodity changes does not necessarily mean that the land is being used for a different purpose.*

He concluded that the justices had correctly quashed the enforcement notice that alleged material change of use.

Lord Bridge in *Westminster City Council* v *British Waterways Board* [1984] emphasised that the identity of the occupier was irrelevant; it was the activities carried on at the premises in issue that were relevant in planning law.

In *Williams* v *Minister of Housing and Local Government* [1967], Widgery J said that there was a significant difference in the character of a use which involves selling produce of the land itself, and selling goods brought in from elsewhere.

It might be argued that a home occupied by a single family unit and a home in multiple occupation are both in residential use. However, Lord Parker CJ in the *East Barnet* case introduced the idea that it is the purpose of the use that is important, and if this is materially different from the previous use from a planning point of view, there is a material change of use:

> *Birmingham Corp* v *Minister of Housing and Local Government and Habib Ullah* [1964]
> Three houses had previously been occupied by a single family each. They were then used for multiple-paying occupation without planning permission. It was held that such a change of use could be material: the purpose of the new use, letting rooms for gain, is different from the previous use of a house for a private family (see now Class C3 of 1987 UCO).

The courts, in construing the meaning of 'material change of use', seem to have developed three ideas: a change of use is material if it has planning consequences on the neighbourhood, or there is a change in the character of the use, or the purpose of the use is different from the old.

3.5.3. A change in the degree of an existing use

The suggestion here is that an intensification of an existing use can involve a material change of use. In *Guildford Rural District Council* v *Penny* [1959], Lord Evershed MR said:

> ... *increasing intensity of use or occupation may involve a substantial increase in the burden of the services which a local authority has to supply, and that, in truth, might, in some cases at least, be material in considering whether the use of the land had been materially changed.*

In other words, if there are planning consequences, there may be a material change of use. In *Penny*, an increase in the number of caravans on a site from 8 to 26 was not considered to involve a material change of use. In the *Habib Ullah* case (see earlier), it was not the increase in the number of occupants of the houses concerned that involved a material change of use, it was the change of the purpose of the use. A material change of the use occurred in *Peake* v *Secretary of State for Wales* [1971]. The owner of a private garage used it for vehicle repair on a part-time basis. When he lost his job, he then repaired vehicles at the garage on a full-time basis. The change of use from an incidental activity to a use of a more substantial character was material.

In *Marshall* v *Nottingham County Council* [1960], an enforcement notice was served alleging intensification of use: the site had been used for the manufacture of wooden buildings, but from 1957, it was used for the selling of caravans and wooden buildings. It was held that there was no material change of use, merely because the wooden buildings were no longer made on site, or because caravans were not originally sold. Neither was there a material change of use through intensification as the uses after 1957 were so greatly intensified in comparison with the original use. The notion that there is some principle of intensification was criticised by the Court of Appeal in *Kensington & Chelsea RBC* v *Secretary of State for the Environment* [1981]. Donaldson LJ said that the issue was about a material 'change' of use from one condition to another.

The problem of intensification of an existing use was clearly stated by Lawton LJ in *Brooks & Burton* v *Secretary of State for the Environment* [1978]:

> ... *intensification of use can be a material change of use. Whether it is or not depends on the degree of intensification. Matters of degree are for the Secretary of State to decide ...*

What is also clear from this case is that if a use is within one of the classes of the UCO, and even if an intensification of the use amounted to a material change of use on the general principles now being discussed, an intensification of such a use does not involve development: changes of use within a class are excluded by section 55(2)(f) of the 1990 Act. Therefore, an intensification of any shop use, or an industrial process, can never involve development. Intensification problems are therefore confined to uses outside the UCO.

In *Blum* v *Secretary of State for the Environment* [1987], it was said that there must be a material change in the character of a use to amount to a 'material change of use by intensification'.

3.5.4. Principal, ancillary and multiple uses

In deciding whether there has, or will be, a material change of use, the principal or primary use must be identified. A principal use may sometimes involve multiple uses.

Buildings or land may have a principal use. It is the purpose of the use that is relevant. Thus, in *London Residuary Body* v *Secretary of State for the Environment* [1988], it was accepted that a building used as offices for a local government body had a principal use of local government and not general office use.

Where there is a principal use of land or buildings, an ancillary use is regarded as part of the principal use. Two points follow: a change of use of the part used for the ancillary use to the same use as the principal use will not be a material change of use; and if the ancillary use becomes the principal use in its own right, a material change of use may be involved. This idea is true whether or not a use is within the UCO:

> *Percy Trentham Ltd* v *Gloucestershire County Council* [1966] CA
> Farm buildings, which had previously been used by the appellant's predecessor for storing farm machinery and equipment in connection with a 75-acre farm, were then used by the appellant for storing building materials. Although both uses were storage, one was incidental to a farming use. The new storage use was for a different principal activity, that of a building contractor, and a material change of use was involved.

In *Essex Water Co* v *Secretary of State for the Environment* [1989], it was said that the uses of a planning unit cannot be ancillary to a principal use carried on outside that planning unit. In *Allen* v *Secretary of State for the Environment* [1990], it was held that the sale of plants and shrubs grown on land is ancillary unless the sales are abnormal in volume. An ancillary use may become a principal use in its own right where the ancillary use becomes functionally separated from the principal use, as in *Wood* v *Secretary of State for the Environment* [1973]. This particular point may merge within the planning unit question addressed below. Indeed, the functional relation with the principal use of a unit is the critical factor of ancillary status; the ancillary use must 'ordinarily' be incidental: see *Harrods Ltd* v *Secretary of State for the Environment, Transport and the Regions* [2002], where a certificate of lawful use was requested for the use of the roof of the store as a helicopter landing.

In some cases, there are mixed uses and no principal use is apparent. Whether the uses are on separate areas or intermingled, a comparison between the overall use of the land at one time should be made with the overall use at a later time; if that comparison suggests a material change of use, then there is development:

> *Wipperman* v *London Borough of Barking* [1965] DC
> Prior to 1962 land was used for storage of fence material and for car-breaking. After that date the land was used for storage of materials for the building of conservatories. It was held that the two original uses were not ancillary to each other but dissimilar uses; that the new storage use of the whole area had changed the character of the land as a whole and involved a material change of use.

This case also established that the mere cessation of one of a number of uses was not as such a material change of use unless there was some intensification of the remaining use: see also *Philglow Ltd* v *Secretary of State for the Environment* [1984].

Where there is a primary use, such as a retail shop, then the creation within that unit of a specialised shop such as a pharmacy would not involve a material change of use: see *R* v *Maldon DC, ex p Pattani* [1999].

3.5.5. The planning unit

Circular 67 of 1949 anticipated the problem that occurs where a change of use involves only part of a building:

> *The minister takes the view that the question whether there is a material change of use should be decided in relation to the whole premises and not merely in relation to the part, ie the point at issue is whether the character of the whole existing use will be substantially affected by the change which is proposed in a part of the building. He would not for instance regard as constituting a material change of use the use by a professional man – say a doctor or dentist – of one or two rooms in his private dwelling for the purpose of consultation with his patients so long as this use remained ancillary to the main residential use.*

Planning control concerns the change of use of 'any buildings or other land'. That raises the question as to the proper unit for consideration. In *East Barnet Urban District Council* v *British Transport Commission* [1962], Lord Parker CJ said it was a matter of commonsense. A useful set of guidelines is to be found in:

> *Burdle* v *Secretary of State for the Environment* [1972] DC
> An area of land had been used for the business of a scrapyard and car breakers; there was also some evidence of on-site retail sales of car parts; a lean-to had been used as an office in connection with the business.

Burdle bought the business. He started to sell new car parts and camping equipment from the lean-to. The local planning authority served an enforcement notice alleging material change of use of the lean-to to a shop. The Secretary of State's decision that the enforcement notice should only relate to the lean-to and not the whole site was not accepted by the court.

The court had to consider what was the proper planning unit. In other words, had there been a material change of use of the whole site, or of the lean-to? Clearly the answer in each case might be different. This is an important question, because if the local planning authority can select some small part of a person's curtilage, it may be easier to show that there has been a material change of use of that small part rather than some larger area. As Lord Widgery LJ in *De Mulder* v *Secretary of State for the Environment* [1974] put the matter:

> *a planning authority by an arbitrary division of an area into a number of smaller areas each with its own enforcement notice cannot by that means impose more severe restrictions on the landowner than might have been imposed on him by an enforcement notice applicable to the whole area.*

The need for a planning authority to correctly identify the proper planning unit is illustrated by *Kensington and Chelsea Royal London Borough Council* v *Secretary of State for the Environment* [1981]. The owners of a restaurant extended the seating area into the garden. It was held that the proper planning unit was the whole premises, and not just the garden, so the question was whether there had been a material change of use of the whole premises, and not just the garden area.

Bridge J in the *Burdle* case propounded some useful tests for determining the proper planning unit:

(a) Where there is a single main purpose of the occupier's use of his land to which secondary activities are incidental or ancillary, the whole unit of occupation should be the planning unit.
(b) Where an occupier carries out a variety of activities and it is not possible to say that one is incidental or ancillary to another, then, again the whole unit of occupation should be the planning unit.
(c) Where, within a single unit of occupation, there are two or more physically separate and distinct areas occupied for substantially different and unrelated purposes, each area used for a different main purpose (together with its incidental and ancillary activities) is a planning unit.

In other words, the unit of occupation is the planning unit unless there are activities that are physically and functionally separate, in which case the planning unit will be a smaller unit. However, in relation to two or more uses of one building, Lord Widgery CJ said in *Wood* v *Secretary of State for the Environment* [1973]:

... it can rarely if ever be right to dissect a single dwellinghouse and to regard one room in isolation as being an appropriate planning unit ...

However, in *Wakelin* v *Secretary of State for the Environment* [1978], the Court of Appeal decided that an inspector could, on the facts before him, conclude that the division of a residential property consisting of a house and a lodge (the latter provided at one time accommodation for a relative), into two separate functional units was a material change of use: planning permission was therefore required for the residential use of the lodge as a separate unit.

This was followed in *Winton* v *Secretary of State for the Environment* [1984]; it was suggested that the division of a single unit of occupation into two or more units might, if there were planning consequences, amount to a material change of use. The effect of this case has been abrogated in its application to the UCO.

In *Church Commissioners for England* v *Secretary of State for the Environment* [1995], it was held that the individual units of a large shopping centre were planning units and not the whole centre in the occupation of the landlord.

Physical separation may deny the existence of a single planning unit. Thus, in *Fuller* v *Secretary of State for the Environment* [1988], an agricultural unit included more than one planning unit.

3.5.6. *Abandonment of a use*

A material change of use suggests a change from use A to use B. Certain problems have arisen where use A has ceased, and, either there has been a period of no use of the land or buildings concerned, to be followed again by use A; or, the change has been use A to use B, and then back to use A. The question in each case is whether a material change of use is involved in reverting to use A. (See part **3.8** of this chapter for certain resumptions of use that do not need planning permission.)

In *Fyson* v *Buckinghamshire County Council* [1958], land that had been used for the storage of scrap was unoccupied for a period of 7 years. It was held that the resumption of scrap storage use was not a material change of use. However in *Hartley* v *Minister of Housing and Local Government* [1970], Lord Denning MR said:

I think that when a man ceases to use a site for a particular purpose and lets it remain unused for a considerable time, then the proper inference may be that he has abandoned the former use. Once abandoned, he cannot start to use the site again, unless he gets planning permission: and this is so, even though the new use is the same as the previous one.

And Widgery LJ, as he then was, said:

It has been suggested in the courts before, and it seems to me that it is now time to reach a view upon it, that it is perfectly feasible in this context to describe a use as having been abandoned when one means that it has not

*merely been suspended for a short and determined period, but has ceased
with no intention to resume it at anytime. It is perfectly true... that the word
'abandonment' does not appear in the legislation. We are not concerned
with the legislation at this stage, but merely with the facts of the matter.*

Lord Denning MR then considered a suitable test to determine whether a use
has been abandoned:

*Has the cessation of use (followed by non-use) been merely temporary, or
did it amount to abandonment? ... Abandonment depends on the circumstances. If the land has remained unused for a considerable time, in such
circumstances that a reasonable man might conclude that the previous use
had been abandoned, then the [planning authority or the Secretary of State]
may hold it to have been abandoned.*

The *Hartley* test, if it may be so called, was applied in the following case:

Ratcliffe v *Secretary of State for the Environment* [1975] DC
    A quarry had been used for tipping waste from 1920 until 1961; apart
from some minor tipping from trespassers, the site was not used for this
purpose thereafter. The appellants made an application under, what is now
section 192 of the 1990 Act, to have determined whether a resumption of
tipping needed planning permission. It was held that the Secretary of State
was entitled on the evidence, to conclude that the tipping use had been abandoned.

The occupier's intention to abandon or otherwise is more important in applying
the *Hartley* test than the length of time of a cessation of use. If land is not being
used merely because a tenant cannot be found for it, it is submitted that the use
right is not lost so long as there remains an intention to find a tenant, no matter
how abandoned the property may appear physically. However, the intention of
the occupier cannot be paramount as the real test must be the view that an ordinary reasonable man would reach on the facts: see *Hughes* v *Secretary of State
for the Environment, Transport and the Regions* [1999].
    In a ministerial decision (Ref. APP/5289/G/79/20), reported at [1980] JPL
p759, the Secretary of State decided that there was no abandonment of a use of
certain residential property, that had been vandalised and left empty for 13 years
and uninhabitable for five years, because the owner had shown an intention not
to abandon the use by taking steps to prevent the vandalism.
    In *Trustees of Castell-y-Mynach Estate* v *Secretary of State for Wales* [1985],
the following four factors were said to be relevant in deciding whether a use had
been abandoned:

(a) The physical condition of the building.
(b) The period of non-use.

(c)   Whether there has been any intervening use.
(d)   The evidence, if any, of the owner's intentions.

Seasonal resumptions of use do not involve development:

> *Webber* v *Minister of Housing and Local Government* [1968] CA
> A four-acre field was used for camping in summer and grazing in winter. It was held there was no material change of use each summer when the camping use was resumed.

In *Philglow Ltd* v *Secretary of State for the Environment* [1984], the Court of Appeal decided that the mere cessation of a use could not itself be a material change of use. Any other decision would mean that a landowner would require planning permission to cease a use. However, the resumption of one of a number of uses, after that use has been abandoned for a time, may be a material change of use:

> *Hartley* v *Minister of Housing and Local Government* [1970] CA
> A site was used until 1961 as a petrol-filling station and for the sale of cars. Then for four years the sale of cars ceased. It was held that the minister was entitled on the evidence to decide that the resumption of the car sales use in 1965 involved a material change of use.

The concept of abandonment was criticised by Glidewell J in *Balco Transport Services Ltd* v *Secretary of State for the Environment* [1985] who pointed out that it was not a concept that is apt to cover a change from one use to another. Indeed, section 57(4) of the 1990 Act provides that planning permission is not required to revert to a previous lawful use following an enforcement notice. The House of Lords held in *Pioneer Aggregates (UK) Ltd* v *Secretary of State for the Environment* [1984] that where the activity of mining operations is authorised by planning permission, a commercial decision to stop those operations for a period of time cannot extinguish the planning permission. Planning permission is granted under statutory powers and enures for the benefit of the land and successive owners (section 75). In the *Pioneer* case, Lord Scarman said:

> ... the introduction into planning law of a doctrine of abandonment by elec-tion of the landowner (or occupier) cannot, in my judgment, be justified. It would lead to uncertainty and confusion in the law, and there is no need for it. There is nothing in the legislation to encourage the view that the courts should import into the planning law such a rule ...

However, Lord Scarman did acknowledge that an existing use, such as a use existing before 1 July 1948, and not requiring planning permissions, could be abandoned; he mentioned the *Hartley* case as a good example.

In *Bramall* v *Secretary of State for Communities and Local Government* [2011], a planning inspector had been entitled to conclude that a prior right to use a cottage on the expiry of a limited period planning permission had been abandoned; a substantial number of years had passed since a period of continued residential use.

The Planning and Compensation Act 1991 introduced provisions relating to the grant of certificates of lawful use; such a certificate may be issued where a use has continued for not less than 10 years without any enforcement action being taken and without contravention of the requirements of any enforcement notice: see sections 171B and 191 of the 1990 Act. It was held in *M & M (Land) Ltd* v *Secretary of State for Communities and Local Government* [2007] that a use the subject of a certificate of lawful use issued under section 191 of the 1990 could, on the facts, be abandoned.

However, immunity against enforcement for material changes of use occurring before 1 July 1948 (existing uses) or 1 January 1964 (established uses) is not lost by the provisions relating to certificates of lawful use. Uses that commenced prior to either of those two dates or before the commencement of the 10-year period for acquiring immunity from enforcement action may be lost by operation of law in one of three ways; by abandonment, by the formation of a new planning unit, or by way of a material change of use: see *Panton* v *Secretary of State for the Environment, Transport and the Regions* [1999]. Accordingly, the concept of abandonment will apply to the cessation of an existing use, that is, one that could otherwise continue without planning permission: see the *Hartley* case in the preceding text.

3.5.7. *A planning permission may extinguish a use*

If there is a right to use land or buildings for a particular purpose, is that right lost if planning permission for some other purpose is then obtained?

In *Prosser* v *Minister of Housing and Local Government* [1968], where planning permission was granted for the rebuilding of a petrol-filling station, a condition was attached to prevent retail sales on the site. It was argued for the appellant that if he had a pre-existing right to use the site for sales, he would not need planning permission to display second-hand cars for sale. The court held that if a planning permission is acted upon, any conditions in it are then binding, and any pre-existing uses are destroyed: planning history recommences with the new building.

In the case of *Petticoat Lane Rentals Ltd* v *Minister of Housing and Local Government* [1971] DC, a building was constructed on a site previously used for market trading: a condition of the permission restricted the future market trading to Sundays. The building was erected on pillars and market trading continued after the completion of the building on weekdays. In upholding the Secretary of State's decision that the continuance of the weekday market trading involved a material change of use, Lord Widgery CJ applied *Prosser*'s case on the ground that the erection of a building over an area of land creates a new planning unit:

The land as such is merged in that new building and a new planning unit with no planning history is achieved. The new planning unit, the new building, starts with a new use, that is to say immediately after it was completed it was used for nothing, and therefore any use to which it is put is a change of use, and if that use is not authorised by the planning permission, that use is a use which can be restrained by planning control.

These cases were considered by the House of Lords in *Newbury District Council* v *Secretary of State for the Environment* [1981]. Viscount Dilhorne said that the taking up of planning permission should not prevent an owner relying on existing use rights. He said existing use rights would only be extinguished if the implementation of a planning permission led to the creation of a new planning unit. For example, the complete rebuilding, as in the *Prosser* case, was to be regarded as the creation of a new planning unit. However, in the *Newbury* case, the grant of planning permission for a storage use did not create a new planning unit; the buildings in question had existing use rights for storage purposes, and those rights could be relied on in preference to the conditions attached to the planning permission.

Although the *Pioneer Aggregates* case would seem to suggest that a planning permission cannot be abandoned, this requires a qualification in the light of the decision of the Court of Appeal in *Cynon Valley Borough Council* v *Secretary of State for Wales* [1986]. The planning permission in *Pioneer Aggregates* was for mining operations; and that in *Cynon Valley* was for a change of use. In the latter case, Balcombe LJ in *Cynon Valley* said that where the development for which planning permission is required is a material change of use, the permission is to change from use A to use B, and is not merely permission to use the property for use B for the indefinite future. If planning permission is obtained and implemented for use C, then the previous planning permission for use B would not survive. He reached this decision after carefully considering the House of Lords case of *Young* v *Secretary of State for the Environment* [1983]: once a planning permission is implemented, it is then spent, and cannot be revived to authorise the reversion to an earlier permitted use.

If planning permission is granted for the erection of a building, the permission enables the building to be used for the purpose for which it is designed (section 75(3) of the 1990 Act). It is only if the intended use is inconsistent with the permission that the *Prosser* case applies. However, if a building is erected without planning permission, not only is this provision authorising its designed use not available, the erection of the building may extinguish any lawful use previously enjoyed by the site covered by the building, a question considered in the next case:

Jennings Motors Ltd v *Secretary of State for the Environment* CA [1982]
A site was lawfully used for a taxi, car and coach hire business, and for vehicle repairs and car sales. A building was erected on part of the site without planning permission and was used for the same purpose as the rest

of the site. It was held in the Divisional Court that the erection of the building destroyed any previous lawful use of the site now covered by the building so that the building had no lawful use. Any use to which it was put, including the original use of the site, was a material change of use, involving development, and requiring planning permission. However, the Court of Appeal allowed an appeal by Jennings Motors against this decision. It was said that although some physical alteration to part of a site, such as by the erection of a new building or the alteration of an existing building, is one of the factors to be taken into account in considering whether there has been a break in the planning history of the site, that break had not occurred in this case; the occupiers had continued their existing use right, and were entitled so to do.

The question was formulated by Lord Denning MR in this case as to whether there was a 'new chapter in planning history': physical alterations to the site or part of it may constitute a break in that planning history only if sufficiently radical. He particularly preferred the theory of a new chapter in planning history to that of a new planning unit in the circumstances of the case. Oliver and Watkins LJJ considered that the theory of the planning unit should be preserved for the geographical problems within the guidance given by Bridge J in the *Burdle* case (see p. 25). All this is some way from the simple proposition in the 1990 Act that a material change in the use of any buildings or other land constitutes development.

The effect of, what is now, the General Permitted Development Order, and development permitted by the order (see Chapter 4), on a pre-existing planning permission, was considered in:

Cynon Valley Borough Council v *Secretary of State for Wales* CA [1986]
    Planning permission was granted in 1958 for a fish-and-chip shop, but a change of use to an antique shop took place in 1978 as development permitted by the then general development order. It was held that the 1958 permission was spent and did not survive the change to an antique shop. However, section 23(8) of the 1971 Act provided that where development was permitted (subject to limitations) by a development order, permission was not required to revert to the normal use of the land. As the permitted change to an antique shop was subject to the general limitation that excludes certain types of shops from the general class of shops, the reversion to the previous shop use was permitted and not lost (see now section 57(3) of the 1990 Act).

In *Durham County Council* v *Secretary of State for the Environment* [1990], it was held that a change of use did not nullify an earlier planning permission for operational development provided that earlier permission could still be implemented. The same point arose in *Camden Borough Council* v *McDonald's Restaurants* [1993], where planning permission was granted for an extension to premises used as a restaurant for a number of years; following the cessation of restaurant use, the premises were used as a bookshop. It was held that the change of use to a bookshop did not nullify the permission to alter the premises; the extension could still be built and used in connection with a restaurant use.

3.6. Uses that commenced before 1964

Until the Planning and Compensation Act 1991 came into effect, a distinction was made between uses that commenced before 1 July 1948 and before 1964. If the use existed in 1948, planning permission was not required for the continuance of that use. Accordingly such continuance was lawful. If a material change of use occurred after 1 July 1948 and before 1964 without planning permission, the continuance of that use was unlawful, although immune from enforcement action.

The amendments made to the 1990 Act by the 1991 Act now provide that any change of use requiring planning permission which takes place without that permission can only be enforced within 10 years of the breach of planning control (see Chapter 15). However, if a change of use occurs and has become immune from enforcement action because of the expiration of the 10-year period, such use is lawful.

3.6.1. Uses existing on 1 July 1948 – 'the appointed day'

Planning permission is not required for any change of use that occurred before 1 July 1948: see Schedule 24 to the 1971 Act as continued by the Planning (Consequential Provisions) Act 1990. Section 57 and Schedule 4 to the 1990 Act also excludes from the need to obtain planning permission, certain uses of a temporary or intermittent nature in 1948:

- Where land was used on the appointed day for a temporary use, planning permission is not required to resume the normal use before 6 December 1968 (Schedule 4).
- Where land was used on the appointed day for one purpose, and also, on occasions, whether at regular intervals or not, for another purpose, planning permission is not required for the occasional use before 6 December 1968 nor is it required for the occasional use after that date provided the occasional use took place at least once between the appointed day and the beginning of 1968 (section 55).
- Where land was unoccupied on the appointed day, but had been occupied at some time between 7 January 1937 and the appointed day, planning permission is not required for the resumption of any use of the land begun before 6 December 1968, if the use was for a purpose for which the land was last used before the appointed day.

Land includes any building (section 336 of the 1990 Act).

3.6.2. *Material change of use commencing between 1 July 1948 and before 1964 without planning permission*

Any such change of use was immune from enforcement action. Until 1992, an established use certificate could be obtained to give protection against enforcement in the following cases:

- Use commenced before 1964 without planning permission and continued since.
- Use commenced before 1964 with planning permission but subject to conditions or limitations that have not been complied with since 1963.
- Use commenced after 1963 as the result of a change of use not requiring planning permission (a change of use may not require planning permission either because it is not a material change of use, and is not therefore development, such as a change of use within one of the classes of the Use Classes Order, or because it is a material change of use that is permitted without need of planning permission.

A certificate could not be granted if the established use was no longer continuing at the time of the application. Nor could it be granted in respect of the use of land as a single dwellinghouse. Where there has been an intensification in the established use by the date of the application for a certificate, the planning authority was entitled to refuse to grant the certificate: *Hipsey* v *Secretary for the Environment* [1984]. It would seem that a certificate cannot be granted for the lesser use before intensification as this use is no longer continuing.

In *Broxbourne Borough Council* v *Secretary of State for the Environment* [1979], Goff J had this to say about a certificate of established use:

> ... *the purpose of an established use certificate is clear. It does not render a use lawful. To that extent it is unlike a grant of planning permission. Therefore, if, for example, the use specified in an established use certificate is abandoned, it cannot lawfully be resumed. Its function is to render the specified use, as long as it persists, immune from an enforcement notice.*

3.7. Use continuing for 10 years after material change

A material change of use without planning permission, which is a breach of planning control, can only be enforced within a period of 10 years from the date of the breach or of four years in the case of a change of use to a single dwellinghouse: see section 171B of the 1990 Act. Any material change of use without planning permission which has not been the subject of enforcement proceedings within these time limits is lawful. A certificate of lawfulness of existing use can be obtained under section 191 of the 1990 Act. However, as discussed earlier, any such use can be abandoned. Breach of planning control is considered in Chapter 15.

3.8. Change of use not requiring planning permission

Section 57 of the 1990 Act specifies certain changes of use in respect of which planning permission is not required:

- Planning permission is not required to resume the previous normal use of land following the expiration of a planning permission for a temporary use (section 57(2)).
- Where planning permission to develop land has been granted, subject to limitations, by a development order, planning permission is not required for the use of that land which is the previous *normal* use of land following the expiration of a planning permission for a temporary use that is granted by the General Permitted Development Order (see Chapter 4) (section 57(3)). The meaning of permission granted by a development order 'subject to limitations' includes permission to use premises as a shop other than shops excluded from Class I of the old Use Classes Order: *Cynon Valley Borough Council* v *Secretary of State for Wales* [1986].
- Planning permission is not required for the resumption of a previous *lawful* use of land following an enforcement notice (section 57(4)).

In considering the *normal* use of land or a building, any use in breach of planning control must be disregarded: see section 57(5).

4 Permitted development

4.1. Introduction

Planning permission is required for development; the meaning of development was explained in the preceding two chapters. Although an actual application must be made to the local planning authority for planning permission for most forms of development, there are a number of classes of development of a minor nature for which planning permission is automatically granted in the Town and Country Planning (General Permitted Development) Order 1995 (in this chapter, referred to as the GPDO: reproduced as Appendix B); the GPDO has been amended from time to time. In most cases, this permitted development can be carried out without any further express consent of, or notice to, the local planning authority: see section 60 of the Town and Country Planning Act 1990.

In special circumstances, the planning permission deemed granted by the GPDO can be withdrawn by what is called an article 4 direction. An express application for planning permission is then necessary for any development that would otherwise have been permitted but for the direction.

There would appear to be several inter-related reasons for permitted development rights. Many of the classes of permitted development concern minor operations such as external painting and temporary uses of land, which, although strictly within the meaning of development, are not likely to involve planning considerations of any consequence. To require express planning applications for these activities would be an unnecessary burden on owners as well as planning authorities.

There is also an element of ministerial policy in the classes of permitted development, as the greater their scope, the less control that can be exercised by local planning authorities. The cases include the enlargement of dwellinghouses or industrial buildings, within certain limitations, and the erection of agricultural buildings; by permitting these developments, the minister is very firmly taking such development out of local control.

The final justification for permitted development concerns the activities of many of the statutory undertakers. Although much of their work will be development, the minister has decided that certain classes of such development shall be permitted: the statutory undertakers may proceed without need of any express planning permission.

In 1986, the concept of Simplified Planning Zones was introduced. Subject to the designation of such zones, development is permitted in accordance with the schemes for such areas.

4.2. The classes of permitted development

The classes of permitted development are set out in Parts 1–38 of Schedule 2 to the GPDO:

1. Development within the curtilage of a dwellinghouse
2. Minor operations
3. Changes of use
4. Temporary buildings and uses
5. Caravan sites
6. Agricultural buildings and operations
7. Forestry buildings and operations
8. Industrial and warehouse development
9. Repairs to unadopted streets and private ways
10. Repairs to services
11. Development under local or private Acts
12. Development by local authorities
13. Development by local highway authorities
14. Development by drainage bodies
15. Development by the Environment Agency
16. Development by or on behalf of sewerage undertakers
17. Development by statutory undertakers
18. Aviation development
19. Development ancillary to mining operations
20. Coal mining development by the Coal Authority and licensed operators
21. Waste tipping at a mine
22. Mineral exploration
23. Removal of material from mineral-working deposits
24. Development by telecommunications code system operators
25. Other telecommunications development
26. Development by Historic Buildings and Monuments Commission for England (English Heritage)
27. Use by members of certain recreational organisations
28. Development at amusement parks
29. Driver information systems
30. Toll road facilities
31. Demolition of buildings
32. Schools, colleges, universities and hospitals
33. Close circuit television cameras
34. Development by the Crown
35. Aviation development by the Crown

36.   Crown railways, dockyards, etc., and lighthouses
37.   Emergency development by the Crown
38.   Development for national security purposes
39.   Temporary protection of poultry and other captive birds
40.   Installation of domestic microgeneration equipment.

Article 3 of the GPDO grants planning permission for the classes of development in Schedule 2 subject to any relevant exception, limitation or condition specified in relation to those classes. By article 3(4), nothing in the GPDO permits development that would be contrary to any condition imposed by any planning permission that has been granted following an express application. The deemed planning permission does not apply in connection with any building, if the construction of that building was itself in breach of planning control, or in connection with an existing use, if that use was in breach of planning control. Thus, for example, one could not erect a dwellinghouse in breach of planning control (see Chapter 15) and then rely upon the rights in the GPDO to extend that dwellinghouse within the limitations of the permitted development.

Many of the classes impose some limitations on the dimensions of permitted development. In *Fayrewood Fish Farms* v *Secretary of State for the Environment* [1984], it was said that if any part of the development exceeded the limitations imposed by the then General Development Order, then the whole of that development would amount to a breach of planning control, and not simply so much of the development as is beyond the limitations. There is no reason to suppose that that would not apply to the new GPDO.

With minor exceptions, the deemed permission does not authorise any development that requires or involves the formation, laying out or material widening of a means of access to an existing highway which is a trunk road or a classified road, or creates an obstruction to the view of persons using any highway used by vehicular traffic, so as to be likely to cause danger to such persons: see article 3(6).

By virtue of the Town and Country Planning (Environmental Impact Assessment) (England and Wales) Regulations 1999, development that requires, or is likely to require an environmental impact assessment cannot be carried out notwithstanding that it may otherwise be permitted development under the GPDO. These regulations are considered in more detail in Chapter 8. Where development is likely to have a significant effect on a site protected by the European directives (Habitats and Birds), then regulation 73 of the Conservation (Natural Habitats, etc.), Regulations 2010 imposes a condition on the grant of planning permission under the GPDO that the development should not commence until approval has been given by the LPA under regulation 68.

The tolerances and limitations in respect of some of the classes of permitted development are more restricted in certain areas. These areas are referred to as article 1(4) land, article 1(5) land or article 1(6) land. These three categories of land are set out, respectively, in Parts 1, 2 and 3 of Schedule 1 to the GPDO. In article 1(4) land, the permitted development rights for the installation of satellite antennae are more restricted. Article 1(5) land is land within a national park, an area of outstanding natural beauty, a conservation area, a specified area under section 41(3) of the Wildlife and Countryside Act 1981, the Broads or a World Heritage Site. The development tolerances in Part 1 (dwellinghouse development), Part 8 (industrial development), Part 17 (statutory undertakers – electricity) and Parts 24 and 25 (telecommunications) are more limited in article 1(5) land. Article 1(6) land is land within a national park or within certain areas that fringe national parks. In relation to agricultural and forestry development in Parts 6 and 7 of Schedule 2, these may involve what is called a prior notification requirement in article 1(6) land. For example, under Part 6 of Schedule 2 to the GPDO the erection of buildings that are reasonably necessary for the purposes of agriculture is permitted, within certain size tolerances. However, in the case of article 1(6) land, where the erection of a new agricultural building is intended, the owner must first apply to the local planning authority for a determination as to whether the prior approval of the authority will be required to the siting, design and external appearance of the building. The development of the building cannot then be begun until the receipt from the local planning authority of a written notice of their determination that such prior approval is not required, the expiry of 28 days without the local planning authority making any determination, or where the local planning authority has given notice that their prior approval is required, the giving of such approval.

Some of the classes of permitted development are summarised in the following text (the position is that in England; there are some slight differences in Wales).

4.2.1. Part 1 – Development within the curtilage of a dwellinghouse

Several classes of permitted development are set out under Part 1. Class A permits the enlargement, improvement or other alteration of a dwellinghouse within certain tolerances. These are now quite complex and largely concern heights, distances from the boundary, form of side walls on the boundary and similarity with the existing building. No part of the building enlarged, improved or altered must exceed the height of the highest part of the roof of the original dwellinghouse; the same applies to eaves' height. Where the building fronts a highway, the enlargement must not extend beyond the existing walls. The total area of ground covered by buildings within the curtilage (other than the original dwellinghouse) must not exceed 50 per cent of the total area of the curtilage.

Class B permits the enlargement of a dwellinghouse by way of an addition or alteration to its roof. Again there are various tolerances and the works must not exceed the height of the highest part of the existing roof or extend beyond the plane of any existing roof slope which fronts any highway.

Class D permits the erection or construction of a porch outside any external door of a dwellinghouse provided the ground area does not exceed three square metres, the height does not exceed three metres above ground level and no part of the structure would be within two metres of the boundary of the curtilage of the dwellinghouse with the highway.

Class E permits the provision within the curtilage of a dwellinghouse of any building or enclosure, swimming or other pool required for a purpose incidental to the enjoyment of the dwellinghouse as such. There are limits on the heights of any such building, and the total area of buildings cannot exceed 50 per cent of the curtilage.

Classes F, G and H permit, respectively, a hard surface, flues and chimneys, and the installation of a satellite antenna, within certain dimensions.

The meaning of 'curtilage' is important in the application of several of the classes in Part 1. Although 'curtilage' was said to mean a small area in which a building is situated in *Dyer* v *Dorset County Council* [1989], the Court of Appeal in *Skerritts of Nottingham Ltd* v *Secretary of State for the Environment, Transport and the Regions (No 1)* [2000] said that the notion of 'smallness' was not relevant. In *McAlpine* v *Secretary of State for the Environment* [1995], it was held that the inspector was entitled to make a current and visual appraisal, in reaching his conclusion as to the extent of the 'curtilage'. Although curtilage can equate with ownership, ownership is only a factor, but is not a determinative factor of the extent of a building's curtilage; see *Lowe* v *First Secretary of State* [2003]. In its review of the meaning of 'curtilage' in the *Skerritts* case, the Court of Appeal did not refer to the test considered in the earlier decision of the House of Lords in *Sinclair Lockart's Trustees* v *Central Land Board* [1950], where it was said that the ground which is used for the comfortable enjoyment of a house or other building, and which serves the purposes of the house or the building in some necessary or reasonably useful way, is the curtilage. The true test in *Sinclair Lockart's Trustees* v *Central Land Board* is that there is no requirement that a subsidiary building like an annex must serve the house or building in some necessary or useful way, it is enough that such a building serves the purposes of the principal house or building in some necessary or reasonably useful way: see *Wheeler* v *First Secretary of State* [2003]. In *McAlpine* v *Secretary of State for the Environment* [1995], an enforcement notice was upheld that required the removal of a swimming pool constructed in the substantial grounds of a listed building. In *Peche d'or Investments* v *Secretary of State for the Environment* [1997], it was held that one cannot erect a building under Class E, as a building incidental to the enjoyment of a dwellinghouse, and then immediately use that building for a primary residential use.

4.2.2. Part 2 – Minor operations

This Part contains three classes. Class A permits the erection, construction, maintenance, improvement or alteration of a gate, fence, wall or other means of enclosure. Class B permits the formation, laying out and construction of a means of access to a highway which is not a trunk road or a classified road, where that access is required in connection with development permitted by any class in the GPDO. Class C permits the painting of the exterior of any building or work.

4.2.3. Part 3 – Changes of use

In Chapter 3, it was pointed out that changes of use within a Use Class are not development. The purpose of Part 3 is to permit, largely by reference to the Use

Classes in the Town and Country Planning (Use Classes) Order 1987 certain changes of use. The effect of this Part is illustrated by Table 4.1. In relation to Classes D, F(c) and G(c) and the references to a display window at ground level, the bay windows in a double-fronted Edwardian house were accepted as being display windows at ground level: see *North Cornwall District Council* v *Secretary of State for Transport, Local Government and the Regions* [2003].

Table 4.1 Table of permitted changes of use

Class in the GPDO	From and to uses in the following classes in Use Classes Order
Class A	From Class A3 (restaurants and cafes), Class A4 (drinking establishments) or A5 (hot food takeaways) to Class A1
Class AA	From Class A4 (drinking establishments) or Class A5 (hot food takeaways) to Class A3 (restaurants and cafes)
Class B(a)	From Class B2 (general industrial) or B8 (storage and distribution) to B1 (business)
Class B(b)	From Class B1 (business) or B2 (general industrial) to Class B8 (storage and distribution) (In the case of Classes B(a) and B(b) of the GPDO, there is a limit of 235 square metres in the case of changes of use to or from a use in Class B8)
Class C	From Class A3 (restaurants and cafes), Class A4 (drinking establishments) or A5 (hot food takeaways) to Class A2 (financial and professional services)
Class D	From Class A2 (financial and professional services) where the premises have a display window at ground level to Class A1 (shops)
Class E	From a use permitted by a planning permission granted on an application to another use which that planning permission would have specifically have authorised, provided the change of use takes place within 10 years from the grant of the permission
Class F(a)	From any purpose within Class A1 (shops) to a mixed use within Class A1 and as a single flat
Class F(b)	From a use for any purpose within Class A2 (financial and professional services) to a mixed use within Class A2 and as a single flat
Class F(c)	From Class A2 (financial and professional services) where the building has a display window at ground level to a mixed use within Class A1 and as a single flat
Class G(a)	From a mixed use for any purpose within Class A1 (shops) and as a single flat to Class A1
Class G(b)	From a mixed use for any purpose within Class A2 (financial and professional services) and as a single flat to Class A2
Class G(c)	From a mixed use for any purpose within Class A2 (financial and professional services) and as a single flat where that building has a display window at ground level to Class A2
Class H	From use as a casino to Class D (assembly and leisure)
Class I(a)	From a use within Class C4 (houses in multiple occupation) to a use in Class C3 (dwellinghouse)
Class I(b)	From a use within Class C3 to a use within Class C4

4.2.4. Part 4 – Temporary buildings and uses

There are two classes in this Part. Class A permits the provision on land of build-ings, movable structures, works, plant or machinery required temporarily in connection with the carrying out of operations for which planning permission has otherwise been granted. In other words, site buildings, cranes and other structures in connection with the erection of a building. In R (*Hall Hunter Partnership*) v *First Secretary of State* [2006], it was held that the erection of many hectares of large, linked, walk-in polytunnels for growing soft fruit on a farm was a building operation, but it was also held that although the polytunnels were erected in connection with an agricultural use, they did not have permitted development rights.

Class B permits the use of any land for any purpose for not more than 28 days in total in any calendar year, of which not more than 14 days in total may be for the holding of a market and for certain motor sports activities.

4.2.5. Part 6 – Agricultural buildings and operations

Class A permits the carrying out on agricultural land comprised in an agricultural unit of five hectares or more in area of works for the erection, extension or alteration of a building or any excavation or engineering operations, which are reasonably necessary for the purposes of agriculture within that unit. The class does not permit the erection, extension or alteration of a dwelling; the provision of a building, structure or works not designed for agricultural purposes; buildings works or structures that would exceed 465 square metres in ground area; a height exceeding three metres in respect of any building, structure or works within three kilometres of an aerodrome or 12 metres in any case; any development within 25 metres of the metalled part of a trunk or classified road; or works to a building, structure or an excavation used or to be used for the accommodation of livestock or for the storage of slurry or sewage sludge where the building, structure or excavation is, or would be, within 400 metres of the curtilage of 'a protected building'. A protected building is defined as a residential building other than such a building on any agricultural unit.

The prior notification procedure, applicable in national parks and other special areas, was described in p. 39 earlier.

Class B permits development within limitations on agricultural units of less than five hectares. Class C permits mineral working for agricultural purposes.

4.2.6. Part 7 – Forestry buildings and operations

This Part permits the carrying out on land used for the purposes of forestry, including afforestation, of development reasonably necessary for those purposes consisting of works for the erection, extension or alteration of a building, the formation, alteration or maintenance of private roads, operations to obtain materi-als for those private roads and other operations. There are certain tolerances and the prior notification procedure operates in the special areas.

4.2.7. Part 8 – Industrial and warehouse development

Class A permits the extension or alteration of an industrial building or warehouse within certain tolerances. There are certain limitations as to the use of the building as extended or altered to ensure that it remains in use as an industrial building or warehouse. The height of the building as extended or altered must not exceed the height of the original building. The cubic content of the original building must not be exceeded by 10 per cent in respect of article 1(5) land (national parks and other special areas), or 25 per cent in any other case. The floor space of the original building must not be exceeded by more than 500 square metres in respect of article 1(5) land or 1,000 square metres in any other case. The external appearance of the premises of the undertaking must not be materially affected and no part of the development must be carried out within five metres of any boundary. Certain conditions are attached to the permitted development relating to use and working hours. Classes B, C and D permit the installation of additional or replacement plant and machinery and a number of other matters of a minor nature relating to industrial processes.

4.2.8. Other classes of permitted development

Except for the two Parts explained in the following text, the remaining Parts of Schedule 2 permit development in respect of local authority, statutory and other users of a specialised nature.

Part 24 permits development by electronic code operators (telecommunications). It is under this class that radio masts and other similar installations may be erected by telecommunications operators. There are various height limits by reference to different types of structures and circumstances. A 42-day prior approval procedure was introduced in 1999 (28 days in the case of development within a site of special scientific interest (SSSI)).

Part 31 permits a building operation consisting of the demolition of a building. Demolition of buildings was considered in Chapter 2. Parts 34–38 concern a number of classes of development by the Crown. The Planning and Compulsory Purchase Act 2004 amended the 1990 Act so that that Act applies to the Crown; the Crown requires planning permission for operational and change of use development.

4.3. Article 4 directions

Under article 4 of the GPDO, either the local planning authority or the Secretary of State may give a direction that specified development described in any Part, class or paragraph in Schedule 2 shall not be carried out without the express grant of planning permission. The effect of such a direction is that the automatic planning permission is not available for the classes of development so specified.

Under article 4(2), an article 4 direction may be made in respect of the whole or any part of a conservation area in which case certain development, principally that relating to development within the curtilage of a dwellinghouse, will not enjoy permitted development rights.

The approval of the Secretary of State to the making of an article 4 direction is not required in respect of:

(a) Directions made under article 4(2) relating to certain categories of permitted development rights in a conservation area mainly affecting dwellinghouses.
(b) Directions relating to listed buildings or buildings notified by the Secretary of State as of architectural or historic interest or development within the curtilage of a listed building: see article 5(3).
(c) Directions relating to any development in Parts 1–4 or Part 31 (Development within the curtilage of a dwellinghouse, sundry minor operations, changes of use, temporary buildings and uses and demolition of buildings).

In the case of a direction within (c), it will only remain in force for six months unless approved by the Secretary of State; it will cease to have effect if disallowed by him within that period.

Where a direction is made, notice is to be served on all owners and occupiers of the affected land.

Where an article 4 direction has been made withdrawing permitted development rights, any person with an interest in the land may seek compensation for abortive expenditure or for other loss or damage under sections 107–108 of the 1990 Act. A precondition of the right to claim compensation is that an application has been made for planning permission, and that application has been refused, or only granted subject to conditions other than those previously imposed by the GPDO. The question of compensation is considered further in Chapter 19.

4.4. Special development orders

Under section 59(3)(b) of the 1990 Act, the Secretary of State may make a special development order applicable only to such land or descriptions of land as may be specified in the order. A special development order has a similar effect to the GPDO save that it applies only to a defined area and that the classes of permitted development are such as the Secretary of State decides to include in the order. Special development orders are also made to modify the GPDO in special areas.

Thus, special development orders were made in respect of many of the urban development areas where there were urban development corporations many of which have been wound up.

4.5. Local development orders

Sections 40–41 of the Planning and Compulsory Purchase Act 2004, which inserts additional provisions into the 1990 Act, make provision for local development orders. These orders are made by local planning authorities in relation to their respective areas. As with the GPDO, and special development orders, they permit such development as is specified in the order for the respective area. The scope of LDOs is likely to reflect local circumstances and the need to implement a policy contained in a development plan document, such as a regeneration policy.

4.6. Simplified planning zones

Sections 82–87 of the Town and Country Planning Act 1990, as amended by Schedule 5 to the Planning and Compensation Act 1991, make provision for simplified planning zones. Where a simplified planning zone (hereafter SPZ) scheme is adopted or approved, the effect is that planning permission is deemed granted for development specified in the scheme or for any development of any class that may be specified. Planning permission may be unconditional or subject to such conditions, limitations or exceptions as may be specified in the scheme.

Under section 83, every local planning authority must consider for which part or parts of their area a SPZ would be desirable. This must be kept under review. A planning authority must prepare a scheme for any part of their area they think desirable for such a purpose. Nothing in a SPZ scheme may affect the right of a person to carry out something that is not development, or is development that either has planning permission, or does not need it. Development permitted by the SPZ scheme will prevail over any limitations or restrictions imposed under a planning permission outside of the scheme: see section 84(3).

Under section 85 of the 1990 Act, a SPZ will take effect on the date of its adoption or approval and will cease to have effect at the end of the period of 10 years beginning with that date. When the scheme ceases to have effect planning permission under it will also cease to have effect except in a case where the development authorised by it has begun.

Land of the following descriptions may not be included in a SPZ: national parks, conservation areas, land within the Broads, areas of outstanding natural beauty, land identified as a green belt, and areas of special scientific interest (SSSI). Where a SPZ is made, it cannot have effect to grant planning permission for development that is likely to have a significant effect on a site protected by the European directives (Habitats and Birds): see Conservation (Natural Habitats, etc.) Regulations 2010.

Schedule 7 to the 1990 Act, as amended in 1991, sets out the details of a SPZ, and how it is to be adopted or approved. A SPZ is to consist of a map and written statement, and such diagrams, illustrations and descriptions as thought appropriate. It must specify:

(a) The development or classes of development permitted by the scheme.
(b) The land in relation to which permission is granted.
(c) Any conditions, limitations or exceptions subject to which it is granted.

Any person may request a local planning authority to make a SPZ, and if the authority fails to do so, or to decide to do so within three months, the person concerned may refer the question to the Secretary of State.

Where a local planning authority is considering a proposal to make or alter a SPZ scheme, publicity must be given and representations considered.

4.7. Neighbourhood development orders

These orders were introduced by the Localism Act 2011. A neighbourhood development order is an order which grants planning permission in relation to a particular neighbourhood area specified in the order. The process for making neighbourhood development orders is somewhat bound up with neighbourhood planning and neighbourhood forums. These matters are dealt with in detail in Chapter 6.

However, under section 61J of the 1990 Act, a neighbourhood development order may not provide for the granting of planning permission for what is called 'excluded development', or for any development that planning permission is already granted for. Under section 61K, the following development is excluded development:

(a) Development consisting of a county matter.
(b) Development consisting of the carrying out of operations relating to waste development.
(c) Development falling within council directive 85/337/EEC (environmental assessment) concerned with the assessment of the effects of certain public and private projects in the environment.
(d) Development consisting of a nationally significant infrastructure project within the meaning of the Planning Act 2008.
(e) Prescribed development or development of prescribed descriptions as may be set out regulations.
(f) Development in a prescribed area or an area of prescribed description.

Under section 61L of the 1990 Act, planning permission granted by a neighbourhood development order may be granted unconditionally, or subject to such conditions or limitations as may be specified in the order. The conditions that may be specified can include obtaining the approval of the local planning authority and specifying the period within which applications must be made to the local planning authority for the approval of the authority of any matter specified in the order. This can include a condition that the development begins before the end of a specified period.

The Secretary of State may, and the local planning authority may with the consent of the Secretary of State, revoke a neighbourhood development order under section 61M of the 1990 Act.

Part II

The decision-making process

5 Planning authorities

5.1. Introduction

By now, the reader will have understood the scope of development control, and will know whether a proposal constitutes 'development' needing planning permission. This chapter explains the planning authorities, their power and functions. Although planning and development control are essentially matters for local government, the Secretary of State for Communities and Local Government possesses very wide powers. Until 1994, it was a feature of the administration of planning that there was a division of responsibility between upper and lower tiers of local government. The Local Government Act 1992 empowers the creation of a number of unitary authorities with sole responsibility for planning in their respective areas. There are also certain authorities, without planning functions, that have a right to be consulted, and, in some cases may sometimes issue directions that may affect the processing of development plans or planning applications; examples include parish and community councils and highway authorities. Finally, there are also Regional Planning Boards with responsibilities for the preparation of regional spatial strategies.

Unhappy with the discharge of planning functions by some local planning authorities, the government in recent years has increasingly altered the balance of power away from local government. The powers of the Secretary of State to decide planning appeals, to widen the scope of the General Permitted Development Order, and to initiate policy circulars and new legislation, has given central government a considerable say in planning matters.

5.2. The distribution of functions between local planning authorities

5.2.1. Outside London

Outside London, there are metropolitan district councils in the former metropolitan counties of West Midlands, Greater Manchester, Merseyside, Tyne and Wear,

South Yorkshire and West Yorkshire. In the non-metropolitan areas, there is either a two-tier structure of county and district councils or a unitary council. A number of unitary authorities have been set up under the powers contained in the Local Government Act 1992. A unitary authority is, and exercises the planning functions of, a local planning authority.

5.2.1.1. Parish councils

All areas that had parish councils before 1 April 1974 continue to have parish councils, but other areas may now also have parish councils. Sometimes a parish council is called a town council. In Wales, a parish council is called a community council. Parish councils are not local planning authorities, but are entitled to be consulted on certain planning and development control matters.

In Wales, there is a system of unitary authorities.

5.2.1.2. Allocation of functions outside London

County councils were responsible for the preparation of structure plans and certain other matters. However, since the Planning and Compulsory Purchase Act 2004, structure plans are being phased out; they were being replaced by regional special strategies prepared by Regional Planning Boards, and local development schemes prepared by district authorities see Chapter 6. County councils remain the minerals and waste planning authorities.

Non-metropolitan district councils are responsible for the preparation of local development schemes and development control, including controls relating to listed buildings and advertisements, and other functions allocated under Schedule 1.

A unitary council is the local planning authority for all purposes. In Wales, each unitary council has all the functions of a local planning authority.

Metropolitan district councils are local planning authorities for all planning functions and are also minerals planning authorities.

In national parks, the planning and development control functions are exercised by the National Park Authorities. Where a district council exists in a National Park Authority area, the district council has concurrent powers in regard to limited functions such as tree preservation orders and protection of trees in conservation areas: see section 4A of the 1990 Act. In the Broads, there is a Broads Authority exercising most of the planning functions of a local planning authority.

5.2.2. Greater London

The arrangement consists of a number of borough councils and the Corporation of the City of London. Each of these authorities is both a local planning authority for all planning functions and a mineral planning authority for its area. Even though the Greater London Authority is not the local planning authority, the Mayor of London has certain powers in relation to planning. These include spatial development strategy, and acting as a statutory consultee on certain significant planning applications. Additionally, section 3 of the 1990 Act provides for a joint planning committee for Greater London to:

(a) Consider and advise the London borough councils on matters of common interest relating to the planning and development of Greater London.
(b) Inform the Secretary of State of the views of those authorities, and other matters he has requested their advice on.
(c) Inform local planning authorities and other bodies in the vicinity of London of the views of the London borough councils on planning matters.

The functions of the now long abolished Greater London Council in respect of listed buildings, conservation areas and ancient monuments have been largely transferred to the Historical Buildings and Monuments Commission (Local Government Act 1985, section 6) usually known as English Heritage.

5.2.3. Urban development corporations

Urban development corporations were set up in a number of areas and possessed certain planning and development functions. Some of the original corporations have fulfilled their tasks, such as the London Docklands Urban Development Corporation, others continue, such as those in Thames Gateway and in West Northamptonshire and elsewhere. The Localism Act 2011 made provision for Mayoral development corporations for development areas designated by the Mayor for London.

5.2.4. Legal challenges to allocation of powers

Section 286(1) of the Town and Country Planning Act 1990 has application. It provides that a challenge cannot be made to the validity of any permission, determination or certificate granted, made or issued or purporting to have been granted, made or issued by a local planning authority, on the ground that any such decision ought to have been given by some other local planning authority. Thus, challenges are prevented on the grounds of mistaken allocation of jurisdiction between authorities. It does not prevent challenges where neither authority has jurisdiction: see *A-G ex rel. Co-operative Retail Services Ltd* v *Taff-Ely BC* [1981].

5.3. The role of the Secretary of State

The responsibilities and functions of the Secretary of State are so numerous that they cannot all be outlined in this chapter. His powers to influence planning policies and planning decisions should not be underestimated. He initiates planning legislation and, increasingly reserves to himself, powers to make directions, orders or regulations. General permitted development orders and special development orders are examples of orders made by the Secretary of State that grant planning permission independently of local planning authorities for classes of development chosen by him. The Secretary of State has wide powers in relation to planning documents. The minister prepared the National Planning Policy Framework (March 2012), which is the principal national policy guidance to local planning authorities, planning inspectors and other decision-makers. The Secretary of State can intervene in the preparation of local development documents (see section 27 of the Planning and Compulsory Purchase Act 2004); he has the power to call in any planning application; he (or inspectors appointed by him) decides appeals against certain decisions made by local planning authorities; and he may issue guidance to local authorities in what are called circulars. A circular is not an item of legislation, although it may contain a direction having legal effect; its purpose is to explain government policy or legislation. A circular can be a 'material consideration' which a local planning authority is required to have regard to in the exercise of many of its powers (see Chapter 8). His department also issues guidance on general and specific planning topics, although a raft of guides and policy statements were replaced by the National Planning Policy Framework published in March 2012.

The power to call in an application is conferred on the Secretary of State to be used in circumstances where the local authority is minded to grant planning permission but the Secretary of State is of the view that the application requires closer examination prior to permission being granted. Where the Secretary of State exercises his power to call in an application, the local planning authority is deprived of their jurisdiction to determine the application themselves. In *R* v *Secretary of State for the Environment, Transport and the Regions, ex p Holding and Barnes plc* [2001], it was held that the Secretary of State was not an independent or impartial tribunal for the purposes of Article 6 of the Convention on Human Rights and Fundamental Freedoms, but there were sufficient procedural safeguards governing his decision-making and his planning functions fell within supervisory jurisdiction of the High Court, and this meant that the requirements of Article 6 were met.

Accordingly, there must be planning issues of more than local importance involved before a direction will be made to exercise the call-in power.

The Courts are unwilling to uphold applications for judicial review of the Secretary of State's exercise of or his refusal to use his call-in power save in circumstances where his decision is 'wildly perverse': see *R* v *Secretary of State for the Environment, ex p Newprop* [1983].

The sum total of these responsibilities and functions is to place in the hands of the Secretary of State enormous powers. Whether Parliament has been wise in so doing is certainly questionable.

5.4. Delegation to committees and officers

There is a wide power in section 101 of the Local Government Act 1972 for a local authority to arrange for the discharge of any of their functions to a committee, a subcommittee or an officer of the authority, or indeed, within limits, to another authority. In practice, many planning functions are delegated to the planning committee of an authority. Where delegation has properly taken place, the council is then bound by the decisions of its committee or officer acting within the delegated powers. Most local authorities now have an executive cabinet that makes a wide range of decisions.

Circumstances may sometimes arise where an officer, without any express authority, but acting within the scope of his ostensible authority, makes a representation or statement upon which another person acts. This occurred in *Lever Finance Ltd* v *Westminster City Council* [1970] where a planning officer told a developer that he did not need planning permission for a variation to an already approved scheme. The Court of Appeal decided that, as the developer had acted upon the officer's representation by completing a house, the council was bound by that officer's representation, and 'estopped' from taking enforcement proceedings against the developer. The word 'estopped' comes from the doctrine of estoppel. This can be applied in a situation where party A believes in what is represented by party B, and then acts to his detriment; estoppel then prevents party B from denying the truth or effect of his representation. In *Camden London Borough Council* v *Secretary of State for the Environment* [1993], an enforcement notice was quashed on appeal on the ground of estoppel, because the planning officer had approved a number of minor variations which, when completed, were made the subject of an enforcement notice.

The application of the doctrine of estoppel in planning cases was considered to be limited in *Western Fish Products Ltd* v *Penwith District Council* [1981], where a differently constituted Court of Appeal considered that before estoppel arose, the mere fact that the officer making the representation was a planning officer was not sufficient to assume he was acting within the scope of his ostensible authority; there must be some evidence that the authority had been delegated to the officer; in any event estoppel cannot prevent a council from exercising its statutory discretion or performing its statutory duty. In the case, it was held that the council was entitled to refuse planning permission for development which the planning officer had represented was within an existing use right if the extent of that use was established.

However, the doctrine of estoppel must now be regarded as having limited or no application to planning law. Lord Hoffmann in *Regina (Reprotech (Pebsham) Ltd)* v *East Sussex County Council* [2002], said it was 'unhelpful to introduce private law concepts of estoppel into planning law'. The vaguely similar concept of 'legitimate expectation' may sometimes, but rarely, have application: see Chapter 20.

In *Wells* v *Minister of Housing and Local Government* [1967], it was said that a letter from a planning officer stating that planning permission was not required for a certain activity could be regarded as a determination of the position under section 64 (now section 192) of the Town and Country Planning Act 1990 (see Chapter 15).

5.5. Negligence

The principle in *Hedley Byrne & Co Ltd* v *Heller & Partners* [1964], that there can be liability for negligent misstatements, would seem applicable to planning authorities and their officers. Local authorities owe a duty of care in statements and representations they may make in connection with planning matters. *Caparo Industries plc* v *Dickman* [1990] and *Murphy* v *Brentwood* [1990] have severely limited this principle in relation to the potential liability of local authorities. There was liability in *Coats Patons (Retail) Ltd* v *Birmingham Corporation* [1971] where a local authority clerk gave an incorrect answer to a question put to him about his council's proposals to build a subway. More recently, in *Lambert* v *West Devon Borough Council* [1997], the local authority was held liable for negligence when one of their senior officers wrongly indicated to a developer that he could go ahead with a development. It was held that a duty of care was owed by the council to the plaintiff, because the officer knew or ought to have known that the plaintiff was relying on him. In *Lam* v *Brennan* [1997], restaurateurs contended that the planning authority had negligently granted planning consent to the adjoining occupier of premises and were liable to them for nuisances and personal injuries suffered by them and the loss of business resulting from fumes emanating from the premises. The Court of Appeal held that no statutory or common law duty of care arose in relation to the process of deciding whether planning permission should or should not be granted. In the Australian case of *Shaddock & Associated Pty Ltd* v *Parramatta City Council* [1981], the *Hedley Byrne* principle was said to apply to public authorities that follow a practice of supplying information, even where the official supplying the information did not know of the precise use to which the information would be put.

In *Kane* v *New Forest District Council* [2002], it was held that the council were liable for negligence, as they had created the danger. The council had required the provision of a footpath, in connection with the approval of a planning application, which was the cause of a danger; the appellant was injured in consequence. The Court of Appeal concluded that a local planning authority did not have blanket immunity in exercising a statutory function under planning legislation. Such immunity was not to be consonant with the recent developments of the law both in England and in Strasbourg.

In *Co-operative Retail Services Ltd* v *Taff-Ely Borough Council* [1983], where the town clerk issued a planning consent neither he nor his authority had power to issue, the judge said:

> ... *a careful and prudent chief executive in such circumstances would have realised that to issue a notice of consent which might be invalid could cause loss or damage even to the person to whom it is issued ... it would clearly be foreseeable that some damage might result even to the applicant himself if a notice of planning consent was issued which was of doubtful validity. So viewed, I have no doubt that the action of [the clerk] would be regarded as negligent.*

The judge continued by saying that had an invalid planning consent come into the hands of a *bona fide* purchaser for value of the affected property, without notice of the circumstances, the local authority would have been liable if the purchaser relied upon it and suffered damage. In fact, the claim failed as the party seeking damages knew, or was deemed to know that the planning consent was invalid.

5.6. Nuisance

In *Allen* v *Gulf Oil Refining Ltd* [1981], it was held that there was no liability in nuisance in respect of fumes and discharges from an oil refinery, because the construction and use of the oil refinery was authorised by an Act of Parliament. Nevertheless, in *Wheeler* v *JJ Saunders* [1995], a pig farmer could not rely on the grant of planning permission as a defence to an action in nuisance. The Court of Appeal in *Barr* v *Biffa Waste Services Ltd* [2012] said that there was no basis for using a statutory scheme (waste licensing) to cut down private law rights, such as nuisance, protective of neighbours. In addition, in *Hunter* v *Canary Wharf Ltd* [1997], the Court of Appeal concluded that the implementation of a planning permission could not, as such, confer immunity to a nuisance action.

6 Development plans, documents and neighbourhood planning

6.1. Introduction

Section 38(2) of the Planning and Compulsory Purchase Act 2004 defines a development plan as:

- Outside Greater London, the regional strategy for the region, if there is one for that region (the Localism Act 2011, Schedule 8, para 13, confusingly amends this point and requires it to be omitted).
- In London, the spatial development strategy.
- The development plan documents (DPDs).
- The neighbourhood development plans for the area.

Land use planning involves two tasks: planning and development control. The developer is mainly concerned with development control, for that affects his decision-making in the most direct way. The landowner will be concerned with both; planning may affect his land, beneficially or injuriously, and development control may be the final confirmation of those effects. In exercising development control, the local planning authority is required to have regard to the contents of the development plan (section 70 of the 1990 Act), the development plan being the document which describes the culmination of the planning task. Further, section 38(6) of the Planning and Compulsory Purchase Act 2004 provides that in making any determination under the planning Acts, the determination shall be made in accordance with the plans unless material considerations indicate otherwise.

The Localism Act 2011 introduced many new features. First, there is now a duty to co-operate in relation to planning of sustainable development. This duty falls on every local planning authority, county council or other prescribed body, to engage constructively, actively and on an ongoing basis in relation to a range of activities including the preparation of development plan and other local development documents (LDDs): see s33A of the Planning and Compulsory Purchase Act 2004. Second, there are new provisions about neighbourhood planning, which include neighbourhood development orders and neighbourhood development plans, in Schedule 9 to the 2011 Act. Third, changes are made to the grant of development consent orders for Nationally Significant Infrastructure Projects.

Therefore, planning and development control are interrelated; simply put, planning eventually allocates land uses, and development control is the detailed day-to-day implementation of the planning task. Although this book concentrates on development control, some attention must be given to the planning process and the legal consequence of development plans, development plan documents, neighbourhood planning and neighbourhood planning documents, as they impinge on development activities.

Planning throws up a number of contradictions which the reader may care to think about. Although the right to develop was nationalised in 1948, the initiative to develop was left with the private landowner and developer to a very large extent. The inclinations and decisions of these parties are clearly crucial to development. However, it is a very different group of people who decide, through the planning process, where and what sort of development should take place. This group – the planning authorities, the Secretary of State, their technical advisers, authorised neighbourhood forum and the members of the public who are interested – make technical and political decisions about land use, and decisions about land not in their ownership. The landowner will usually make economic decisions with regard to his land, and it is therefore not surprising that these economic decisions are so often in conflict with the technical and political decisions of others. This conflict is reduced if planning and landownership are combined in one party, either in the state, or in the individual in co-operation with the state (see D. Denman's *Co-operative Planning*, published by Geographia, 1974). The former is not universally favoured; the latter, although accepted in some countries, has never been tried in the United Kingdom.

Until the enactment of the Planning and Compulsory Purchase Act 2004, the development plan generally consisted of either two plans, a structure plan and a local plan, and then later on increasingly only a unitary development plan. However, the 2004 Act made radical changes to the concept of development plans. A development plan now consists of a series of documents.

Something must first be said about the pre-2004 Act development plans. However, the Localism Act 2011 has introduced a significant role for authorised neighbourhood forums where neighbourhood development orders are made. Accordingly, this chapter considers neighbourhood planning after an explanation of the 2004 Act development plans.

Finally, the publication of the National Planning Policy Framework in March 2012, with its principal objective of achieving sustainable development, will have a profound influence on planning decisions.

6.2. Pre-2004 Act development plans

6.2.1. *Development plans*

Under the law prior to the 2004 Act, development plans comprised unitary plans or structure plans and local plans. Substantial changes were made to the system of development plans and their preparation by amendments made to the 1990 Act by the Planning and Compensation Act 1991.

6.2.2. Structure plans

Structure plans were prepared by county councils. A structure plan is a written statement formulating the authority's general policies in respect of the development and use of the land in their area. It must contain policies in respect of the conservation of the natural beauty and amenity of the land, the improvement of the physical environment and management of traffic. The structure plan therefore sets out key, strategic polices as a framework for local planning by the district councils. The strategic nature of the structure plan is exercised by the limitation that the plan must not be map-based (see section 31 of the 1990 Act, the Town and Country Planning (Development Plan) Regulations 1991 and PPG 12 – *Development Plans*).

The explanatory memorandum to a structure plan must set out the regard the authority had to the following:

- Regional or strategic guidance given by the Secretary of State.
- Current national policies.
- The resources likely to be available.
- Other matters the Secretary of State directs shall be taken into account.
- Social and economic considerations.
- Environmental considerations.
- Policies and proposals of any urban development corporation.

6.2.3. Local plans

These plans were district-wide. Local plans must contain a written statement of the authority's detailed policies for the development and use of land in their area. A local plan must include policies in respect of the conservation of the natural beauty and amenity of the land, the improvement of the physical environment and management of the traffic (see section 36(3) of the 1990 Act).

A local plan must not contain any policies in respect of minerals (unless it is a plan for a national park) or in respect of disposal of refuse or waste. A local plan must contain a map and other diagrams. It may designate any area as an action area for comprehensive treatment by development, redevelopment or improvement (section 36(7)). A local plan has to be in general conformity with the structure plan.

6.2.4. Unitary development plans

Unitary development plans (UDPs) comprised two parts: Part I of a UDP consists of a written statement formulating the authority's general policies in respect of the development and use of the land in their area. It will include policies in respect of the conservation of the natural beauty and amenity of the land, the improvement of the physical environment and management of the traffic (see section 12 of the 1990 Act).

Part II of a UDP consists of a written statement formulating in such detail as the authority thinks appropriate their proposals for the development and use of land in the authority's area based on a map, together with a reasoned justification of the general policies in both Parts I and II (section 12(4) of the 1990 Act).

In formulating the policies in Part I, the authority is required to have regard to:

- Regional or strategic guidance given by the Secretary of State.
- Current national policies.
- The resources likely to be available.
- Other matters the Secretary of State directs shall be taken into account.
- Social and economic considerations.
- Environmental considerations.
- Policies and proposals of any urban development corporation.
- Any waste disposal plans.

UDPs applied to the metropolitan areas, and to those unitary and national park authorities designated by the Secretary of State.

6.2.5. Joint unitary development plans

Section 23 of the 1990 Act authorises the making of joint UDPs by two or more local planning authorities in Greater London or in the metropolitan areas.

Following the reforms of local government in Wales under the Local Government (Wales) Act 1994, which created a system of unitary authorities, section 23A of the 1990 Act enables joint UDPs to be made by two or more local planning authorities.

6.2.6. Unitary development plans for national parks

We have seen in the preceding text that the Secretary of State could have designated a national park authority for the purpose of preparing a UDP. In Wales, a UDP has to be prepared for each national park (see section 23B).

6.2.7. Minerals local plans

These plans were prepared by those authorities designated as minerals planning authorities. They set out policies for the supply of minerals and for the required degree of environmental protection associated with the mineral development. They indicate those areas where provision is made for mineral working and the disposal of mineral waste.

6.2.8. Waste local plans

Outside the metropolitan areas, separate waste local plans must be prepared by the county planning or national park authorities: see section 38. The authorities may choose to prepare a separate waste local plan or combine it with their minerals local plans. The waste local plans set out detailed land use policies for the treatment and disposal of waste. They must be complementary with the waste disposal plans required to be prepared under the Environmental Protection Act 1990: PPG 23 – *Planning and Pollution Control*.

6.3. Post-2004 Act development plans

The Planning and Compulsory Purchase Act 2004 introduced substantial changes to the planning process. Further substantial changes were made by the Localism Act 2011 and its arrangements for neighbourhood planning. The expression 'the development plan', in any enactment, now has a much extended meaning: see section 38 of the 2004 Act. For the purposes of any area in Greater London, the development plan is the development plan documents (taken as a whole) which have been adopted or approved in relation to that area, including neighbourhood development plans.

For the purpose of any other area in England, the development plan is the development plan documents (taken as a whole) which have been adopted or approved in relation to the relevant area including neighbourhood development plans.

For the purposes of any area in Wales, the development plan is the local development plan adopted or approved in relation to that area including neighbourhood development plans.

These documents are fully explained and discussed in the following text. The introduction of a number of documents, which as a whole constitute a development plan, will inevitably give rise to inconsistencies where documents are made at different times. Section 38(5) of the 2004 Act provides that where there is a conflict, it must be resolved in favour of the policy which is contained in the last document, to be adopted, approved or published.

It is beyond the scope of this work to identify all the various matters and considerations that should be addressed by the development documents. However, to give one example, directive 79/409/EEC (Birds Directive) on the conservation of wild birds requires member states to establish special protection areas. Under directive 92/43/EEC (Habitats Directive) on the conservation of natural habitat areas, provision is made for the establishment of special areas of conservation. The requirements of these Directives have been transposed into British law by the Conservation of Habitats and Species Regulations 2010. However, the Regulations set out matters that should be considered in the preparation of developments plans.

6.3.1. Regional spatial strategy

Under the 2004 Act, each region would have had a regional spatial strategy (RSS).The purpose of a RSS was to provide a spatial framework to inform the preparation of local development documents, local transport plans (LTPs) and regional and sub-regional strategies and programmes that have a bearing on land use activities. Regional spatial strategies are no longer part of the development plan for most purposes in consequence of the limitations imposed by the Localism Act 2011.

6.4. Local development schemes and documents

The 2004 Act abolishes structure plans and makes provision for surveys, the preparation of local development schemes, and local development documents. These schemes and documents are referred to by the generic title of local development plans in the National Planning Policy Framework (March 2012).

6.4.1. Surveys

Under section 13 of the 2004 Act, a local planning authority must keep under review a number of matters expected to affect the development of the authority's area or the planning of its development. These include:

- The principal physical, economic, social and environmental characteristics of the area of the authority.
- The principal purposes for which land is used in the area.
- The size, composition and distribution of the population of the area.
- The communications, transport system and traffic of the area.
- Any other considerations which may be expected to affect those matters.
- Such other matters as may be prescribed or as the Secretary of State (in a particular case) may direct.

Because of the original shift towards regional strategic planning, structure plans are abolished, and this removes from county councils a significant role that they have so far had in structural planning. Section 14 of the 2004 Act requires the county councils to be the principal authority for the purposes of conducting surveys under section 13, subject to directions that may be given by the Secretary of State. In effect, there may be a certain amount of flexibility between the two tiers of local planning authorities where there is no unitary authority.

6.4.2. Local development scheme

A local development scheme sets out a programme for, and the documents that will constitute, the local development documents. In effect, the scheme is a programme and timetable for the production and revision of the documents. The schemes are prepared by the local planning authorities (not the county councils, other than unitary authorities): see section 15 of the 2004 Act.

County councils retain responsibility for minerals and waste development. Under section 16 of the 2004 Act, a county council must prepare a minerals and waste development scheme (save where there is a unitary authority).

6.4.3. *Local development documents*

Section 17 of the 2004 Act makes provision for local development documents (LDDs).

LDDs are documents of such descriptions as may be prescribed in the local development scheme and also in the local planning authority's statement of community involvement. The local planning authority may also specify other documents, as they think appropriate. The purpose of local development documents is to set out the authorities' policies (however expressed) relating to the development and use of land in their area. If any policy set out in a local development document conflicts with any other statement or information in the document, the conflict must be resolved in favour of the policy: see section 17(5). The local planning authority must keep under review their local development documents having regard to the results of any surveys that they are required to carry out. A document is only a local development document if it is adopted by resolution of the local planning authority as a local development document, and is approved by the Secretary of State: see section 17(8). Regulation 6 of the Town and Country Planning (Local Development) (England) Regulations 2004 provides descriptions of documents that must be specified as LDDs in a local development scheme. Those referred to as *core strategies* are:

- Documents relating to development and use of land which the local planning authority wishes to encourage during any specified period.
- Documents containing statements of objectives relating to design and access that are to be encouraged during any specified period.
- Statements concerning any environmental, social and economic objectives which are relevant to the attainment of development and use of land.
- The authority's general policies in respect of the matters referred to in the preceding text.

The documents referred to as a *submission proposals map* are:

- Other documents that deal with policies relating to identified sites or areas and refer to an Ordnance Survey or other map.

In connection with specific sites or areas, the following documents will also be LDDs, and are referred to as *area action plans*:

- Policies relevant to any areas of a local planning authority identified as areas of significant change or special conservation, and
- Documents relating to site allocation policies.

Under regulation 7 of the 2004 Regulations, the documents that must be development plan documents (DPDs) are core strategies, area action plans and any other document that includes a site allocation policy.

In the preparation of the development plan, and therefore any of the preceding documents, planning authorities must take into account the terms of the National Planning Policy Framework (NPPF) (March 2012) issue by the Secretary of State for Communities and Local Government: see sections 19(2)(a) and 38(6) of the Planning and Compulsory Purchase Act 2004. The NPPF says that the purpose of the planning system is to contribute to the achievement of sustainable development, and that this has three dimensions – an economic, social and environmental role.

6.4.4. Independent examination of development plan documents

The local planning authority must submit every development plan document to the Secretary of State for independent examination: see section 20(1) of the 2004 Act. The examination is carried out by an examiner appointed by the Secretary of State: see section 20(4). Representations can be made, and the examiner must make recommendations, and give his reasons: see section 20(7). There is a right to be heard at the examination, although that must be considered in the context of an examination at which cross-examination might not be allowed, and topics for consideration might be kept to those of general and particular importance.

6.5. Neighbourhood planning

Section 116 of, and Schedule 9 to, the Localism Act 2011 make provision for neighbourhood development orders and neighbourhood development plans. Schedule 9 to the 2011 Act contains a number of provisions concerning neighbourhood development orders that are inserted as sections 61E to 61Q of the Town and Country Planning Act 1990. The same Schedule makes amendments to the Planning and Compulsory Purchase Act 2004 to provide for neighbourhood development plans.

A neighbourhood development plan is a plan which sets out policies (however expressed) in relation to the development and use of land in the whole or any part of the particular neighbourhood area specified in the plan: see the new section 38A(2) of the 2004 Act. A neighbourhood development order is an order which grants planning permission in relation to a particular neighbourhood area specified in the order for development of any class specified in the order: see section 61E of the 1990 Act. The process for making neighbourhood development orders is contained in a new Schedule 4B to the 1990 Act.

Because the preparation of neighbourhood development orders and neighbourhood development plans are somewhat interrelated, both are considered in this Chapter.

6.5.1. Neighbourhood development plans

Under section 38A of the Planning and Compulsory Act 2004, any qualifying body is entitled to initiate a process for the purpose of requiring a local planning authority in England to make a neighbourhood development plan. A qualifying body means a parish council, or an organisation or body designated as a neighbourhood forum (see the following text). A neighbourhood plan is a plan which sets out policies (however expressed) in relation to the development and use of land and the whole or any part of a particular neighbourhood area specified in the plan. Schedule 4B to the 1990 Act makes provision for the independent examination of orders proposed by qualifying bodies. Under section 38A(4), a local planning authority to whom a proposal for the making of a neighbourhood development plan has been made, must make a neighbourhood development plan to which the proposal related if, in each of the referendums required under Schedule 4B, more than half of those voting have voted in favour of the plan, and must make the plan as soon as reasonably practicable after the referendum is held. In the case of a neighbourhood area designated as a business area, the obligation on the authority is discretionary. The authority is not obliged to make a neighbourhood development plan if they consider that it would breach or would be incompatible with any EU obligation or any convention rights under the Human Rights Act 1998. Regulations will provide for the holding of an examination and for the procedure for the making and consideration of a neighbourhood plan.

Under section 38B of the 2004 Act, a neighbourhood development plan must specify the period for which it is to have effect, may not include provision about development that is excluded development, and may not relate to more than one neighbourhood area. Only one neighbourhood development plan may be made for each neighbourhood area.

The terms of the National Planning Policy Framework (NPPF) (March 2012) issue by the Secretary of State for Communities and Local Government, must be taken into account in the preparation of neighbourhood development plans.

6.5.2. *Designation of neighbourhood forums*

Neighbourhood development plans are dependent on a local forum called a 'qualifying body' and designated neighbourhood forums. A qualifying body is entitled to submit a proposal to a local planning authority for the making of a neighbourhood development order by the authority in relation to a neighbourhood area. A qualifying body means a parish council, or an organisation or body designated as a neighbourhood forum, authorised for the purposes of a neighbourhood development order in relation to the relevant neighbourhood area. In the case of an area that does not consist of or include the whole of any part of the area of a parish council, a local planning authority may designate an organisation or body as a neighbourhood forum if it meets certain conditions: see section 61F(5). First, the body is established for the express purpose of promoting or improving the social, economic and environmental well-being of an area consisting of including the neighbourhood area concerned. Second, its membership is open to individuals who work there, and individuals who are elected members of the county council, district council or London borough council whose area falls within the neighbourhood area concerned. Third, its membership includes a minimum of 21 individuals each of whom live in the neighbourhood area concerned, work there, or is an elected member of the local authorities mentioned in the preceding text. Fourth, it has a written constitution, fifth, subject to any other conditions that may be prescribed.

The local planning authority may also designate an organisation or body as a neighbourhood forum if they are satisfied that the organisation or body meets certain prescribed conditions. Under section 61F(7), a local planning authority must, in deciding whether to designate an organisation or body, have regard to the desirability of designating an organisation or body which has secured that its membership includes at least one individual within the categories mentioned in the preceding text, whose membership is drawn from different places in the neighbourhood area concerned and from different sections of the community in that area, and whose purpose reflects (in general terms) the character of that area. The local planning authority may designate only one organisation or body as a neighbourhood forum for each neighbourhood area, and only if the organisation or body has made an application to be designated. It must give reasons to an organisation or body applying to be designated as a neighbourhood forum where the local planning authority refuses the application. A designation ceases to have effect at the end of a period of five years: see section 61F(8). And a local planning authority may withdraw an organisation or body's designation as a neighbourhood forum if they consider that that body is no longer meeting the conditions by which it was designated or any other criteria that the authority was required to have regard to when making the designation. Each local planning authority must make such arrangements, as they consider appropriate for making people aware as to the times when organisations or bodies would make applications to be designated as neighbourhood forums for neighbourhood areas.

Under section 61H of the 1990 Act, whenever a local planning authority exercises their powers to designate an area as a neighbourhood area, they must consider whether they should designate the area concerned as a business area. The power to designate a neighbourhood area as a business area is exercisable by the authority only if, having regard to matters that may be prescribed, they consider that the area is wholly or predominately business in nature.

6.5.3. *Process for making neighbourhood development orders*

As mentioned in the preceding text, a neighbourhood development order is an order which grants planning permission in relation to a particular neighbourhood area specified in the order. The new Schedule 4B to the Town and Country Planning Act 1990 sets out the process for making neighbourhood development orders. Regulations are yet to be made setting out the details, the prescribed forms, and the standards, that will be expected. However, a qualifying body (parish council or neighbourhood forum) is entitled to submit a proposal to a local planning authority for the making of a neighbourhood development order by that authority in relation to a neighbourhood area. The proposal must be accompanied by a draft of the order and a statement which contains a summary of the proposals and sets out the reasons why an order should be made in the proposed terms. A local planning authority must give such advice and assistance to qualifying bodies as, in all the circumstances, they consider appropriate for the purpose of, or in connection with, facilitating the making of proposals for neighbourhood development orders in relation to neighbourhood areas, within their area: see paragraph 3 of Schedule 4B. Regulations will make provision for the requirements that must be complied with, and these will include the giving of notice and publicity, the information and documents to be made available, the consultation and participation by the public, the consideration of representations, and for prescribing steps to be taken. Under paragraph 5 of Schedule 4B, a local planning authority may decline to consider a proposal submitted to them if they consider it a repeat proposal. A repeat proposal is one made within two years of the date the authority refuses an earlier proposal or a referendum on an order has been held and half or less than half of those voting in favour of the order, and that there has been no significant change in relevant consideration since the refusal of the proposal or the holding of the referendum. Paragraph 6 of Schedule 4B sets out the arrangements where the local planning authority is prepared to consider a proposal. They must consider whether the qualifying body is authorised for the purposes of a neighbourhood development order to act in relation to the neighbourhood area concerned, whether the proposal complies with provisions under section 61F, as well as those under section 61E, 61J and 61L.

Paragraph 7 of Schedule 4B sets out arrangements for an independent examination. If the local planning authority considers that the proposal satisfies the formal requirements of section 61, it must submit for independent examination the draft neighbourhood development order, and such other documents as may be prescribed. The authority may appoint a person to carry out the examination, but only if the qualifying body consents to the appointment. If there is no consent, then the Secretary of State may appoint a person to carry out the examination. The examiner must be independent of the qualifying body and the authority, must not have an interest in any land affected by the draft order, and must have appropriate qualifications and experience. Although one assumes that the planning inspectorate might be involved here, it also appears that the Secretary of State or another local planning authority may enter into arrangements with the authority for the provision of the services of any of their employees as examiners.

Paragraph 8 of Schedule 4B sets out the matters that the examiner must consider:

- Whether the draft neighbourhood development order meets the basic conditions.
- Whether the draft order complies with the provision made by or under sections 61E, 61J and 61L.
- Whether any period specified in section 61L is appropriate.
- Whether the area for any referendum should extend beyond the neighbourhood area to which the draft order relates.
- Such other matters as may be prescribed.

A draft order meets the basic conditions if:

- Having regard to national policies and advice containing guidance issued by the Secretary of State, it is appropriate to make the order.
- Having special regard to the desirability of preserving any listed building or its setting or any features of special architectural historic interest that it possesses, it is appropriate to make the order.
- Having special regard to the desirability of preserving or enhancing the character or appearance of any conservation area, it is appropriate to make the order.
- The making of the order contributes to the achievement of sustainable development.
- The making of the order is in general conformity with the strategic policies contained in the development plan for the area of the authority (or any part of that area).
- The making of the order does not breach, and is otherwise compatible with, EU obligations.
- Prescribed conditions are met in relation to the order and prescribed matters have been complied with in connection with the proposal for the order.

Schedule 4B (paragraph 9) directs that the general rule is that the examination of the issues by the examiner is to take the form of the consideration of written representations. Although the examiner must cause a hearing to be held for the purpose of receiving oral representations about a particular issue at a hearing in a case where he considers that the consideration of oral representations is necessary to ensure adequate examination of the issue or a person has a fair chance to put a case. Only certain persons are entitled to make oral representations about an issue at a hearing, namely, the qualifying body, the local planning authority, where the hearing is held to give a person a fair chance to put a case, that person, and such other persons as may be prescribed. The hearing will be in public. It is for the examiner to decide whether a person making order representations may be questioned by another person.

The examiner must make a report on the draft order containing recommendations either that the order is submitted to a referendum, or that modifications specified in the report are made to the draft order, and the draft order as modified is submitted to a referendum, or that the proposal for the order is refused. There is a fairly narrow list of matters that can be the subject of modifications set out in paragraph 10 to Schedule 4B. The report must give reasons for each of its recommendations and contain a summary of its main findings.

On receipt of the report of the examiner, the local planning authority must consider each of the recommendations made by the report and decide what action to take in response to them. Nevertheless, if the authority is satisfied that the draft order meets the basic conditions, is compatible with the Convention rights, or would meet those conditions or be compatible with modifications, then a referendum must be held on the making by the authority of a neighbourhood development order: see paragraph 12 of Schedule 4B.

6.5.4. *Referendum*

Paragraphs 14–16 of Schedule 4B to the 1990 Act make provision for the holding of a referendum on the making of a neighbourhood development order. The relevant council must make arrangements for the referendum to take place in so much of their area as falls in what is called the referendum area. A person is entitled to vote in the referendum if on the prescribed date, that person is entitled to vote in an election of any councillors of the relevant council and the persons qualifying address for the election is in the referendum area. Regulations will be made about the holding of referendums, such as, the arrangements, prescribing the dates by which the referendum must be held, the questions to be asked in the referendum and any explanatory material to be provided, the publicity, the limitation of expenditure, the conduct of the referendum, and other detailed matters as to where and how voting is to take place and the votes cast are to be counted.

6.5.5. *Development granted by neighbourhood development order*

As mentioned in the preceding text, a neighbourhood development order is an order which grants planning permission in relation to a particular neighbourhood area specified in the order. However, under section 61J of the 1990 Act, a neighbourhood development order may not provide for the granting or planning permission for what is called 'excluded development', or for any development that planning permission is already granted for. Under section 61K, the following development is excluded development:

(a) Development consisting of a county matter.
(b) Development consisting of the carrying out of operations relating to waste development.
(c) Development falling within council directive 85/337/EEC (Environmental Assessment) concerned with the assessment of the effects of certain public and private projects in the environment.
(d) Development consisting of a nationally significant infrastructure project within the meaning of the Planning Act 2008.
(e) Prescribed development or development of prescribed descriptions as may be set out regulations.
(f) Development in a prescribed area or an area of prescribed description.

Under section 61L of the 1990 Act, planning permission granted by a neighbourhood development order may be granted unconditionally, or subject to such conditions or limitations as may be specified in the order.

The conditions that may be specified can include obtaining the approval of the local planning authority and specifying the period within which applications must be made to the local planning authority for the approval of the authority of any matter specified in the order. This can include a condition that the development begins before the end of a specified period.

The Secretary of State may, and the local planning authority may with the consent of the Secretary of State, revoke a neighbourhood development order under section 61M of the 1990 Act.

6.6. The relationships and legal consequences of development plans

6.6.1. The status of development plans

Section 38(6) of the 2004 Act provides that if regard is to be had to the development plan for the purposes of any determination to be made under the planning Acts, the determination must be made in accordance with the plan unless material considerations indicate otherwise. The duty arising under this provision will apply in relation to the determination of planning applications, planning appeals, imposition of planning conditions, revoking or modifying a planning permission, making a discontinuance order, enforcement notices and several other provisions. It must be remembered that neighbourhood development plans are now part of the development plan.

However, decision-makers (planning authorities and inspectors) must take into account the terms of the National Planning Policy Framework (NPPF) (March 2012) issue by the Secretary of State for Communities and Local Government when considering planning applications and appeals: see sections 19(2)(a) and 38(6) of the Planning and Compulsory Purchase Act 2004. The NPPF says that the purpose of the planning system is to contribute to the achievement of sustainable development, and that this has three dimensions: an economic, social and environmental role. It details a list of objectives from building a strong, competitive economy to conserving and enhancing the natural and historic environments.

A planning permission issued without compliance with the duty under section 54A of the 1990 Act, the predecessor provision to section 38(6) of the 2004 Act, which was not in quite the same terms, established a presumption in favour of the development plan, was held to be *ultra vires* in *R* v *Canterbury City Council, ex p Springimage Ltd* [1993]. An inspector is not required to make express mention of section 54A in his decision, and presumably that must include section 38(6) of the 2004 Act, but it must contain a clear finding of whether the proposal accorded with the plan or not: see *Jones* v *Secretary of State for the Environment* [1998].

6.6.2. Blight and compulsory purchase

One of the less attractive consequences of the development plan is the blighting it can cause to properties affected by the possibility of unattractive public authority schemes. The provisions of the development plan that directly cause blight – making property difficult to sell, except at a depressed price – may enable certain owner–occupiers to serve blight notices to compel the appropriate authorities to acquire the property concerned.

The provisions of a development plan are important in the assessment of compensation for a compulsory purchase of land. Certain assumptions as to planning permission, based upon the development plan, may be made by a claimant to establish the value of his land had no compulsory purchase taken place (see Barry Denyer-Green, *Compulsory Purchase and Compensation*, Tenth Edition, Estates Gazette, 2013).

6.6.3. Challenging a plan

Any person aggrieved by what is called a relevant document may bring proceedings to challenge the same on the grounds that the document is not within the appropriate power, or some procedural requirement has not been complied with: see section 113 of the Planning and Compulsory Purchase Act 2004 and *Taylor Wimpey UK Ltd* v *Crawley BC* [2008]. Each of the following is a relevant document:

- The Wales Spatial Plan, where this remains.
- A development plan document.
- A local development plan.
- A revision of any of the preceding three documents.
- The Mayor of London's spatial development strategy.
- An alteration of or replacement of the spatial development strategy.

The procedures concerning judicial proceedings are treated in more detail in Chapter 20; there is a strict six-week time limit. If a challenge may only be brought once the relevant document has been approved, then an early procedural error could mean that the document is quashed at a much later stage, which is unfortunate. If an early challenge could be lawfully brought, much time and expense could be saved: see *South Northamptonshire DC* v *Charles Church Ltd* [2000].

Examples of challenges under the predecessor provision in section 87(1) of the 1990 Act include the following. In *Great Portland Estates plc* v *Westminster City Council* [1984], a district plan stipulated that office development would not normally be permitted outside the central activities zone of the city. The circumstances when office development might be permitted outside that zone were set out in a guide which was not part of the district plan. This part of the plan was quashed as it did not set out the local planning authority's 'proposals of the use and development' of the land – these were partly in the guide. In *Fourth Investments* v *Bury Metropolitan Borough Council* [1985], part of a local plan was quashed, as the inspector should not have included certain land as part of the green belt had he properly considered the housing needs envisaged by the structure plan. In *Thames Water Utilities* v *East Hertfordshire District Council* [1995], a policy in a local plan was quashed, because the local authority had changed their mind about the boundary of the green belt during the local planning inquiry.

6.6.4. Interpretation of the plan's policies

In *Tesco Stores Ltd* v *Dundee City Council* [2012], the Supreme Court gave guidance on the approach that should be taken in interpreting a development plan. Policy statements should be interpreted objectively in accordance with the language used, with a measure of flexibility. They should be read in the appropriate context. However, a development plan was not to be interpreted like a statute or a contract. Many development plan provisions required a measure of judgment to be applied to a mix of facts, and fell within the jurisdiction of local authorities. Where, as in the *Tesco* case, the development plan contained a provision that called for judgment as to whether a site was 'suitable' for a particular purpose, that was not a matter of planning judgment but involved a question of textual interpretation, which could only be answered by construing the language in its context.

7   The planning application

7.1. Introduction

The scope of development control was set out in Chapters 2 and 3; in Chapter 4, certain developments for which planning permission is automatically granted was outlined; this chapter considers the planning application for development that needs the express consent of the local planning authority.

7.2. Pre-application consultation

By amendments made to the Town and Country Planning Act 1990, where a person proposes to make an application for planning permission for development of a description in a development order, that person must carry out certain required publicity and pre-application consultation (section 61W). That person must take into account the responses to the consultation in deciding whether the application should be on the same terms as originally conceived (section 61X).

7.3. The form of the application

It is important to the applicant for planning permission that he uses the correct form, completes it satisfactorily, informs the parties entitled to be told of an application, and sends the form, appropriate certificates, plans and fee to the right authority. If he is successful thus far, he may be excused any impatience he may later experience in the processing of his application.

The Town and Country Planning (Applications) Regulations 1988 sets out the formal requirements for a planning application. Additionally, the Town and Country Planning (Development Management Procedure) Order 2010 directs that a design and access statement should accompany every application save for those relating to material change of use, engineering or mining operations, development to, or in the curtilage of, a dwellinghouse (save in certain designated areas): see Article 8. A design and access statement, which is also required in connection with listed buildings, explains the design principles in a structured way, and enables applicants to demonstrate an integrated approach that will deliver design: see para 58 of Circular 1/2006 – *Guidance on Changes to the Development Control System*.

In respect of applications likely to involve the need for an environmental impact assessment, the Town and Country Planning (Environmental Impact Assessment) (England and Wales) Regulations 1999, which give effect to Directive 85/337/EEC, may have effect. These regulations are explained further in Chapter 8.

Most application forms ask for brief particulars of the proposed development, including the purpose for which the land and or buildings are to be used. These should be provided as fully and as accurately as is possible. This is because a planning permission may be granted 'in accordance with the plan and application submitted ...'. The description that is contained in the application form, particularly of the proposed use, may therefore be the only form of words available for an inspector or court to consider if an enforcement notice is served alleging development without planning permission. In *Kwik Save Discount Group Ltd* v *Secretary of State for Wales* [1981], the Court of Appeal decided that the description 'retail showroom including new display windows', in the application form, had to be construed within the context of the whole application, which was for alterations and extensions to a service station, and could not be isolated out to enable a supermarket use to take place. This point is further considered in Chapter 9.

In *Geall* v *Secretary of State for the Environment, Transport and the Regions* [1999], it was decided that the Secretary of State must decide whether a valid application has been made on the materials provided to him on an appeal. If a local planning authority rejects a purported application, on the grounds that it is invalid, that decision can be challenged by an application for judicial review (see Chapter 20). Alternatively, the applicant may appeal to the Secretary of State under section 78: see *R* v *Secretary of State for the Environment, Transport and the Regions, ex p Bath North East Somerset District Council* [1999].

It is no longer possible for planning permission to operate as a listed building consent; accordingly, any work to a listed building must be the subject of a separate listed building consent application (section 10 of the Planning (Listed Buildings and Conservation Areas) 1990 Act).

There is no express requirement that a planning application be made by the owner or person with an interest in the land to which the application relates. In *Hanily* v *Minister of Housing and Local Government* [1952], it was suggested that a person 'who genuinely hopes to acquire the interest in the land' may make an application.

In the past, developers frequently made duplicate applications with a view to appealing one of the applications if a decision was not made timely, whilst continuing to discuss the second application with the local planning authority. This practice is now less frequent as an authority has power to reject a duplicate or repeat application, in certain circumstances: see section 70B of the 1990 Act.

Where development has taken place in breach of planning control, and an enforcement notice has been served, a local planning authority now has power to decline to determine a retrospective application for planning permission for development that constitutes the breach of planning control: see the new section 70C of the 1990 Act introduced by the Localism Act 2011.

7.4. Fees

Section 87 of the Local Government, Planning and Land Act 1980 empowers the Secretary of State to make regulations providing for the payment of fees to local planning authorities in respect of any application for planning permission, consent, approval, determination or certificate under the planning Acts. Provision is also made for the payment of a fee to the Secretary of State where he decides a deemed planning application in connection with an appeal against enforcement proceedings (see Chapter 14).

The present regulations are the Town and Country Planning (Fees for Applications and Deemed Applications) Regulations 1989. The fees are increased from time to time.

The fees are payable upon an application for planning permission; for the approval of reserved matters; for the consent for the display of advertisements; and for deemed applications for planning permission (under section 303 of the 1990 Act). In the case of deemed applications, there are exemptions (regulation 8).

The fee must be paid at the time of the application, which will otherwise be invalid. There is no refund if the application is not consented to!

7.4.1. No fee payable

A fee is not payable in a number of circumstances.

No fee is payable if the local planning authority is satisfied that the application relates:

> *solely to the carrying out of operations for the alteration or extension of an existing dwellinghouse or the carrying out of operations (other than the erection of a dwellinghouse) in the curtilage of an existing dwelling for the purpose, in either case, of providing means of access to or within the dwellinghouse for a disabled person who is resident or proposes to reside in that dwellinghouse, or of providing facilities designed to secure his greater safety, health or comfort.* (regulation 4)

This regulation now includes alterations to public buildings.

No fee is payable where an application is made solely for development within Schedule 1 to the General Permitted Development Order 1995 (development for which planning permission is normally automatically available without application) and express planning permission is necessary for such development (regulation 5). This may happen where an article 4 direction is in force withdrawing such automatic consent (see Chapter 4), or because of the requirements of a condition limiting the permitted development (see Chapter 9; particularly *City of London Corporation* v *Secretary of State for the Environment* [1972]). Regulation 5 includes applications for the retention of a building or the continuance of a use without complying with a condition that limits development otherwise permitted by the GPDO. The purpose of regulation 5 is to recognise that, in the ordinary case, development specified by the General Permitted Development Order needs no application, and therefore no fee would ordinarily become payable.

No fee is payable where an application is made for planning permission or the approval of reserved matters for a modified proposal within 12 months of the grant of planning permission to the same applicant (regulation 7).

A fee paid in connection with an enforcement notice – a deemed planning application – is refunded if the appeal succeeds (regulation 10).

The scale of fees is set out in Schedule 2 to the Regulations.

7.5. Application for outline planning permission

Outline planning permission is defined in the 1990 Act as:

> *... planning permission granted ... with the reservation for subsequent approval by the local planning authority ... of matters (referred to ... as 'reserved matters') not particularised in the application.* (section 92(1))

Outline planning permission is also defined in article 2 of the Town and Country Planning (Development Management Procedure) Order 2010 as a planning permission for the erection of a building, which is granted subject to a condition requiring the subsequent approval of the local planning authority with respect to one or more reserved matters.

Reserved matters are defined in article 2 of the Development Management Procedure Order as any of the following in respect of which details have not been given in an application, namely: access, appearance, landscaping, layout and scale, within the upper and lower limit for the height, width and length of each building stated in the application. Under article 4(3), the application should state the approximate location of buildings, routes and open spaces.

Outline planning permissions are not otherwise separately considered or provided for in the 1990 Act (other than in relation to a time limit for the application for approval of reserved matters). An outline permission is in all senses and for all purposes the planning permission; the only difference from a planning permission, that is not outline, is that it is granted subject to conditions that require the approval of the local planning authority as to one or more of the reserved matters. For other conditions that may be imposed, see Chapter 9.

Although an application for outline can therefore save design time and costs, the local planning authority may notify the applicant within one month of the application that further details must be submitted before it can be considered (see article 4 of the Development Management Procedure Order 2010).

When submitting an application for the approval of the reserved matters, it is important that a planning application form is not used, as otherwise, the authority may be entitled to consider the application *de novo*. However, in *Etheridge* v *Secretary of State for the Environment* [1983], it was said that an application for full planning permission is at the same time an application for the approval of reserved matters: what matters is not the form of the application, but the objective effect. Article 5 of the Development Management Procedure Order specifies that an application for the approval of reserved matters should be in writing, identify the outline planning permission, and include such particulars, plans and drawings as are necessary to deal with the reserved matters. The further certificates as to ownership are not therefore required (*R* v *Bradford-on-Avon Urban District Council, ex p Boulton* [1964]).

The application for the approval of reserved matters should not radically differ from the development proposed by the outline. If it does, the local planning authority will be entitled to consider the application as one for planning permission, and refuse it (see *Shemara* v *Luton Corporation* [1967]). In *Calcaria Construction Co Ltd* v *Secretary of State for the Environment* [1974], where outline planning permission had been granted for a warehouse of 55,000 square feet for wholesale and retail distribution, it was decided that the local planning authority, and the Secretary of State, had rightly refused consent to the detailed application which showed a supermarket and ancillary storage and parking space; a warehouse for wholesale and retail distribution could not mean a supermarket.

Matters took a different course in *Chalgray Ltd* v *Secretary of State for the Environment* [1976], where the application for approval of the reserved matters included a new access which was not intended in the outline. Following the refusal of consent by the local planning authority, the Secretary of State actually declined to decide the appeal as he said the details submitted were *ultra vires* the original planning permission. An appeal against his decision was dismissed. In *R* v *Castle Point District Council, ex p Brooks* [1985], outline planning permission was granted for a bungalow, but the plans and details approved as the 'reserved matters' were of a house; a neighbour was successful in having the decision quashed by the procedure known as judicial review (see Chapter 19).

Where outline planning permission is granted for the carrying out of building or other operations, it shall be granted subject to conditions that the application for approval of the reserved matters is made within three years of the grant of the outline planning permission, and that the development itself must be begun within five years of the grant of the outline planning permission, or within two years of the approval of the reserved matters, whichever is the later date: see section 92(2) of the 1990 Act. Any outline planning permission not containing these conditions is deemed to contain them: see section 92(3) of the 1990 Act. The local planning authority may substitute for these periods of time such other periods as they may direct (section 92(4) of the 1990 Act).

These conditions can cause problems if the local planning authority refuses consent to an application for approval of the reserved matters and the time limit has expired for making a further application. In practice, it is advisable to submit more than one such application, each slightly different in detail, and all in good time. A right of appeal against a refusal to approve the reserved matters required by a planning condition is now possible under section 78(1) of the 1990 Act.

The planning authority cannot decide that no approval shall be given to reserved matters, because the original outline planning permission would no longer be so decided (*Hamilton* v *West Sussex County Council* [1958]). However, if they have already given approval to the reserved matters, can they refuse to consider a further application for the approval of reserved matters? The Court of Appeal decided in *Heron Corporation Ltd* v *Manchester City Council* [1978], where the first approved detailed scheme was rendered abortive by the subsequent listing of certain buildings, that the second application for the reserved matters was within the scope of the original outline planning permission and must be considered by the local planning authority. In *R* v *Hammersmith and Fulham LBC, ex p Greater London Council* [1986], the Court of Appeal applied the *Heron* case, and decided that where an application for the approval of reserved matters omitted a bus garage and public library from a large scheme of development that had an outline planning permission, approval could be validly given: the development approved was not something different from that in the outline.

Conditions may be imposed on the grant of approval of reserved matters to an outline planning permission; a condition cannot be imposed that modifies the grant of the outline permission itself: *Redrow Homes Ltd* v *First Secretary of State* [2003]. A condition cannot relate to matters not reserved: *R* v *Elmbridge BC, ex p Health Care Corp* [1991].

7.6. Notices to owners and tenants

Section 65 of the 1990 Act makes provision for the service of notices and for publicity in relation to making of an application. The detailed requirements for notices are in the Development Management Procedure Order 2010 (DMPO). A local planning authority shall not entertain an application unless these requirements are satisfied. Under article 11 of the DMPO, an applicant must give notice to any person who is the owner of the land to which the application relates or is an agricultural tenant. Under article 12, the applicant for planning permission is required to sign one or more of the appropriate certificates about ownership and the giving of notices.

The term 'owner' means the freeholder and the owner of a lease of which not less than seven years remain unexpired (section 65(8)). This means that tenants with lesser interests are not entitled to be notified of a planning application affecting their property. To this, there is an exception in the case of an agricultural holding within the meaning of the Agricultural Holdings Act 1986; the applicant must certify that he has notified any tenant of the agricultural holding. This exception exists, because an agricultural tenant may lose security of tenure if planning permission is granted.

It is an offence to give a certificate which a person knows is false or misleading in a material particular or to recklessly issue a certificate which is false or misleading (section 65(7)). If no certificate is issued, the local planning authority will have no jurisdiction to consider the application. Even if a certificate is issued that is incorrect, the decision of the authority will remain valid (*R* v *Bradford-on-Avon Urban District Council, ex p Boulton* [1964] where the applicant mistakenly certified that he owned all the land). In *Main* v *Swansea City Council* [1984] CA, the certificate was wrong as part of the land the subject of the application was owned by someone who had not been given notice. This omission did not render the subsequent grant of planning permission a complete nullity as the court had a discretion, which it exercised, to refuse to quash the permission because of the lapse of time (three years). Nevertheless, in *R (Pridmore)* v *Salisbury DC* [2005], permission was quashed where a certificate was found to be manifestly false. Where an applicant completes an application, and the appropriate certificate, purporting to do so as agent of the true owner, he will be required to account for any profits he may obtain if he later acquires the land from an unsuspecting owner at a price below market value (*English* v *Dedham Vale Properties Ltd* [1978]).

The date for determining whether a person is an owner is the date 21 days before the date of the application.

7.7. Publicity for applications

Every planning application must be entered on a register (section 69(1) of the 1990 Act). The Development Management Procedure Order 2010 makes detailed provision for these registers, and for the details, they must contain (article 36). They are kept at the offices of the local planning authority, and are available for public inspection. They contain not only new applications yet to be decided but also a permanent record of all applications and decisions.

Under article 13, a local planning authority must publicise an application in the following manner:

(1) Applications for development which are required to be accompanied by an environmental statement, or which depart from the development plan, or affect a public right of way must be publicised by a site notice for not less than 21 days and by local advertisement.

(2) Applications for major development must be publicised by a site notice for not less than 21 days or by notice on adjoining owners or occupiers and by local advertisement. Major development includes winning and working of minerals, waste development, erection of more than ten dwellinghouses, erection of a building of more than 1,000 metres or development on a site exceeding one hectare.

In any other case, an application must be publicised by site notice for 21 days or by notice on any adjoining owners or occupiers. In all cases, applications shall be listed on the local planning authority's web site,

The publicity forms are specified in the Development Management Procedure Order. The forms of advertisement and site notice invite any person to make a representation to the local planning authority.

Where an application has not been properly publicised in accordance with article 11 of the DMPO, an adjoining owner prejudiced by that failure might be awarded damages as in *R (on the application of Gavin)* v *Haringey London Borough Council and another* [2004]. In *Rubin* v *First Secretary of State* [2004], there was procedural unfairness where a planning application was made by a neighbour and there was a failure to inform the complainant of both this application and a subsequent planning inquiry.

7.8. Development affecting conservation areas and listed buildings

Conservation areas and listed buildings are considered more fully in Chapter 10. However, any application for development which would, in the opinion of the local planning authority, affect the character or appearance of a conservation area, or affect the setting of a listed building must be given publicity by the authority (sections 67 and 73 of the Planning (Listed Buildings and Conservation Areas) Act 1990). The publicity consists of a newspaper advertisement and site notice inviting any person to make representations to the authority. The application must not then be determined within 21 days of the notices appearing. The difficulty for the developer that arises from this publicity requirement is that it concerns not simply development in a conservation area, or of a listed building, but also development that merely affects such an area or building.

A notice that fails to state the date by which objections and representations are to be made is not a proper notice, and a subsequent planning permission may be invalid: *R* v *Lambeth London Borough Council, ex p Sharp* [1986]. Strict compliance with these requirements is expected where a local planning authority exercises a power to grant planning permission to themselves and is not exercising a duty.

There are additional requirements for the notification of applications to English Heritage: see Circular 01/01 – *Arrangements for Handling Heritage Applications – Notification and Directions by the Secretary of State*. In *R* v *South Hereford District Council, ex p Felton* [1990], a planning permission, which had been granted without being properly advertised or notified, was quashed.

7.9. Development not in accordance with the development plan

Where the local planning authority proposes to grant permission for development that is not in accordance with the development plan, it must comply with the *Town and Country Planning (Development Plans and Consultation) (Departures) Direction* 1999 (Circular 07/99). Article 27 of the Development Management Procedure Order 2010 also provides for the granting of planning permission in accordance with directions where the development is not in accordance with a development plan. Certain applications must be notified to the Secretary of State; these will be those raising planning issues of more than local importance. They are departure applications which the authority proposes to grant for development consisting of more than 150 dwellings, more than 5,000 square metres of gross retail, leisure, office or mixed commercial floor space, development of land of a planning authority, or any other development that would significantly prejudice the implementation of the development plans polices and proposals.

It has been held that a failure to comply with a direction does not restrict the jurisdiction of a local planning authority to determine a planning application: see *Co-operative Retail Services* v *Taff-Ely Borough Council* [1979]. Thus, in *R* v *St Edmundsbury Borough Council, ex p Investors in Industry Commercial Properties Ltd* [1985], it was said that the omission to advertise was breach of a directory rather than a mandatory requirement. In the circumstances, there had been widespread publicity in local newspapers.

8 How a local planning authority determines a planning application

8.1. Introduction

It may be thought that a local planning authority, as an element in the democratic processes of local government, has a fair degree of discretion in deciding a planning application. However, not only are there a number of statutory directions and procedures to be followed, and other authorities and bodies to be consulted, there are also a considerable number of legal decisions which have circumscribed the limits of discretion available to the authority. A local authority is of statutory creation exercising statutory powers and duties: it is subject to judicial control and some degree of ministerial supervision.

Much of the detailed guidance to local planning authorities was found in the Town and Country Planning (General Development Procedure) Order 1995 (GDPO). That Order has been revoked and the Town and Country Planning (Development Management Procedure) (England) Order 2010 (DMPO) sets out a number of requirements and details that must be submitted with a planning application, and of the consultations that the local planning authority should carry out. The DMPO also deal with the rights to appeal planning decisions, or a failure to make a timely decision. One important provision is that, unless the applicant agrees, a decision on an application must be made within eight weeks, or 13 weeks where the application is for a major development (article 29 of the DMPO). A major development includes mining operations and one involving more than ten houses. In recent years, ministerial circulars have been more direct and specific in giving guidance to local planning authorities. There is also the series of Planning Policy Statements that address a number of specific policy matters.

The local planning authority makes their decision on an application by virtue of section 70 of the 1990 Act: they may grant planning permission unconditionally, or subject to such conditions as they think fit; or they may refuse planning permission. In dealing with a planning application, section 70(2) provides that an authority shall have regard to the provisions of the development plan, so far as material to the application, and to any other material considerations. The Secretary of State is similarly bound when determining an appeal. However, section 38(6) of the Planning and Compulsory Purchase Act 2004 provides that where regard is to be had to the development plan, the determination must be made in accordance with the plan unless material considerations indicate otherwise.

Although in practice local planning authorities may inform neighbours about planning applications, there is no requirement or duty, and it would be difficult to decide the scope of persons likely to be affected: see *R v Secretary of State for the Environment, ex p Kent* [1988].

8.2. Consultations with other bodies

Before planning permission can be granted by the local planning authority, certain authorities or persons must be consulted (see the table at Schedule 5 to the DMPO). For example, where any development is likely to affect land in the area of another local planning authority, or land in a national park, the consultee will be the respective local planning authority or the county planning authority, as the case may be. The Health and Safety Executive may notify areas that contain risks and must be consulted about development in those areas. Development likely to result in a material increase in the volume or a material change in the character of traffic affecting trunk roads, or classified roads must be the subject of consultation with the appropriate highway authority. Development affecting the site of scheduled monuments must be referred to the Historic Buildings and Monuments Commission. The Environment Agency must be consulted on a wide range of development involving proposals likely to affect rivers or give rise to pollution risks.

Certain development is specified as being a county matter (see Chapter 5), and if an application is received that appears to the district planning authority to relate to such a matter, the county planning authority decides the application. The county planning authority is required to afford an opportunity to the district planning authority to make recommendations; and is bound to take these into account (article 16 of the DMPO).

8.3. Highways: directions and consultations

Where an application is for development consisting of the making or altering of an access to a trunk road with a speed limit exceeding 40 mph, or to a special road; or other development within 67 metres of an existing or proposed trunk road, the Secretary of State must be consulted. The application cannot be determined except in accordance with a direction. However, the application can be determined if a direction is not received within 28 days (article 26 of the DMPO).

The highways authority (local or the Secretary of State for Transport, as appropriate) must be consulted in the following cases (Schedule 5 to the DMPO):

- Development likely to result in a material increase in the volume or a material change in the character of traffic entering or leaving a trunk road, or using a level crossing (the Secretary of State for Transport is the consultee);

- Development likely to result in a material increase in the volume or a material change in the character of traffic entering or leaving a classified road or proposed highway;
- Development involving the formulation, laying or altering of any means of access to a highway (other than trunk road) or to a toll road (in the latter case, the consultee is the concessionaire);
- Development which consists of or includes the laying out or construction of a new street.

8.4. Directions and policy guidance given by the Secretary of State

Sections 74 and 77 of the 1990 Act contain wide powers for the Secretary of State to issue directions in relation to planning applications. There is also a power in article 25 of the DMPO.

8.4.1. The general power of direction

The Secretary of State may give directions restricting the grant of planning permission by a local planning authority, either indefinitely or during such period as may be specified, in respect of a particular development or any specified class of development. The local planning authority must then comply with such a direction (article 25 of the DMPO).

8.4.2. Development not in accordance with the development plan for the area

Subject to the significance of the development plan in all planning determinations, and considered more fully in the following text, the Secretary of State may issue directions governing the grant of planning permission that is not in accordance with the development plan (article 27 of the DMPO).

The present direction is the *Town and Country Planning (Development Plans and Consultation) (Departure) Direction* 1999 (in Circular 07/99). A failure to comply with the Development Plans Direction does not render the planning permission null and void; the directive is not mandatory: see *R v St Edmundsbury Borough Council, ex parte Investors in Industry Commercial Properties Ltd* [1985]. Where county matters are concerned, county planning authorities decide if an application involves a departure from the development plan. Otherwise, it is for the district planning authorities to make the decision. If a departure application is so identified, and the respective planning authorities are not going to refuse it, the publicity requirements of section 71 of the 1990 Act and the DMPO apply. This gives an opportunity for representations to be made; the planning authority, when making their decision, must take these representations into account. Certain forms of development, such as development consisting of the provision of more than 150 houses or 5,000 square metres of gross retail, leisure, office or mixed commercial floor space must be notified to the Secretary of State.

8.4.3. The call-in procedure

The Secretary of State may give directions requiring planning applications or applications for the approval of reserved matters to be referred to him for decision. A direction may be given to a local planning authority, or to all such authorities; it may relate to a particular application or to a class of applications as specified (section 77 of the 1990 Act). This power appears to confer on the Secretary of State an unqualified direction which is subject only to the *Wednesbury* principle of judicial review (see Chapter 20). The Secretary of State, in deciding to call in applications, can only be challenged if his decision is perverse or totally unreasonable; his decision is one of policy or value judgment for him: *Rhys-Williams* v *Secretary of State for Wales* [1985].

As there is no right of appeal from a decision of the Secretary of State, he will, if requested by either the applicant or the local planning authority, appoint an inspector before whom the parties may appear (section 77(5)).

8.4.4. Policy guidance

The current principal policy guidance is the National Planning Policy Framework (NPPF) (March 2012) issued by the Secretary of State for Communities and Local Government. This must be taken into account when considering planning applications and appeals: see sections 19(2)(a) and 38(6) of the Planning and Compulsory Purchase Act 2004. The NPPF says that the purpose of the planning system is to contribute to the achievement of sustainable development, and that this has three dimensions: an economic, social and environmental role. It details a list of objectives from building a strong, competitive economy to conserving and enhancing the natural and historic environments.

The Department of Communities and Local Government also issues circulars and, occasionally, consultation papers. A raft of Planning Policy Statements (PPSs) have now been cancelled (March 2012). The PPSs contained policy guidance on a variety of specific areas of planning and development control. Circulars very often contain advice which is less durable in that it may reflect the policies of a particular Secretary of State.

Circulars have been held to be 'material considerations', and therefore regard may have to be paid to any relevant policies they contain: see *Pye (Oxford) Estates Ltd* v *West Oxfordshire District Council* [1982]. Nevertheless, provided the local planning authority has considered the policy in a relevant circular, they may depart from the policy if they have reasons for doing so: see *R* v *Camden London Borough Council, ex parte Comyn Ching & Co (London) Ltd* [1983]. It would be wrong for a local planning authority to allow a circular to decide an application without the full exercise of their own proper discretion: see *R* v *Worthing Borough Council, ex parte Burch* [1985].

The same applied to PPSs (and before them the PPGs). They were capable of being material considerations. The terms of the National Planning Policy Framework (March 2012) must be taken into account when preparing development plans, determining planning applications and appeals. Circulars and PPSs do not have to be interpreted like statutes. A decision-maker will only be wrong if an interpretation is adopted that is perverse and not one the words are properly capable of bearing: see *R* v *Derbyshire County Council, ex parte Woods* [1997].

8.5. The effect of the development plan

There is a significant administrative, democratic and public participation effort behind the preparation or amendment of the development plans. Under section 38(6) of the 2004 Act, a determination of a planning application shall be made in accordance with the development plan unless material considerations indicate otherwise: see *Persimmon Homes (North West) Ltd* v *First Secretary of State* [2006]. This statutory direction, first introduced in 1991, gives a significant dominance to the development plan in decision-making. In *R* v *Leominster District Council, ex parte Pothecary* [1997], the Court of Appeal gave guidance on the application of section 54A, the statutory predecessor to section 38(6) of the 2004 Act, in those difficult cases where there are factors that make a bald application of the section difficult. Otherwise, the section establishes a presumption in favour of the development plan: see *St Albans District Council* v *Secretary of State for the Environment* [1993] and *City of Edinburgh* v *Secretary of State for Scotland* [1997].

The Secretary of State is in no different position with regard to the effect of the development plan when deciding appeals. Thus, in *Jones* v *Secretary of State for the Environment* [1998], the decision of a planning inspector, deciding an appeal on behalf of the Minister, was required to determine the appeal on the basis of section 54A. The court considered that the inspector should have shown in his decision letter that he had properly considered the section.

If a local planning authority decides to grant planning permission that is not in accordance with the development plan, they must follow the departure procedure outlined in Chapter 7, in part **7.8**, and take into account any representations they receive. If this occurs, the policy that contradicts a provision of a development plan is not subject to the public participation process that applies to the making or modification of the local or structure plans. This occurred in *Covent Garden Community Association Ltd* v *Greater London Council* [1981], where the GLC granted planning permission to themselves that was contrary to the local action area plan. However, because they had followed the statutory procedures, and were not regarded as having been in breach of the rules of natural justice (see Chapter 20), the Community Association were left without any legal remedy to compel the GLC to submit the new policy to an inquiry.

8.6. Other material considerations

The 1990 Act provides that a local planning authority, in dealing with a planning application, must not only have regard to the development plan, so far as material, but additionally 'to any other material considerations' (section 70(2)); see also section 38(6) of the 2004 Act. This also applies to the Secretary of State or an inspector when a planning appeal is being determined, although it has been decided that, where the decision is made by the Secretary of State, he does have regard to his own circulars even if he does not specifically refer to these in terms: see *Hewlett* v *Secretary of State for the Environment* [1985].

There are a number of cases on this provision: the issue is usually either that some matter is alleged to be a material consideration, and was not considered; or, particular regard was paid to some matter which it is alleged was not a material consideration. It has already been mentioned that circulars containing relevant policies may be material considerations: see *Carpets of Worth Ltd* v *Wyre Forest District Council* [1991]. A quite separate question is, if a particular matter is a material consideration, what weight should be given to it? Whether something is a material consideration, is a question of law, but as explained in *Tesco Stores Ltd* v *Secretary of State for the Environment* [1995], the question of weight is a matter of planning judgment for the decision-maker.

In *Stringer* v *Minister of Housing and Local Government* [1970], Cooke J said:

> *In principle … any consideration which relates to the use and development of land is capable of being a planning consideration. Whether a particular consideration is material in any given case will depend on the circumstances.*

In *R (Hinds)* v *Blackpool Borough Council* [2012], the Court of Appeal had to consider materiality of an emerging government policy to revoke regional spatial policies. It concluded that a 'material consideration' was one which had some weight in the decision-making process and was capable of tipping the balance somewhat: see *R(Kides)* v *South Cambridgeshire District Council* [2002]. However, the guidance in *Kides* required a cautious approach to be applied with commonsense on the particular facts, especially where the new matter arises after the resolution to grant planning permission and before the final decision: see *R(Dry)* v *West Oxfordshire District Council* [2010]. The following represent only a selection of a large number of matters that have been held to be material considerations.

8.6.1. Financial considerations

Can economic questions be material considerations? Ackner J in *Murphy (J) & Sons Ltd* v *Secretary of State for the Environment* [1973] considered this problem and said:

> *I have never heard it suggested before that a planning application involves a valuation exercise. … I hold that as a matter of law the Minister [or the local planning authority] is not entitled to have regard to the cost of developing a site in determining whether planning permission … should be granted.*

Later, he was to say that he had stated the proposition too widely (*Hambledon & Chiddingfold Parish Councils* v *Secretary of State for the Environment* [1976]). In *Brighton Borough Council* v *Secretary of State for the Environment* [1978], Sir Douglas Frank upheld the Secretary of State's decision that by granting planning permission to the applicants, certain school trustees, this would generate funds to enable the maintenance and refurbishment of a listed building nearby: a distinct planning benefit. The same judge, in *Walters* v *Secretary of State for Wales* [1978], decided that it would be wrong if a planning authority justified their refusal of planning permission for development on the ground that it would be uneconomic. He said:

> ... *the grant of planning permission cannot depend on the resources and intention of the applicant. It is for the developer to make the economic decision whether to carry out the development and not the local planning authority.*

It seems that economic matters can, in appropriate cases, be material considerations: although the unprofitability of any particular development proposal should not be a reason for refusal. In *Niarchos (London) Ltd* v *Secretary of State for the Environment* [1978], it was held that the planning authority should consider the costs of adaptation of premises to residential use notwithstanding a policy in the development plan in favour of residential use, rather than a continuance of office use under a temporary planning permission: economic viability was considered by the judge to be an important consideration in deciding the applicability of that policy.

The cases present no very clear pattern as to whether economic considerations can be taken into account. It is submitted that the economic viability of a proposed development should never be considered by a planning authority as a ground for refusing planning permission. On the other hand, it would be wrong for a planning authority to ignore economic matters that are obvious and relevant, such as in *Northumberland County Council* v *Secretary of State for the Environment* [1989], where it was held that the economic advantages of an open cast mining development was a material consideration.

8.6.2. Effect on future decisions

In *Collis Radio Ltd* v *Secretary of State for the Environment* [1975], it was held that if planning permission were granted for the development sought, the fact that it would make applications for similar development in the locality difficult to refuse could be a material consideration. The local planning authority may therefore rely on this argument to refuse permission even if they have no other objection to the actual application before them. The authorities were explained in *Poundstretcher Ltd* v *Secretary of State for the Environment* [1988], where it was said that mere fear of a concern about creating a precedent can be a material consideration, but the decision-maker must have the relevant evidence.

8.6.3. Preserving or encouraging uses

The desirability of preserving existing uses of land and buildings is a material consideration even if there is no other objection to an application for change of use. This was upheld in *Clyde & Co* v *Secretary of State for the Environment* [1977], where the local planning authority desired to retain residential accommodation in an area of housing shortage. Although an authority has no powers to compel an occupier to use his land for a permitted use, it is a perfectly proper planning consideration to encourage that use by refusing permission for other uses. However, if planning permission is refused for, say, offices, on the ground that another use such as residential would be preferred, consideration must be given to whether the residential use was a realistic expectation: *Finn & Co* v *Secretary of State for the Environment* [1984].

The matter was considered further by the House of Lords in *Westminster City Council* v *British Waterways Board* [1984]. It was accepted that planning permission could be refused if the ground of refusal was the desirability of protecting the existing use of premises, but that planning permission could not be refused if the ground of refusal was to protect the occupation of the present occupier of the premises. Lord Bridge explained that it was necessary to show on a balance of probability that if permission was refused for use B, the land in disputed would effectively be put to use A.

8.6.4. Non-statutory policies

The *Clyde & Co* case is also important for another aspect of this problem. It had been argued that a planning consideration could only be such, and therefore a material consideration, if set out as a policy in the development plan. This was rejected in the Court of Appeal. Planning authorities are entitled to have regard to other policies than those formally within the development plans.

8.6.5. Planning gain

In *R* v *Westminster City Council, ex parte Monahan* [1989], it was held that if a proposed development can finance planning benefits, this is capable of being a material consideration. In *Tesco Stores Ltd* v *Secretary of State for the Environment* [1995], the House of Lords gave guidance on planning gain. An offer of planning gain which has nothing to do with the proposed development will not be a material consideration. If the offer has some connection with the proposed development, then regard must be had to it. Nevertheless, in *R* v *South Holland DC, ex p Lincoln Co-operative Society Ltd* [2000], where permission was first refused for a retail scheme, but later granted when the developer offered £100,000, the decision was quashed; there were no rational grounds for supporting the conclusion that the sum could redress the harm that the proposal would cause.

8.6.6. Other material considerations

The list of material considerations is never closed, as any relevant fact or issue of a planning nature may be pertinent. Accordingly, such matters as the planning history of the site, and safety and environmental hazards are capable of being material considerations.

In directive 79/409/EEC (Birds Directive) on the conservation of wild birds, member states are required to establish special protection areas. Under directive 92/43/EEC (Habitats Directive) on the conservation of natural habitat areas, provision is made for the establishment of special areas of conservation. The requirements of these Directives have been transposed into British law by the Conservation of Habitats and Species Regulations 2010. However, the Regulations contain controls over decisions relating to developments plans and planning permissions, including a review of existing consents, such as planning permissions. In relation to a specific site, where the matters sought to be protected by the Directives are in issue, such matters will be material considerations.

8.7. Environmental impact assessment

The Town and Country Planning (Environmental Impact Assessment) (England and Wales) Regulations 1999 give effect to Directive 85/337/EEC on the assessment of the effects of certain public and private projects on the environment. Directive 97/111/EEC increased the number of categories subject to environmental impact assessment (EIA).

All development in Schedule 1 to the regulations requires EIA. The development in Schedule 1 includes industrial development such as power stations; it also includes installations for intensive livestock rearing, where the numbers of livestock exceed certain threshold limits. Quarries and opencast mining where the surface of the site exceeds 25 hectares are within the Schedule.

Schedule 2 to the regulations lists development where an EIA is required if the development is either to be carried out in a sensitive area or satisfies a specified threshold or criterion.

The regulations prohibit the grant of planning permission for EIA development unless the decision-maker has first taken account of the environmental information. In the case of Schedule 2 development, there is a screening procedure to determine whether the development is an EIA development. The regulations also enable a person to obtain 'a scoping opinion or direction'; this assists the applicant as to the types of information that will be required in an environmental statement that must accompany an application for EIA development.

Schedule 4 to the regulations sets out the information that is to be included in the environmental statement. This includes the expected residues and emissions resulting from the development and the description of the aspects of the environment likely to be significantly affected by the development.

In *Berkeley* v *Secretary of State for the Environment* [2001], the House of Lords decided that a planning permission was *ultra vires* where the application had not been accompanied by an EIA, as affected individuals had the right to have the need for an EIA considered before the grant of a permission. Following this and other cases, the Department issued guidance to the effect that an application for an outline permission was unlikely to comply with the EIA regulations unless accompanied by an EIA; an outline permission should be tied to the environmental details provided, although there should not be too much rigidity; and development could be unlawful if carried out under a reserved matters approval outside the scope of the outline permission.

8.8. Development affecting conservation areas and listed buildings

In considering whether to grant planning permission for development that affects a listed building or its setting, or a building in a conservation area or its setting, the local planning authority, or the Secretary of State, shall have special regard to the desirability of preserving the building or its setting or any features of special architectural or historic interest which it possesses (section 16(2) of the Planning (Listed Buildings and Conservation Areas) Act 1990).

Decision-makers (planning authorities and inspectors) must take into account the terms of the National Planning Policy Framework (NPPF) (March 2012), issued by the Secretary of State for Communities and Local Government, when considering planning applications and appeals: see sections 19(2)(a) and 38(6) of the Planning and Compulsory Purchase Act 2004. In relation to conserving and enhancing the historic environment, local planning authorities should take into account the desirability of sustaining and enhancing the significance of heritage assets, the positive contribution they make, and the desirability of new development making a positive contribution to local character and distinctiveness.

The further controls that apply to conservation areas and Listed Buildings are considered in Chapter 10.

8.9. Planning applications in which a local planning authority has an interest

The Town and Country Planning General Regulations 1992 make special provisions for applications made by the local planning authority itself. In general, these are to be determined on the same basis as all other applications.

Where a local planning authority is proposing to grant planning permission to itself, there is always a risk of bias. In *Steeples* v *Derbyshire County Council* [1981], where the local planning authority was said to have a financial interest in granting planning permission for a large project they had an interest in, the permission was quashed on the ground of a breach of the rules of natural justice. The case was distinguished in *R* v *Amber Valley District Council, ex parte Jackson* [1984], where the majority of the councillors were predisposed towards the proposed development. Because the authority had no contractual or financial interest, unlike in the *Steeples* case, there was said to be no reason why they could not make a fair decision.

This problem was considered further in *R* v *St Edmundsbury Borough Council, ex parte Investors in Industry Commercial Properties Ltd* [1985]. It was acknowledged that there are many cases where a local planning authority will have an interest in the outcome of their own planning decisions. There is no presumption that because they have an interest there must be bias. The test, considered in *Steeples*, whether the reasonable man, knowing all the facts, would conclude that the planning decision was fairly reached, is more suited to judicial rather than administrative decisions. The test in *R* v *Sevenoaks District Council, ex parte Terry* [1985] was whether the authority took into account what was relevant, and disregarded the irrelevant. Today, the test of bias in *R* v *Gough* [1993] applies to decision-making bodies. A person is disqualified from participation in a decision if there is a real danger that that person will be influenced by a pecuniary or personal interest: see *R* v *Secretary of State for the Environment, ex parte Kirkstall Valley Campaign Ltd* [1997].

8.10. Judicial supervision

When a planning authority is exercising its statutory discretion in determining a planning application, it is subject to the judicial supervision exercised by the Administrative Court of the Queen's Bench Division.

The purpose of this supervision is, in simple terms, to ensure fairness. However, because this supervision is applicable to more circumstances than a planning application, the matter is separately considered in Chapter 20.

9 Planning permissions, conditions and agreements

9.1. Introduction

In the preceding chapter, a number of matters relevant to the submission of a planning application were considered. We now move forward to the decisions a local planning authority may make on an application. This chapter is concerned with the nature and effect of a planning permission, and of the conditions that may be attached to a permission. It also considers planning agreements that may be made between landowner and local planning authority. An applicant may also offer unilateral planning obligations. If a decision involves refusal of planning permission, the reader should turn on to Chapter 17. The reader should bear in mind that what is said here in relation to a local planning authority also applies in those cases where the Secretary of State grants or is deemed to grant planning permission.

9.2. Planning permission

9.2.1. What is the planning permission?

This important question has arisen on a number of occasions where there has been a disparity between the decision of the planning committee and the formal notification; or the committee re-decide an earlier decision.

Where on an application for planning permission or for approval of reserved matters, and a permission or approval is granted subject to conditions or the application is refused, under article 22 of the Town and Country Planning (General Development Procedure) Order 1995, the local planning authority shall give notice of the reasons for refusal or for any condition imposed. However, there is no duty to give reasons when granting planning permission: see *R* v *Aylesbury Vale District Council, ex p Chaplin* [1997].

In *R* v *Yeovil Borough Council, ex p Trustees of Elim Pentecostal Church* [1971], the planning committee resolved that their town clerk be authorised to approve the application for planning permission subject to certain conditions. At a later meeting, the committee decided to refuse planning permission. The applicants appealed to the Divisional Court, contending that the committee had granted a conditional permission at their first meeting. The court decided that there existed no planning permission, as no notice in writing had been issued as was required by the then General Development Order. In this decision, Lord Widgery CJ relied on *dicta* of Lord Denning MR in *Slough Estates Ltd v Slough Borough Council (No 2)* [1969] CA where he said that the grant of a planning permission has to be in writing.

Lord Widgery CJ himself came to a different conclusion in *Norfolk County Council v Secretary of State for the Environment* [1973], where it was decided that the decision of a planning committee to refuse an application was the decision, and the written notification, which incorrectly granted planning permission, could not be regarded as having any validity.

The *Norfolk* case was followed in *Co-operative Retail Services Ltd v Taff-Ely Borough Council* [1979]. Here, a resolution of the district planning authority favoured the grant of planning permission, but subject to a consultation procedure with the county planning authority as it involved a county matter: the Court of Appeal decided the resolution was not a grant of planning permission. The letter issued by the clerk of the district planning authority, of a planning consent, was held to be void and of no legal effect. In *R v West Oxfordshire Borough Council, ex p CH Pearce Homes Ltd* [1986], Woolf J fully reviewed the case law and decided that the decision of the planning committee could not by itself be the planning permission.

These cases suggest that the written planning consent cannot always be relied upon if there has been a mistake in its issue, a forged signature, or it is issued without authority. A purchaser paid £500,000 for certain land in the *Co-operative Retail Services* case on the sight of the written planning consent that turned out to be of no legal effect. A justification for declaring the permission void in that case was given by Lord Denning MR: a wrong planning permission may result in damage to the public interest and this must prevail over any harm to the private interest of individuals who may be able to resort to private remedies.

Reference has already been made in Chapter 7 to the importance of the planning application. The details it contains may be considered in construing the meaning of the planning permission where that permission expressly refers to the application and therefore incorporates it into the permission (see *Slough Estates Ltd v Slough Borough Council (No 2)* [1971] HL). The application and any other relevant material may be considered in construing an ambiguously worded permission: see *Staffordshire Moorlands District Council v Cartwright* [1992]. Further guidance on the construction of a planning permission is found in *R v Ashford BC, ex p Shepway DC* [1999].

However, in *Kwik Save Discount Group Ltd v Secretary of State for Wales* [1981], where there was no express reference to the application for planning permission, the Court of Appeal concluded that an application for planning permission, taken as a whole, was implicitly limited to the use of the subject premises for the sale of motorcars; the permission could not be construed as including a supermarket.

9.2.2. The effect of planning permission

Unless a planning permission otherwise provides, any grant of planning permission lasts for the benefit of the land and of all persons with an interest in that land (section 75 of the 1990 Act). Planning permission for mining operations is therefore not abandoned during a period when the operations are suspended: *Pioneer Aggregates* v *Secretary of State for the Environment* [1984]. However, where planning permission has been granted for a material change of use, it is 'spent' once the authorised change has taken place and cannot be used to revert to the same use on a second occasion: *Cynon Valley Borough Council* v *Secretary of State for Wales* [1986]. Where land has the benefit of more than one planning permission, can any permission, or any part of a permission, be utilised?

> *Lucas & Sons* v *Dorking and Horley Rural District Council* [1964]
>   Planning permission was granted in 1952 to erect 28 houses on a plot of land. A further permission was granted in 1957 to erect six detached houses on the same plot. Lucas built two detached houses, under the 1957 permission, and 14 houses under the 1952 permission. Winn J decided that, in the case where a planning permission covered a scheme of development, there could not be a breach of planning control by the completion of only some of the houses within the scheme, because of the obvious injustice to the purchasers of those houses were the law otherwise. Accordingly, the 1952 permission was valid as to the 14 houses built, and so also was that of 1957.

The problem may arise rather differently where a subsequent planning permission is inconsistent with an earlier:

> *Pilkington* v *Secretary State for the Environment* [1973] DC
>   In 1953, permission was granted to erect a bungalow in the northern part of a plot of land, with the rest of the plot to be used for a smallholding. In 1954, permission was granted to build a bungalow in the middle of the plot, and a further permission was granted to build another bungalow in the southern part of the plot. The two bungalows authorised by the later two permissions were erected. It was held that the permission of 1953 could then no longer be implemented as intended, as it was inconsistent with the later permissions which now prevented the use of the plot as a smallholding.

Planning permission may authorise a number of separate buildings, as in the *Lucas* case, or just one. Moreover, while not all the development need be carried out in the circumstances that arose in the *Lucas* case, is there a breach of the planning permission if part only of one building is not completed, or not completed in accordance with the permission? The following case deals with this point in the context of enforcement. However, the same question could arise under section 94 of the 1990 Act where, following a completion notice, planning permission may cease to have effect if the development authorised by the planning permission is not completed. This section is dealt with at page 96 below.

> *Copeland Borough Council* v *Secretary of State for the Environment* [1976] DC
> A house was built in accordance with a planning permission with the exception that the roof tiles were different in colour from those specified. It was held that there was a breach of that planning permission, notwithstanding that most of what it authorised was carried out, the part not so authorised or carried out meant that the development was not as authorised.

Obvious difficulties still remain. What if a planning permission is for development consisting of a house and detached garage, and the garage is not built. Is the development that is carried out, in this case only the house, development authorised by the planning permission? In *Sheppard* v *Secretary of State for the Environment* [1975], Willis J divided a complex planning permission which in part related to a change of use that was not, in the end, carried out. Certain conditions in the planning permission were not complied with, but by dividing the planning permission, the conditions were found to relate to the change of use and therefore no breach of these conditions occurred by carrying out the development that did not involve the change of use. Conditions are not binding if the permission to which they relate is not exercised: *Handoll* v *Warner Goodman & Streat* [1995].

9.2.3. Planning permission differs from application

Questions have arisen as to whether the local planning authority may grant permission for part of the development sought in the application. For example, if an application is for the erection of ten separate houses on ten separately identifiable plots, can the authority grant permission for only five of the houses? Sir Douglas Frank thought so in *Kent County Council* v *Secretary of State for the Environment* [1976] provided the application contained separate and divisible elements; it was then lawful to deal with each separately.

A different view was taken in *Bernard Wheatcroft Ltd* v *Secretary of State for the Environment* [1980] where permission had been sought for the residential development of some 35 acres. Forbes J considered that it would be unlawful to grant planning permission for development which was substantially different from that sought in the application, as to do so would deprive those entitled to be consulted of their consultation rights. He thought that a test of substantial difference was to be preferred to the test of severability; but if the latter test was used, it should only be in connection with the test of substantial difference. The correct test, in his view, was whether the development to be permitted was substantially different from that in the application, and the difference in substance was to be determined in relation to the rights of those to be consulted. Forbes J considered that a planning condition which had the effect of reducing the permitted development to 25 acres was therefore valid; residential development on 25 acres was not substantially different from residential development on 35 acres. This planning condition is considered further on p. 101.

It is submitted that as many planning applications contain separate and divisible elements, the test of severability should be applied. Nevertheless, where the application is not severable, then the test of substantial difference would be appropriate.

If planning permission is granted for, say a single building, and the building is not completed, it may be possible to secure the completion of the development by the service of a completion notice under section 94 of the 1990 Act. However, a condition requiring the completion of development in a planning permission relating to a large estate would probably be unenforceable.

In *Breckland District Council* v *Secretary of State for the Environment* [1992], it was held that the submission of an amended plan enlarging the application site was not *ipso facto* invalid, but it must be looked at with special care. The decision to approve the amended plan in that case was invalid having regard to the *Bernard Wheatcroft* case.

9.2.4. *Planning permission for buildings*

Where planning permission is granted for the erection of a building, then, if no purpose is specified, the permission will authorise the building to be used for the purpose for which it is designed (section 75(3) of the 1990 Act). A planning permission for the use of land will not, however, authorise the carrying out of building operations: *Sunbury-on-Thames Urban District Council* v *Mann* [1958]; therefore planning permission for 'warehousing' is not permission for building operations to erect warehouses: *Wivenhoe Port Ltd* v *Colchester Borough Council* [1985]. Strictly, planning permission is for a material change of use, not simply a use: see the *Cynon Valley* case (at **9.2.2** above).

The use of a building can be prescribed by the description used in the permission, rather then by way of a condition. In *Wilson* v *West Sussex County Council* [1963], the planning permission described the development as 'an agricultural cottage'.

9.2.5. *Planning permission for existing buildings,*
works or use of land

Planning permission may be sought, and granted, for the retention on land of buildings or works, or the continuance of a use of land, where these matters exist at the time of the application and do so in breach of planning control (section 73A of the 1990 Act).

9.2.6. *Planning permission and time limits*

Every planning permission is to be granted subject to a condition that development must be begun within three years of the grant. In the absence of an express condition, every planning permission is deemed granted subject to such a condition.

The local planning authority may grant, in appropriate cases, planning permission subject to a longer or shorter time limit, having regard to the provisions of the development plan and to any other material considerations (section 91 of the 1990 Act). The three-year period is extended by one year where there are any proceedings challenging the validity of the permission: see section 91(3A).

Where outline planning permission is granted for building or other operations, it shall be granted subject to a condition that application for approval of the 'reserved matters' must be made within three years from the outline planning permission. The development must begin within five years of the outline planning permission or two years from the final approval of the reserved matters, whichever is the later date (section 92(2)).

A development is begun on the date when the following 'material operations' have been carried out: any work of construction or erection of any work, of demolition of a building; digging of a foundation trench; laying of mains or pipes or part of the foundations; laying or constructing a road or part of one; and change of use of any land which constitutes material change of use (section 56 of the 1990 Act). The laying of pipes and a road constitute 'material operations' even if the pipes are not quite in accordance with the plans: *Spackman* v *Secretary of State for the Environment* [1977]; and in *Malvern Hills District Council* v *Secretary of State for the Environment* [1982], the pegging out of an estate road amounted to a specified operation. In *East Dunbartonshire Council* v *Secretary of State for Scotland* [1999], it was held that it is irrelevant that a specified operation might be undertaken as a mere token or pretence; an objective test must be applied as to whether what was done is a specified operation. In *R (Connaught Quarries Ltd)* v *Secretary of State for the Environment, Transport and the Regions* [2001], the grubbing out of a hedge preparatory to creating an access road was an insufficient operation to implement a permission.

If the development authorised by a planning permission is begun before the expiry of the time limit, but the time limit has expired before the development has been completed, the local planning authority may serve a completion notice requiring the development to be completed within a stated period of not less than 12 months. Failure to comply with the notice means that the planning permission ceases to be valid. There are certain rights of appeal (section 94 of the 1990 Act).

Thus, development permitted by a planning permission must commence before the expiration of the time limit. However, where the permission was subject to planning conditions, it was held in *F G Whitley & Sons* v *Secretary of State for Wales* [1992] that a permission is controlled by its conditions, and that therefore if operations commence in breach of the conditions, the operations are not lawfully commenced and there has been a breach of planning control. The effect of conditions in this context was considered in *R(Hart Aggregates Ltd)* v *Hartlepool Borough Council* [2005] and in *Greyfort Properties Ltd* v *Secretary of State for Communities and Local Government* [2011].

9.2.7. *Planning permission unnecessarily obtained*

Interesting questions arise if planning permission is unnecessarily obtained in the belief that a proposed activity is development; this situation is most likely to involve a change of use. Are the previous existing use rights thereby lost, and is one bound by conditions attached to the planning permission? In *Newbury District Council* v *Secretary of State for the Environment* [1981] HL, where, as it turned out, planning permission was not necessary to use some former hangars as warehouses, it was decided that the owners could rely on their previous existing use rights and ignore any planning conditions (although for the validity of these, see the following text) attached to a planning permission.

9.2.8. *Outline planning permission*

For applications for outline planning permission, and for the approval of reserved matters, see Chapter 7; for time limits, see at **9.2.6** above on duration of planning permission.

9.3. Planning conditions

9.3.1. *The statutory power to impose conditions*

A local planning authority 'may grant planning permission ... subject to such conditions as they think fit' (section 70(1)(a) of the 1990 Act).

Apart from this apparently wide power, a local planning authority may impose conditions:

> ... *for regulating the development or use of any land under the control of the applicant (whether or not it is land in respect of which the application was made) or requiring the carrying out of works on any such land, so far as appears to the local planning authority to be expedient for the purposes of or in connection with the development authorised by the permission...* (section 72(1)(a) of the 1990 Act)

Section 72(1)(a) can only be used if the land made subject to the conditions is under the 'control' of the applicant. Such conditions may affect land not the subject of the planning application. 'Control' does not necessarily mean ownership, and what amounts to control is a question of fact and degree: *George Wimpey & Co Ltd* v *New Forest District Council* [1979].

There is a further power:

> ... *for requiring the removal of any buildings or works authorised by the permission, or the discontinuance of any use of land so authorised, at the end of a specified period, and the carrying out of any works required for the reinstatement of land at the end of that period.* (section 72(1)(b) of the 1990 Act)

The conditions referred to in section 72(1)(b) in the preceding text would apply to a temporary or limited period permission.

Where planning conditions are imposed in a planning permission, the conditions only become effective if the permission is implemented: *Handoll* v *Warner Goodman & Streat* [1995]. However, where a planning condition required certain matters to be done before any development commenced, the commencement of that development without compliance with the condition was unauthorised, and there could be no legitimate expectation that the permission could be treated as having been implemented, as that would bypass statutory safeguards for the public: *Henry Boot Homes Ltd* v *Bassetlaw DC* [2002].

To the extent that the conditions are otherwise valid, they will bind successors in title to the affected land. The existence of planning conditions will normally be discovered, as those imposed after July 1977 in express planning permissions are registrable as local land charges. A failure to register does not render the conditions unenforceable. In *Atkinson* v *Secretary of State for the Environment* [1983], where land was sold between the application for, and the grant of, planning permission, a condition affecting the land sold was triggered by the implementation of the permission on the vendor's retained land. The purchaser was bound by the condition.

9.3.2. The limit on the statutory power

At an early stage of the history of planning law, the courts, in considering the legality of conditions, established a set of useful criteria. The following cases show how some of these criteria have developed.

Pyx Granite Co v *Ministry of Housing and Local Government* [1958] CA
In this case, conditions governing the crushing and screening of granite were imposed in a permission to quarry granite. If complied with the conditions would have retracted a previous lawful use of certain plant. Lord Denning MR said that despite the wide power to impose conditions 'to be valid [they] must fairly and reasonably relate to the permitted development … [and] the planning authority are not at liberty to use their powers for an ulterior object, however desirable that object may seem to them to be in the public interest'. The conditions were held to satisfy these requirements.

Fawcett Properties Ltd v *Buckingham County Council* [1961] HL
In a planning permission for a cottage in the green belt, a condition was imposed restricting its occupation to persons employed or last employed in agriculture. Lord Denning (again) added two further criteria, to that contained in *Pyx*: the condition must not be so uncertain as to have no sensible or ascertainable meaning (ambiguity is not uncertainty); and an authority entrusted with a discretion must not impose a condition that is so unreasonable no reasonable council could have imposed it. In this case, the condition was also upheld, as it satisfied the objects of the green belt policy.

Hall & Co v Shoreham-by-Sea Urban District Council [1964] CA

A planning condition required the developer to construct an ancillary access road and provide rights of access over it to other persons. The Court of Appeal decided the condition was void for two reasons: it required the developer to provide extensive rights of access to other persons over his own private land, and there was an alternative procedure under the Highways Acts for the acquisition at public expense and on the payment of compensation. Wilmer LJ referred to *Colonial Sugar Refining Co Ltd v Melborne Harbour Trust Commissioners* [1927] where Lord Warrington said: 'a statute should not be held to take away private rights of property without compensation unless the intention to do so is expressed in clear and unambiguous terms'.

The whole question of legal validity was reconsidered by the House of Lords in *Newbury District Council v Secretary of State for the Environment* [1981], which approved some broad principles: a condition must not be totally unreasonable that no reasonable authority would impose it; it must relate to the permitted development; it must serve a planning purpose and not some other ulterior motive; and it must not be uncertain as to meaning. Over the years, the cases have collected other problems that are considered in the following text.

The decisions in these and other cases, together with the additional guidance of the Secretary of State, are now usefully summarised in paragraph 14 of Circular 11/95 – *The use of conditions in planning permissions* – to the effect that conditions must be:

(a) Necessary;
(b) Relevant to planning;
(c) Relevant to the development to be permitted;
(d) Enforceable;
(e) Precise;
(f) Reasonable in all other respects.

9.3.3. Unreasonableness

This criterion is really about the application of the *Wednesbury* principles (see Chapter 20), and restrains an authority from abusing a power granted by Parliament. The question is whether there has been a perverse use of the power to impose conditions. It is arguable that the other criteria that are considered in the following text, are but more specific applications of the 'unreasonableness' criterion. A condition requiring an applicant to provide land, construct a road upon it and grant rights of passage over it was held to be totally unreasonable in *Hall & Co Ltd v Shoreham-by-Sea Urban District Council* [1964] CA, on the basis that the local authority would obtain the benefit of a public road without having to pay for it. They had alternative powers under the Highways Acts to achieve a similar object, but at their cost.

In *R* v *Hillingdon London Borough Council, ex p Royco Homes Ltd* [1974] DC, two conditions imposed in a planning permission for residential development required the completed dwellings to be first occupied by persons on the authority's housing waiting list, and that they should have security of tenure. These were held to be invalid, as they went beyond anything Parliament could have intended, or a reasonable authority could have imposed, as they required the developer to discharge at his expense the duties of the local housing authority.

A condition that reduces the permitted development below the development applied for is not totally unreasonable provided the permitted development is not substantially different from that in the application so as to deprive those entitled to be consulted of that opportunity: *Bernard Wheatcroft Ltd* v *Secretary of State for the Environment* [1980] (where the condition reduced the area for residential development from 35 acres to 25 acres – see also p. 95).

In *Mirai Networks Ltd* v *Secretary of State for the Environment* [1994], a condition preventing the creation of additional interior floor space was not invalid, notwithstanding that such work was not development, as the condition had a legitimate relationship to the permitted development.

9.3.4. Relate to the permitted development

A simple example may illustrate this criterion. If an applicant submits a planning application to redevelop his petrol-filling station in Queenstown, the authority cannot impose a condition that affects another of his filling stations in Kingstown. The cases involve more difficult questions than this in practice.

In *Penwith District Council* v *Secretary of State for the Environment* [1977], planning permission had been granted for a factory extension subject to conditions that restricted the use of, and emission of noise from, the old factory as well as the extension. The extension was the permitted development; but the conditions also affected the old factory. They were held to be lawful as the new extension enabled the old factory to be more intensively used, and without the conditions, the extension would not have been permitted.

The House of Lords had the opportunity of considering this problem in *Newbury District Council* v *Secretary of State for the Environment* [1981]. A rubber company had acquired an interest in some existing hangars, and applied for planning permission to use them for warehouses. This was granted subject to their removal at the end of a specified period. It was held that as the permitted development was a permission for a change of use, a condition requiring the removal of the hangars was invalid as removal of the hangars had nothing to do with the use of them as warehouses. (This case involved other issues, see Chapter 3.)

One way of deciding whether a condition relates to the permitted development is to consider whether the condition helps to eliminate the detrimental consequences of the proposed development: *Gill* v *Secretary of State for the Environment* [1985]. Frequently, a local planning authority may wish to see certain matters achieved before granting planning permission for a scheme of development.

For example, they might want road improvement works carried out to the local highway network before granting planning permission for residential development. The imposition of a planning condition requiring the developer to either carry out or pay for such works might well be invalid on a number of grounds. The works may have to be carried out on land not controlled by the developer, or the works may not be strictly necessary as a consequence of redevelopment. It is here that the imposition of a negative condition preventing the commencement, or use, of the development may be considered (see part **9.3.9** below – *negative conditions*).

In *Tarmac Heavy Building Materials Ltd* v *Secretary of State for the Environment, Transport and the Regions* [2000], where planning permission was granted for sand and gravel extraction, a condition requiring the removal of existing buildings on the cessation of the extraction workings was not fairly and reasonably related to the planning permission.

9.3.5. Serve a planning purpose

The condition in the *Fawcett* case (see the preceding text) served a valid planning purpose, as it furthered the maintenance of green belt policies. It was accepted that the condition requiring the removal of the hangars in the *Newbury* case (see the preceding text) served a valid planning purpose – the improvement of the amenity of the area. However, the conditions in the *Royco* case (see the preceding text), although void for unreasonableness, could be said to have served a housing purpose rather than a planning purpose. This criterion may therefore often overlap with that concerning unreasonableness, as a condition imposed for some ulterior motive may be struck down as unreasonable. An ulterior motive may involve a non-planning object. The courts are more familiar with considering the question of unreasonableness, rather than deciding what is a planning purpose.

Some guidance as to what is a planning purpose may, it is submitted, be found in the matters which should be the subject of policy and proposal statements in development plans (see Chapter 6). There is also helpful guidance in Circular 11/95 on the relevance of a condition to planning.

9.3.6. Certainty as to meaning

Lord Denning MR, in the *Fawcett* case (see the preceding text) said that a planning condition is only void for uncertainty if it can be given no meaning or no sensible or ascertainable meaning: ambiguity or a meaning that may lead to absurd results can be construed by the courts. A condition in *Britannia Ltd* v *Secretary of State for the Environment* [1978] required the provision of a small village-type social/shopping centre, and this condition the Secretary of State decided was void for uncertainty as the words were incapable of definition and therefore impossible to give effect to. Although the judge thought this condition could be given a meaning, he, unfortunately, did not actually decide whether it was valid or not because he quashed the Secretary of State's decision on another matter.

A condition reserving for future approval the density of the proposed development in an outline planning permission is not within the definition of reserved matters. Such a condition is not regarded as uncertain: *Inverclyde District Council v Inverkip Building Co Ltd* [1983].

Circular 11/95 also gives guidance as to the need for precision. However, in *Chichester District Council v Secretary of State for the Environment* [1992] the court accepted that although a condition restricting the use of chalets to use for holiday accommodation only might be difficult to enforce, the situation would not be impossible and the condition was held to be valid.

9.3.7. Taking away existing rights

The *Hall* case (see the preceding text) was decided on the basis that the condition requiring the provision of land and its dedication as a public road did involve taking away property without compensation, and was therefore invalid as being totally unreasonable. This was applied in *Bradford City Metropolitan Council v Secretary of State for the Environment* [1986] where a road improvement was required on land outside the applicant's control. However, the trend of recent cases is to allow conditions that interfere with or take away existing proprietary rights provided the conditions are otherwise valid; the argument is that the applicant only loses his existing rights if he makes use of the planning permission.

A condition in *City of London Corporation v Secretary of State for the Environment* [1972] restricted the use of office premises to those of an employment agency. This prevented a change to any other office use within the same class of the Use Classes Order, a change which would not have been development (see section 55 of the 1990 Act). The condition was upheld as valid.

In *Kingston-upon-Thames London Borough Council v Secretary of State for the Environment* [1974], the effect of a condition imposed with a planning permission to redevelop a railway station required certain other land of British Rail to be made available for car parking. This would have interfered with their existing lawful use of that other land without payment of compensation. Relying on the earlier case of *Prossor v Minister of Housing and Local Government* [1968], the court upheld the validity of the condition. In any event, the provisions of section 72(1)(a) of the 1990 Act (see earlier) seems to confirm this use of a condition.

A planning condition may restrict development that would otherwise be permitted by a development order (see *Gill v Secretary of State for the Environment* [1985]). As the condition is only triggered by the implementation of the express planning permission, it will be in the applicant's interest to initiate the permitted development first. In *Mirai Networks Ltd v Secretary of State* [1994], it was held that a condition may prevent activities that are excluded from the definition of development.

9.3.8. Conditions dependent on matters outside control of the applicant

Conditions imposed under section 72(1)(a), for regulating the development or use of land, whether or not it is land in respect of which the application was made, which are necessary for the permitted development, can only be imposed in respect of land under the control of the applicant (see p. 98).

In *Augier* v *Secretary of State for the Environment* [1978], the applicant gave an undertaking that he would secure rights over land not in his ownership to secure visibility splays to an access road. On the strength of that undertaking, planning permission was granted subject to a condition that the undertaking would be fulfilled. It was held that the condition was valid and binding as the applicant could not later be heard to suggest that the undertaking could not be complied with, having encouraged the grant of permission on the strength of it. However, in *Bradford City Metropolitan Council* v S*ecretary of State for the Environment* [1986], a condition requiring road improvements on land not in the applicant's ownership or control was held to be manifestly unreasonable and *ultra vires*.

The problems associated with undertakings are better secured by a planning agreement considered in the next section of this chapter.

In *British Railways Board* v *Secretary of State for the Environment* [1993], the House of Lords had to consider the validity of imposing a negative condition relating to the provision of an access over land not controlled by the applicant, although the land was within the application site. A planning application can relate to land not controlled by the applicant. Although it was said in the case that the development might have no reasonable prospect of being achieved without the consent of the owner of the access land, it is still valid to impose a negative condition that is justified and appropriate on planning principles.

9.3.9. Negative conditions

It can be seen so far that a condition requiring the applicant to provide land for a highway at his expense, or to do something on land not in his control, may be struck down as unreasonable and *ultra vires*. In *Grampian Regional Council* v *City of Aberdeen* [1984], Lord Keith in the House of Lords said that there are many uncertainties in the development process, and a condition worded to prevent the commencement of development until matters (including those not within the applicant's control) necessary for the development were achieved would be valid. In the *City of Bradford* case, it was also suggested that the road improvements could also have been validly provided for by a suitably worded negative condition.

The use of negative conditions was also considered in *British Railways Board* v *Secretary of State for the Environment* [1993] in a situation where there was no reasonable prospect of development being implemented in the absence of consent of an owner of land who was not the applicant. The House of Lords saw no reason why a desirable negative condition could not be imposed.

9.3.10. The effect of an unlawful condition

If a condition is unlawful, what is the effect on the permission to which it is attached?

In *Kent County Council* v *Kingsway Investments* [1970], the House of Lords suggested that if a condition was fundamental to the permission, both should be quashed if the condition was held to be invalid. That was the case in the *Pyx Granite*, *Shoreham* and the *Royco* cases (see the preceding text). There is therefore a danger to the applicant who successfully challenges a condition that he will lose the whole planning permission.

Although there is *dicta* in the *Kingsway* case suggesting circumstances where a void condition may alone be severed, because it is not fundamental to the development permitted, there is little reliable case authority to support this. In *Allnatt London Properties* v *Middlesex County Council* [1964], a condition was found void, because it affected existing use rights, and the planning permission remained severed of the condition; but that case has been doubted following the *Kingston* case (see at **9.3.7** above). In *Mouchell Superannuation Fund Trustees* v *Oxfordshire County Council* [1992], the Court of Appeal accepted that a positive condition relating to access over land not in the control of the applicant was invalid and because it was of fundamental importance to the planning permission as a whole, the permission could not stand.

9.3.11. The circumstances in which conditions may be imposed

So far we have been considering the legal validity of some doubtful conditions. In practice, a great many conditions are imposed, and their legal validity is never in question. They are imposed in cases where the application would otherwise be refused, and the details of the application do not, or cannot, make reference to the matters which are the subject of the conditions. For example, in a planning permission to redevelop a petrol-filling station, conditions may be attached which limit the total lumens of light onto the forecourt, or which restrict the range of products sold at the cash point to motor accessories such as tyres, batteries and light bulbs. Such matters would not otherwise be the subject of planning controls.

A condition may also be imposed to reinforce some matter of particular importance to the planning authority which may anyway be in the application. The fact that it is highlighted in this way may further the possibility that it will be secured: a breach of a condition is clearly a breach of planning control.

Circular 11/95 sets out in its Annex very full guidance on the use of planning conditions. Apart from general considerations, it also gives guidance relating to specific problems such as contaminated sites and personal conditions.

9.4. Challenging planning conditions

There are various ways of challenging planning conditions. They may be invalid in law, for the reasons considered in the preceding section of this chapter, or they may be unnecessary or irrelevant on planning grounds.

A planning condition may be challenged by appeal to the Secretary of State where the applicant is aggrieved by the decision to impose conditions (section 78 of the 1990 Act). The planning merits of the conditions can then be considered by the inspector appointed by the Secretary of State (this is more fully considered in Chapter 17). A further appeal from the decision of the Secretary of State, or his inspector, may be taken to the High Court, but only to question the legal validity of a planning decision. Because the Secretary of State, in determining this type of appeal, may reconsider the whole of the original planning permission, it is often preferable to use the alternative procedure under section 73 of the 1990 Act.

Under section 73, an application may be made for planning permission for the development of land without complying with conditions subject to which a previous planning permission was granted. On such an application, the local planning authority shall consider only the question of the conditions. Thus, whatever view the authority might take, the original permission must be left intact: see *Knott* v *Secretary of State for the Environment* [1996].

An alternative procedure is available if the applicant believes the planning condition is legally invalid: he may apply direct to the court for a declaration to that effect (*Pyx Granite Co* v *Ministry of Housing and Local Government* [1958]); or for a quashing order to quash the decision of the authority (*R* v *Hillingdon London Borough Council, ex p Royco Homes Ltd* [1974]). (See further on this in Chapter 20.)

A further possibility is to proceed with the development without complying with the condition, and then in any appeal to the Secretary of State against an enforcement notice that may be served in respect of the non-compliance, the legal and planning merits can be raised. This procedure has some obvious difficulties. A local planning authority may well decide to issue a breach of condition notice under section 187A of the 1990 Act (see Chapter 15). Where a breach of condition notice is served, there is no procedure to challenge the condition on the ground that it cannot be justified by planning criteria.

9.5. Planning agreements and obligations

9.5.1. The statutory basis

Agreements between local authorities and developers and landowners are common. They are used for a number of purposes. There is a general power for local authorities to make or receive payments or enter into agreements 'calculated to facilitate, or is conducive or incidental to, the discharge of their functions': section 111, Local Government Act 1972. There is a specific power in section 38 of the Highways Act 1980 for a developer to make an agreement for the provision and adoption of a road as a public highway; and power in section 278 of the same Act for a highway authority to agree with a developer to carry out road works in return for a contribution by the developer towards the costs.

In relation to planning agreements, especially those containing continuing obligations on the part of the landowner, it is obviously necessary that such agreements are enforceable against successors in title to the affected land.

Until 1991, there were powers in the old section 106 of the 1990 Act (re-enacting section 52 of the Town and Country Planning Act 1971) enabling a local planning authority to make an agreement with any person interested in land in their area for the purpose of restricting or regulating the development or use of the land, either permanently or during such period as may be prescribed by the agreement. These agreements were enforceable as if they were restrictive covenants. However, new provisions for what are now called planning obligations were introduced in 1991. The new section 106 provides that any person interested in land may, by agreement or otherwise, enter into what is called a planning obligation.

Planning obligations may therefore be offered unilaterally and do not always have to be made with the agreement of a local planning authority. The power of a landowner to bind land unilaterally enables a developer to overcome any reluctance of a local planning authority to enter into an agreement in a case where, for example, the Secretary of State or the inspector on an appeal is prepared to grant planning permission if some planning objection is first addressed. This could be the lack of off-site works that cannot be addressed by a mere planning condition. The developer can offer these works by way of a unilateral planning obligation, and planning permission can then be granted on appeal.

9.5.2. *Agreements cannot fetter statutory powers*

There is a general principle that a body with statutory powers cannot by contract agree not to use its statutory powers (*Ayr Harbour Trustees* v *Oswald* [1883]). It is also wrong for a local planning authority to decide a planning application in order to fulfil an agreement. In *Stringer* v *Minister of Housing and Local Government* [1970], the planning authority had an agreement with the owners of Jodrell Bank radio telescope to restrict development that might interfere with its operations; the court decided that proper consideration must be given to every planning application, and a decision could not be given merely to fulfil such an agreement.

A planning agreement may contain terms that the local planning authority will not exercise one or more of its powers under the 1990 Act (i.e., not take enforcement action, or not refuse permission for a certain application). These terms would ordinarily be *ultra vires* as an authority cannot contract not to exercise a statutory power: *Ransom and Luck Ltd* v *Surbiton Borough Council* [1949]. Under the pre-1991 law, the statutory provisions provided that an agreement could not restrict the use of powers exercised in accordance with the provisions of the development plan. This, and the principle that an agreement cannot fetter the exercise of statutory powers, was considered in *Windsor and Maidenhead Royal Borough* v *Brandrose Investments Ltd* [1983]. The planning authority made an agreement allowing the redevelopment of certain buildings. They later extended a conservation area and sought to exercise their statutory powers to prevent the demolition of the buildings, the exercise of those powers being contrary to the agreement. It was decided that the authority were free to use their powers to extend the conservation area, and the exercise of that power could not be fettered by the agreement. The position relating to the power of an authority to override contractual obligations is probably now decided in the following two cases: *Dowty Boulton Ltd* v *Wolverhampton Corporation* [1971] and *R* v *City of London Corporation, ex p the Barbers of London* [1996]. The effect of these cases is that while an authority will be bound by the terms of an agreement for most purposes, if there is some new scheme or policy the authority may be entitled to rely on statutory powers to override an agreement in the furtherance of the new project or scheme.

Apart from the enabling power in the 1990 Act, there are numerous local Acts, usually of county councils, that contain analogous powers. In the past, many planning agreements are made under these.

9.5.3. The scope of planning obligations

Section 106(1) and (2) sets out the matters that can be contained in a planning obligation:

(a)   Restricting the development or use of the land in any specified way;
(b)   Requiring specified operations or activities to be carried out in, on, under or over the land;
(c)   Requiring the land to be used in any specified way;
(d)   Requiring a sum or sums to be paid to the authority.

9.5.4. Enforcement of planning obligations

The advantage to the local planning authority of a planning obligation over the normal means of development control is that the remedies for breach are contractual and lie in the field of private law, rather than the statutory procedures of the planning Acts which are subject to appeals and the requirements of administrative fairness.

Under section 106(5), a restriction or requirement imposed under a planning obligation may be enforced by an injunction. Further, under subsection (6) the local planning authority may enter land to carry out operations and recover the expenses in so doing where such operations have not been carried out under the terms of a planning obligation. A planning obligation is registrable as a land charge: see subsection (11). It is enforceable against the original party to the obligation and any person deriving title from that person: see subsection (3).

Under the old law, in *Avon County Council* v *Millard* [1986], the Court of Appeal decided that where a breach of a planning agreement also involved a breach of planning control, the local planning authority were entitled to restrain the breach of the agreement by the remedy of injunction; the authority were not obliged to first exhaust the planning remedy of an enforcement notice. The same position probably applies to the new law. However, in *Southampton City Council v Hallyard Ltd* [2008], a section 106 agreement was unenforceable against a successor in title because the agreement failed to identify the interest of the original covenantor contrary to the statutory requirements. The court left open the question whether a person deriving title, but not a successor in title, would be bound. Declaratory relief is not available as to the meaning of an agreement, as this is inconsistent with the authority's enforcement powers: see *Milebush Properties Ltd* v *Tameside MBC* [2012], where judicial review was available. A section 106 agreement may include obligations by the local planning authority. In *Patel* v *Brent London Borough Council* [2005], the landowner had deposited money with the local authority, under the terms of an agreement, for certain road improvements. It was held that the local authority was not entitled to use the money for more extensive works.

9.5.5. The discharge or modification of covenants in pre-1991 planning agreement

Reference has already been made to the fact that old agreements were enforceable against successors in title to the land as if they were restrictive covenants. Restrictive covenants may be discharged or modified by the Lands Tribunal under section 84(1) of the Law of Property Act 1925 if it is satisfied the covenants are obsolete; they restrict the reasonable use of the land, or impede such use as to deny either practical benefits of substantial value or advantage, or are contrary to the public interest, and money will compensate any loss by reason of the discharge or modification; the beneficiary of the restriction agrees; or, no beneficiary will be injured (see now the Upper Tribunal (Lands Chamber)).

The Lands Tribunal allowed a modification of a covenant in a planning agreement in *Re Beecham Group Ltd's Application* [1980]. Although planning permission had been granted following an appeal to the Secretary of State, Beecham's could not develop in breach of the planning agreement. The Lands Tribunal considered the decision of the Secretary of State highly relevant in deciding that the covenant could be modified. The fact that the local planning authority would suffer no injury by the modification, because they owned no land capable of being benefited by the restriction, and that no objection to the development had been put forward by the authority on aesthetic grounds, seemed important to the Tribunal. It is curious that a private law approach is applied to an important matter of public law, in this case planning. Nevertheless, the Tribunal has allowed the modification of an agreement in *Re Cox's Application* [1985], to permit the sale of an old farmhouse.

In *Abbey Homesteads (Developments) Ltd* v *Northampton County Council* [1986], the Court of Appeal decided that the Tribunal was wrong in considering that a restriction reserving land for school purposes could be modified once the land was acquired by the educational authority. Because of surrounding development, that land had to be reserved for a school on a permanent basis, and it would be wrong to regard the restriction as no longer necessary.

9.5.6. Modification and discharge of planning obligations

Under section 106A, planning obligations can be modified or discharged by agreement between the parties bound and the local planning authority. Alternatively, a person against whom a planning obligation is enforceable may apply to the local planning authority for a modification or a discharge; there is a right of appeal to the Secretary of State. No such application may be made before the expiration of a 'relevant period' which is currently five years. In the case of planning obligations made in relation to a development consent granted a Nationally Significant Infrastructure is the relevant authority is the Secretary of State, In London it is the Mayor, and otherwise the local planning authority: see section 106A(11).

Section 84 of the Law of Property Act 1925 does not apply to a planning obligation.

9.5.7. Challenging planning obligations

In *Tesco Stores Ltd* v *Secretary of State for the Environment* [1995], the House of Lords decided that planning obligations are not analogous to planning conditions. Accordingly, the criteria for testing the legal validity of a planning condition (considered in the preceding text) do not apply to a planning obligation.

Although planning obligations are frequently entered into 'behind closed doors', Circular 05/2005 set out guidelines for their use; this was cancelled in March 2012 and replaced by the National Planning Policy Framework (2012).

9.6. Community infrastructure levy

Section 205 of the Planning Act 2008 authorises the making of regulations for the community infrastructure levy (CIL). The purpose of this is to ensure that the costs incurred in supporting the development of an area can be funded (wholly or partly) by the owners or developers of land in a way that makes development of the area economically viable: see section 205(2) as amended by section 115 of the Localism Act 2011. Local authorities, and similar bodies, will be charging authorities entitled to charge CIL for their respective areas: see section 206. Any person may assume liability to pay the CIL before development begins, such as an intending purchaser, but in the absence of such an assumption, the regulations will provide that the owner or developer shall be liable. A charging authority must publish a charging schedule setting rates or other criteria for the imposition of CIL. The regulations must require the charging authority to apply the levy for supporting development by to funding the provision, improvement, replacement, operation or maintenance of infrastructure that will include: roads and other transport facilities, flood defences, schools and other educational facilities, medical facilities, sporting and recreational facilities, open spaces, and affordable housing: see section 216(2). The Community Infrastructure Levy Regulations 2010 are now in force. The regulations deal with collection, enforcement and procedures. They also contain exemptions and reliefs.

The Localism Act 2011 makes some changes to the CIL. Section 211 of the Planning Act 2008 is amended to include an obligation on a charging authority to use appropriate evidence to inform their preparation of a charging schedule, and to provide for regulations to make provision about such evidence, its preparation and use. A new section 212A of the 2008 Act deals with the examination of a charging schedule by an examiner, and sets out a number of details, requirements and obligations in the consideration of a charging schedule; in particular, a charging authority may not approve the charging schedule if the examiner recommends its rejection. CIL regulations may require CIL paid to a charging authority to be passed to another person, and that where such money is passed to another person, it is used to support the development of the area by funding the provision, improvement, replacement, operation or maintenance of infrastructure, or anything else that is concerned with addressing demands that development places on an area: see the new section 216A of the 2008 Act.

9.7. Development consents under the Planning Act 2008

The Planning Act 2008 made provision for the making of applications to, and the issue by, the Infrastructure Planning Commission of development, consents for nationally significant infrastructure projects. Where issued, a development consent has the effect of granting planning permission in accordance with its terms. The Infrastructure Planning Commission has been abolished, as such, and incorporated in the Planning Inspectorate. These powers, and the effect of the changes made by the Localism Act 2011, are more fully described in Chapter 16.

Part III
Special controls

10 Listed buildings, conservation areas and other special areas

10.1. Introduction

There is a special control, additional to the development controls dealt with in the earlier part of this book, that applies to buildings of special architectural or historic interest. This special control has a more detailed relevance to the problem of the preservation of such buildings. This chapter also deals with the related policy concerning conservation areas: the object here is the preservation and enhancement of areas of special architectural or historic interest. A number of special areas that give further protection to the countryside, and the protection to ancient monuments and archaeological areas, are more briefly described.

Decision-makers (planning authorities and inspectors) must take into account the terms of the National Planning Policy Framework (NPPF) (March 2012), issued by the Secretary of State for Communities and Local Government, when considering planning applications, listed building consent applications, and appeals: see sections 19(2)(a) and 38(6) of the Planning and Compulsory Purchase Act 2004. In relation to conserving and enhancing the natural and historic environments, local planning authorities should take into account the desirability of sustaining and enhancing the significance of heritage assets, the positive contribution they make, and the desirability of new development making a positive contribution to local character and distinctiveness.

The Historic Buildings and Monuments Commission was established by the National Heritage Act 1983. Known simply as 'English Heritage', it has an advisory role, but is also concerned more generally with the preservation of, and the promotion of interest in, ancient monuments and historic buildings. The commission has more specific powers of control in relation to buildings in London.

10.2. Control over listed buildings

10.2.1. Listing

Apart from buildings temporarily protected by a building preservation notice, and discussed in the following text, a building must be listed by the Secretary of State to be protected by the controls considered in this part of this chapter. Guidance is contained in *Principles of Selection for Listed Buildings* (March 2010) – and in Circular 07/09 – *Protection of World Heritage Sites*. In deciding to list a building, consideration is given not only to the building itself, but the contribution its exterior makes to the architectural or historic interest of any group of buildings; and the desirability of preserving any feature of the building on the ground of its architectural or historic interest, any feature of the building or on the land within the curtilage of the building. The meaning of 'curtilage' was considered in *Dyer* v *Dorset County Council* [1989] and *Skerritts of Nottingham Ltd* v *Secretary of State for the Environment, Transport and the Regions (No 1)* [2000]. The important criterion is that the building is of special architectural or historic interest (Planning (Listed Buildings and Conservation Areas) Act 1990, section 1(3)).

If a building is listed, notice is served on every owner and occupier, and a copy of the list is registered as a local land charge. The local planning authorities are informed, and copies of the list are available for public inspection (section 2(1) to (5)). There is no appeal against listing at this stage. The consequences of listing can be seen in *Amalgamated Investment and Property Co Ltd* v *John Walker & Sons Ltd* [1976]. The purchaser of a warehouse had contracted to pay its value of £1,700,000. The building was listed after the exchange of contracts, and before completion. The effect of listing was to reduce its value to £200,000; the purchaser was obliged to complete the purchase at the original contract price.

On listing, a building is given a grade to indicate its significance: Grade I is of exceptional interest, Grade II is for important buildings, and Grade II is of special interest warranting preservation. The legal rules described in this chapter do not distinguish between the grades, but the grade has a relevance to decision-making. About 90 per cent of listed buildings are in Grade II, the lowest grade.

Where a building is listed, then every part of that building is listed, including any part, such as the interior, which is not itself of special interest: see *City of Edinburgh* v *Secretary of State for Scotland* [1997]. And anything that is fixed and ancillary to the building after listing is also protected: *Richardson Development Ltd* v *Birmingham City Council* [1999].

10.2.2. Meaning of a building

A listed building means a building included on a list compiled or approved by the Secretary of State. Any object or structure fixed to the building, or any object or structure within the curtilage of the building, which, although not fixed to the building, forms part of the land and has done so since before 1 July 1948, shall be treated as part of the building (section 1(5)). In *Attorney-General* v *Calderdale Borough Council* [1982], a terrace of cottages connected by a bridge to a mill, was regarded as listed, as the mill was a listed building. The House of Lords in *Debenhams plc* v *Westminster City Council* [1987] decided that the reference in the Act to a 'structure' meant something ancillary to the principal building, and would not include an independent building even if connected by a bridge or tunnel. In *R* v *Secretary of State for Wales, ex p Kennedy* [1996], it was held that the definition of 'fixture' was the same for listed buildings legislation as for any other area of the law, and that the inspector was entitled to hold that a clock was a fixture. A 'listed building' does not mean part of a building, but the whole building as listed: see *Shimizu (UK) Ltd* v *Westminster City Council* [1996].

10.2.3. Certificate of immunity from listing

If planning permission has been granted for any development consisting of the alteration, extension or demolition of a building, any person may apply to the Secretary of State for a certificate that the building will not be listed for five years (section 6(1) to (2)). The effect of this is to prevent the Secretary of State from listing the building, or the local planning authority from serving a building preservation notice, for the five years. There must be some doubt about the usefulness of this provision, as the Secretary of State is unlikely to issue his certificate if he believes the building may have the special interest, and in cases where it clearly has no such interest, no one will bother to apply in the first place. In *R (Bancroft)* v *Secretary of State for Culture, Media and Sport* [2005], it was held that the Secretary of State was entitled to take into account a number of factors, such as design flaws, in deciding not to list, and to issue a certificate of immunity, relating to Pimlico School, London.

10.2.4. Building preservation notice

It may appear to a local planning authority that an unlisted building is of special architectural or historic interest, and is in danger of demolition or of alteration in such a way as to affect its characterisation as a building of such interest. The local planning authority may then serve a building preservation notice on the owner and occupier. The effect of such a notice is to give to the building the legal protection it would enjoy were it listed and to inform the parties concerned that the Secretary of State will be asked to list (section 3(1) to (2) of the Listed Buildings Act 1990). If the Secretary of State does so list the building, it will then, of course, enjoy permanent legal protection. If he chooses not to list, the building preservation notice will cease to have effect. It only lasts for six months in any event. Any person who has an interest in the building which has been the subject of a building preservation notice, but which is not ultimately listed, is entitled to compensation for any loss or damage directly attributable to the effect of the notice. This would include sums paid to contractors for any breach of contract caused by the cancelling of work (section 29(1) to (2) of the 1990 Act).

10.2.5. Need for listed building consent

It is a criminal offence to demolish a listed building or to carry out any works of alteration or extension which would affect its character as a building of special interest without listed building consent (section 9(1) and (4)). The erection of a fence under the General Permitted Development Order is not 'demolition, alteration or extension' requiring listed building consent: *Cotswold District Council* v *Secretary of State for the Environment* [1985]. Whether any work will affect the character of a building of special interest is a question of fact and degree for the local planning authority: in one case the painting of a door in yellow was regarded by the Secretary of State as such work (although by the time the appeal was heard the painted door was acceptable, because the yellow paint had meanwhile faded!). However, where there is a proposal to demolish a part of a listed building, the original policy guidance in PPG 15 advised that if the part is a significant part, then the application should be treated as if there was a proposal to demolish the whole: as explained in *R (Sullivan)* v *Warwick DC* [2003], 'significant' must be considered in terms of quality and contribution, and not just scale.

Application for listed building consent is made to the local planning authority. Additionally, in the case of the demolition of a listed building, one month's notice has to be served on English Heritage, and an opportunity given to them to record the building; they may state that either they have recorded the building or they do not wish to do so. Where work has been carried out without listed building consent, then, although that remains a criminal offence, provision is made for written consent to be granted to retain the works of demolition, alteration or extension, as the case may be: the works will then be authorised from the date of the grant of that consent (section 8(3)).

Policy guidance is now set out in the National Planning Policy Framework (March 2012) with regard to the matters that should be taken into account when considering an application relating to a listed building.

If listed building consent is granted subject to conditions, it is similarly a criminal offence not to comply with these conditions (section 9(1)).

In determining the amount of any fine, the court shall in particular have regard to any financial benefit which has accrued or appears likely to accrue to him in consequence of the offence (section 9(5)). It is a defence to a prosecution that the work was urgently necessary in the interests of safety or health, or for the preservation of the building, and that a notice was given to the local planning authority in writing justifying in detail the carrying out of the works (section 9(3)). However, this defence is limited to where it can be shown that temporary support or shelter was not reasonably practicable; the minimum measures necessary were taken; and, the notice contains reasons justifying the works undertaken. Where, as here, there is a statutory defence, the burden of proof lies on the defendant: see *R* v *Alath Construction Ltd* [1991].

The criminal offence is one of strict liability. This means that a magistrates' court would be wrong in dismissing a charge against a building contractor for doing work on a building without listed building consent merely because the builder was ignorant of the listed building status: *R* v *Wells Street Metropolitan Stipendiary Magistrate, ex p Westminster City Council* [1986].

10.2.6. *Application for listed building consent and planning permission*

If the proposed works affect the special interest of a listed building, and also constitute development, both listed building consent and planning permission are required. Clearly, certain work to the interior of a building, or which does not materially affect its external appearance, or a demolition, while requiring listed building consent, may not need planning permission.

If planning permission is required, it may be applied for either before or at the same time as the application for listed building consent. A fee is payable for a planning application, none is payable for a listed building consent application.

The Planning (Listed Buildings and Conservation Areas) Regulations 1990 deal with the form and manner of making applications for listed building consent. Application for a listed building consent is made in the first place to the local planning authority. The application must be accompanied by a certificate of ownership or a certificate that the applicant has notified the owner along the lines of the certificate that must accompany a planning application (section 11 of the Planning (Listed Buildings and Conservation Areas) Act 1990).

The local planning authority must then advertise the application in a local newspaper and by way of a site notice stating where the proposals may be inspected, and inviting representations. Any representations received must be taken into account. Advertisement is not necessary if the application relates only to interior work to Grade II listed buildings (regulation 5 of the 1990 Regulations). The Secretary of State has directed that local planning authorities outside London must notify English Heritage of applications relating to works in respect of a Grade I or II* listed building or works involving demolition, and of their decision on them.

The Secretary of State may give a direction to call in an application (or such applications as he specifies) for his decision: see sections 13 and 15 of the Planning (Listed Buildings and Conservation Areas) Act 1990 and Circulars 01/01 and 09/05. The Secretary of State must be notified of any application which the local planning authority proposes to grant – thus enabling him to call in if he chooses (section 13). In London, it is English Heritage that must be notified. Additionally, any application to demolish a listed building must be notified to a number of specified and interested bodies and societies.

There is an appeal to the Secretary of State against a decision of the local planning authority to refuse consent, or to grant consent subject to conditions. The applicant may, as one of his grounds of appeal, claim that the building should not be listed (sections 20–21 of the Planning (Listed Buildings and Conservation Areas) Act 1990).

Further provisions regarding applications and appeals are found in the Planning (Listed Buildings and in Conservation Areas) Regulations 1990 (as amended).

In deciding whether to grant listed building consent, the local planning authority, or the Secretary of State, are required to have special regard to the desirability of preserving a listed building or its setting or any features of special architectural or historic interest which it possesses (section 16(2)).

10.2.7. Conditions and time limits

Listed building consent may be granted subject to conditions, and in particular, conditions with respect to the preservation of particular features of the building;

the making good of damage caused to the building by the works so authorised; and the reconstruction of the building, or any part, with the use of original materials where practicable (section 17(1)).

Listed building consent for the demolition of a listed building may be granted subject to a condition that the building shall not be demolished before a contract for the carrying out of works of redevelopment of the site has been made, and planning permission for the redevelopment granted (section 17(3)(a) to (b)). It is not clear with whom a contract is to be made nor, indeed, precisely what would suffice as such a contract: would a contract with one's secretary satisfy the condition? The purpose of such a condition is to prevent premature demolition.

Listed building consents are now subject to a condition that the authorised works must be commenced within three years or such other period beginning with that date as the authority granting the consent may direct, being a period which the authority considers appropriate having regard to any material considerations (section 18(1)(a) to (b)).

Provision is now made for an application to vary or modify any condition attached to a listed building consent (section 19(1)).

10.2.8. Exempted buildings

Listed building consent is not required, and a building preservation notice cannot be served, in respect of certain buildings. These exempted buildings are ecclesiastical buildings used for ecclesiastical purposes, and ancient monuments (sections 60(1) and 61(1)). In *Attorney-General* v *Howard United Reformed Church* [1975], the House of Lords decided that if a church were to be demolished, it would not then be used for ecclesiastical purposes and that therefore, if listed, listed building consent would be needed for the demolition.

The Ecclesiastical Exemption (Listed Buildings and Conservation Areas) Order 1994 removes the exemption for all ecclesiastical buildings, save for those listed in the order. The list includes church buildings belonging to the principal Christian churches used for ecclesiastical purposes.

10.2.9. Enforcement

There are certain enforcement provisions available to local planning authorities that apply to listed buildings. These are fully described in Chapter 15. They include powers to require repairs to be carried out. There is no time limit for the service of a listed buildings enforcement notice, so the purchaser of a building may be bound by the unlawful actions of a previous owner, as in *Braun* v *First Secretary of State* [2003]. If an owner of a listed building is refused listed building consent, he may be entitled to some compensation, or to serve a purchase notice compelling the authority to acquire his interest (see further discussion in Chapters 18 and 19).

10.3. Conservation areas

10.3.1. Duty to designate conservation areas

Every local planning authority shall determine which parts of their area are of special architectural or historic interest. These are areas, the character or appearance of which it is desirable to preserve or enhance. The areas identified are then designated as conservation areas (section 69(1) of the Planning (Listed Buildings and Conservation Areas) Act 1990).

Conservation areas are selected for their comprehensive interest, and may include a whole town or part of one, a village, or perhaps only a street or square. The area may contain listed buildings: very often, many of the buildings will not, individually, be worthy of attention, but as a group, they provide the necessary interest. Full advice relating to the use of conservation areas, and the control of development within them, is set out in the National Planning Policy Framework (March 2012) in relation to policies concerned with conserving and enhancing the historic environment.

Designation has certain consequences (see the following text), and therefore the fact of designation has to be notified to the Secretary of State, advertised and registered as a local land charge (sections 69(4) and 70).

The local planning authorities publish their proposals for preservation and enhancement; these are considered at a public meeting (section 71(1) to (3)). A local planning authority may therefore have a particular policy with regard, not only to development in a conservation area, but also to positive steps to enhance the attractiveness of the area such as traffic management and restrictions, parking restrictions and removal of unsightly advertisements and street furniture.

10.3.2. Legal consequences of conservation area designation

There are a number of legal consequences that may or will affect development in or near a conservation area.

(a) Any planning application for development that in the opinion of the local planning authority may affect the character or appearance of a conservation area must be advertised by the authority in a local newspaper and by way of site notice. The authority must then take into account any representation received, and may not determine the application within 21 days of the advertisements (section 67(2) to (7), as applied by section 73 of the Planning (Listed Buildings and Conservation Areas) Act 1990): see *R* v *Lambeth* LBC, *ex parte Sharp* [1986], where an application made by an officer of a local authority was not given the proper consideration as required.

(b) The demolition of buildings, with some exceptions, in conservation areas needs conservation area consent of the local planning authority (section 74(1) and (2)). Conservation area consent follows listed building consent in some respects (see Planning (Listed Buildings and Conservation Areas) Regulations 1990). Consequently, a number of the provisions in the Planning (Listed Buildings and Conservation Areas) Act 1990 that apply to listed buildings also apply to an application to demolish a building in a conservation area (see regulation 12 of 1990 Regulations). A proposal may therefore require consent to demolish, and a planning permission.

(c) Trees in conservation areas are afforded the same degree of protection that would be available under a tree preservation order (section 211 of the 1990 Act); there are a number of exemptions and these are more fully considered in Chapter 12.

(d) Conservation areas are referred to in the Town and Country Planning (General Permitted Development) Order 1995 as 'article 1(5) land'. In such areas, the classes of permitted development may contain limitations not applicable to other areas. For example, the permitted enlargement of a dwellinghouse is limited to 50 cubic metres or 10 per cent, whichever is the greater, in conservation areas, and in other areas the limit is 70 cubic metres or 15 per cent. Other limitations are more fully considered in Chapter 4 where the development orders are discussed.

(e) The local planning authority may make use of their powers to issue an article 4 direction, and thereby withdraw the automatic planning permission available under the General Permitted Development Order 1995 for certain development that would otherwise be permitted in a conservation area. The article 4 direction would specify the development no longer automatically permitted (see further in Chapter 4).

(f) For the purposes of advertisement control (see next chapter), a conservation area may involve the use of discontinuance notices to discontinue advertisements which have deemed consent but which are contrary to the amenity interests; a conservation area may also be an area of special control of advertisements which places greater restrictions on the display of advertisements.

10.4. Ancient monuments and archaeological areas

10.4.1. Monuments

Under the Ancient Monuments and Archaeological Areas Act 1979, the Secretary of State may compile a schedule of monuments; it then becomes an offence to carry out certain works to a 'scheduled monument' without consent. A scheduled monument may include a building, structure or work, or a site comprising the remains of such things. The proscribed work includes work of demolition, damage, destruction, removal, repair, alteration or addition, or flooding or tipping operations (section 1). There are a number of defences (section 2). In *Essex County Council* v *AH Philpott & Sons Ltd* [1987] a farming company was fined £1,500 for laying a drain through the site of a roman villa.

Scheduled monument consent is obtainable from the Secretary of State (section 2). Consent may be granted with or without conditions. Conditions may specify how the work is to be carried out and by whom, i.e., an archaeologist. The Secretary of State may grant scheduled monument consent by an order: this may relate to certain categories of work, and in respect of specified classes of scheduled monuments (section 3). This power is similar to the general development order in general development control.

The Act also makes provision for the general protection of ancient monuments. These include both scheduled monuments and others of great interest and defined by the Secretary of State. Either the Secretary of State or a local authority may become the guardian of an ancient monument, with the consent of the owner (section 12). Responsibility for upkeep passes from the owner. Unless scheduled monument consent exists, it is an offence to destroy or damage any scheduled monument or any ancient monument which is the subject of guardianship (section 28).

10.4.2. Archaeological areas

The 1979 Act contains powers enabling the Secretary of State or a local authority to designate an area of archaeological importance (section 33). It then becomes an offence to carry out operations that disturb the ground, or flooding or tipping that affect the area, without first giving notice to the district council (section 35). There are certain powers of entry for the Secretary of State and others to make investigations before the operations are carried out (sections 38–40).

10.5. Protection of the countryside

There are several special areas that are accorded particular titles in planning legislation and practice. These are very briefly described; some do and some do not have legal consequences for the development of land; they are all significant in their own way on the planning decision process and all are directed to the protection of the countryside.

10.5.1. National parks

These have been created under the National Parks and Access to the Countryside Act 1949. Under this Act, and the Countryside Act 1968, national parks are administered by seven national park authorities in England and three authorities in Wales. There is a special Broads Authority for the Norfolk and Suffolk Broads, which, although not a national park, is administered by the authority with many of the functions of a local planning authority. These authorities have particular obligations with regard to access to the countryside, whether by access agreements or orders; management and maintenance of footpaths and bridleways and, the conservation of natural beauty. Natural England (or the Countryside Council for Wales) is a national body concerned with the provision and improvement of facilities for the enjoyment of the countryside; the conservation and enhancement of natural beauty and amenity; and the need to secure public access.

It has interests and duties in all the countryside and not just in national parks (see the following text).

A national park authority will exercise their planning functions to safeguard the interests of the national park: not only will development inconsistent with national park policies not be allowed, but the authority may also make particular use of its positive planning powers to enforce planning control and discontinue non-conforming uses (see Chapters 17 and 18).

Several of the classes of permitted development are more restricted in national parks and the other special areas: see Chapter 4 and the Town and Country Planning (General Permitted Development) Order 1995. There are therefore certain limitations in respect of permitted development rights regarding the erection of agricultural or forestry buildings, or the construction of private roads: see Appendix B.

10.5.2. Areas of outstanding natural beauty

These are usually smaller areas than national parks, and may be designated following the proposals of Natural England (or the Countryside Council for Wales) or the local planning authority: National Parks and Access to the Countryside Act 1949, sections 87–88.

In an AONB, the development permitted by the Town and Country Planning (General Permitted Development) Order 1995 is subject to the limitations that also apply to national parks and conservation areas. The local planning authority may decide that such an area should be particularly protected from uncontrolled permitted development by the use of an article 4 direction withdrawing one or more classes of permitted development (see Chapter 4).

10.5.3. Sites of special scientific interest

Natural England has a duty in England to notify the local planning authority, and every owner and occupier concerned, of a site of special scientific interest. The notification will specify the flora, fauna, or geological or physiographical interest, and the operations which are likely to damage the features giving rise to that interest (section 28, Wildlife and Countryside Act 1981). In *R v Nature Conservancy Council, ex p Bolton Metropolitan Borough Council* [1996], it was held that the NCC, as Natural England was then called, was in breach of the rules of natural justice, because it had failed to properly consider the objections properly raised by the Council. Provision is now made for the making of representations or objections, and for the confirmation or withdrawal of a notification in consequence. It is an offence to carry out operations proscribed by the notification unless Natural England gives written consent, or the operation is carried out under an agreement made under the National Parks and Access to the Countryside Act 1949, the Countryside Act 1968, or the Natural Environment and Rural Communities Act 2006, or the operation is carried out in accordance with a management scheme or management notice under the 1981 Act itself.

Where it appears to a local planning authority in England that development proposed in a planning application is likely to affect a site of special scientific interest, they must consult with, and take into account the views of, Natural England: see Town and Country Planning (General Development Procedure) Order 1995, article 10.

10.5.4. Management agreements in the countryside

A management agreement may be made between an owner or occupier and the local planning authority (in a national park, the national park authority) for the purpose of conserving or enhancing the natural beauty or amenity of any land (section 39(1) of the Wildlife and Countryside Act 1981).

Such an agreement may impose restrictions on the owner 'as respect of the method of cultivating the land, its use for agricultural purposes or the exercise of rights over the land and may impose obligations on that person to carry out works or agricultural or forestry operations or do other things on the land'. An agreement may also permit the authority to carry out works on the land; and provide for the payment of money by either party to the other (section 39(2)).

A management agreement will bind successors in title to the land, unless it provides otherwise. It is not clear if this means that it is enforceable as if it contains restrictive covenants. If so, they could be the subject of modification upon application to the Upper Tribunal (see Chapter 9).

10.5.5. Green belts

The purpose of a green belt is to prevent urban sprawl by keeping land permanently open. They help protect the countryside. They check the unrestricted sprawl of large built-up areas, and preserve the setting and special character of historic towns. A green belt is a matter of planning policy rather than planning law; although it will have an inevitable effect on the decision-making process. Where a green belt policy is part of the development plan for an area, under section 38(6) of the Planning and Compulsory Purchase Act 2004 a planning decision that requires regard to be had to the development plan must be made in accordance with the plan unless material considerations indicate otherwise (see Chapter 8). National Planning Policy Framework (NPPF) (March 2012), issued by the Secretary of State for Communities and Local Government, contains policy guidance with regard to green belts. Construction of new buildings in a green belt should be regarded as inappropriate except in very special circumstances, although the NPPF does contain a short list of exceptions, such as agricultural and forestry buildings and limited infilling.

10.5.6. European Union directives

In directive 79/409/EEC (Birds Directive) on the conservation of wild birds, member states are required to establish special protection areas. Under directive

92/43/EEC (Habitats Directive) on the conservation of natural habitat areas, provision is made for the establishment of special areas of conservation. The requirements of these Directives have been transposed into British law by the Conservation of Habitats and Species Regulations 2010.

In *R* v *Secretary of State for the Environment, ex p Royal Society Protection of Birds* [1997], the European Court of Justice held that in designating a special protection area under the Birds Directive, no account shall be taken of economic considerations in fixing the boundaries.

10.5.7. *Natural England*

The Countryside Agency was set up in 1999 with the amalgamation of the Countryside Commission and the Rural Enterprise Board. The Nature Conservancy Council was concerned with nature conservation and the notification and protection of sites of special scientific interest and national nature reserves. Natural England is the successor to all these agencies. Its priority areas are promoting social equity and economic opportunity for those who live in the countryside, conserving and enhancing the countryside, notifying and protecting SSSIs and national nature reserves, and helping everyone, wherever they live, enjoy this national asset. Decision-makers (planning authorities and inspectors) must take into account the terms of the National Planning Policy Framework (NPPF) (March 2012), issued by the Secretary of State for Communities and Local Government, when considering planning applications and appeals: see sections 19(2)(a) and 38(6) of the Planning and Compulsory Purchase Act 2004. In relation to conserving and enhancing the natural environment, local planning authorities should take into account a range of considerations listed in the NPPF.

11 Advertisements and other controls

11.1. Introduction

In this chapter, the reader will find the special controls over the display of advertisements.

There is a brief description of enterprise zones for the sake of completeness; additional zones were created in 2011.

11.2. Advertisement controls

The statutory code governing the display of advertisements is contained in the Town and Country Planning Act 1990 and in the Town and Country Planning (Control of Advertisements) (England) Regulations 2007 (SI 2007/783). Circular 03/07 provides advice and explanations. Regulations are made where the Secretary of State considers it is expedient in the interests of amenity and public safety: section 220(1) of the 1990 Act and regulation 3(1) of the 2007 Regulations. The scheme of the regulations is to exclude certain advertisement displays from control; and give deemed consent to a specified list of advertisements. Otherwise it is an offence to display any other advertisement without the consent of the local planning authority: see regulation 30 of the 2007 Regulations. The offence can be committed either by the owner or occupier of the land, or by the person whose goods or business is advertised (section 224(1)). Accordingly, if an advertisement is neither excluded from control nor has deemed consent under the regulations, an express consent is required.

11.3. What constitutes an advertisement

Advertisements are defined in section 336(1) of the 1990 Act, as amended, as:

> *...any word, letter, model, sign, placard, board, notice, awning, blind, device or representation, whether illuminated or not, in the nature of, and employed wholly or partly for the purposes of, advertisement, announcement or direction, and (without prejudice to the previous provisions of this definition), includes any hoarding or similar structure used, or designed or adapted for*

use and anything else principally used, or designed or adapted principally for use, for the display of advertisements ...

The Town and Country Planning Act 1990 section 55(5) provides that:

Without prejudice to any regulations made under the provisions of this Act relating to the control of advertisements, the use for the display of advertisements of any external part of a building which is not normally used for that purpose shall be treated for the purposes of this section as involving a material change in the use of that part of the building.

This section brings the use of external parts of a building for display of advertisements within the definition of development. However, section 222 of the 1990 Act provides that where an advertisement is displayed in accordance with the 2007 Regulations, which involves the development of land, planning permission is deemed to be granted by virtue of that section and no application for planning permission is required. Therefore, a display of an advertisement on the external part of a building may involve a breach of planning control and of the advertisement controls.

Any consent under the regulations for the display of an advertisement also operates as a planning permission to the extent that the display involves development (i.e., building or engineering operation or a material change of use) (section 222).

In *Westminster City Council* v *Secretary of State for the Environment and Bally Group (UK) Ltd* [1990], two separate businesses displayed the names of their respective businesses on a canopy and a blind. Enforcement notices, each alleging a breach of planning control, were issued on the businesses. It was held that, in deciding whether something is an advertisement, one must look at the subject matter in dispute and use commonsense. Then one should look at the regulations to see whether anything flies in the face of commonsense. The court decided that the canopy (or blind) and the words must be conjoined; they were an advertisement and therefore, there was deemed planning permission for the canopy (and blind). The decision of *Glasgow City District Council* v *Secretary of State for Scotland* [1989] was not applied. There the court took the view that the words contained on the canopy were not capable of being included within the ambit of advertisement 'device' and that they did not form an advertisement.

The Planning and Compensation Act 1991 has now amended the definition of 'advertisements' by inserting the words 'awning and blind'. Accordingly, planning permission would now be required where an advertisement is displayed on or by the use of a canopy or blind.

When deciding whether the display of an advertisement falls within the control provisions of the 1990 Act, it is necessary to look at the intention and purpose of the display. In *Great Yarmouth Borough Council* v *Secretary of State for the Environment* [1996], it was held that beams of light projecting skywards to produce a floral image at the base of any cloud could amount to the display of an advertisement within the meaning of section 336(1) of the 1990 Act. The beams were intended to direct members of the public to a specific destination. They were in the nature of a 'signpost'. There was no requirement that the advertisement had to be tangible; accordingly, there was nothing in the definition to exclude a display by means of a hologram.

11.4. Basis of control of advertisements

Under section 220(1) and regulation 3(1) of the 2007 Regulations, the regulations are made, and the powers under them exercised, only in the interests of amenity and public safety. The content of any advertisement cannot be controlled, as such. In respect of amenity, the material factors include the general characteristics of the locality, such as historic, architectural, cultural or similar interest. In the case of public safety, the material factors include the safety of any person using such facilities as roads or railways, and whether any advertisement will obscure or hinder the ready interpretation of any traffic sign, railway signal or navigational aid.

The Secretary of State has power to make different advertisement regulations for different areas. He can make regulations in respect of conservation areas, areas defined as experimental areas and areas of special control. An area of special control may either be a rural area or an area requiring special protection on grounds of amenity. In such an area, there is a greater restriction on the display of advertisements than elsewhere (section 221).

11.5. Advertisements excluded from control

The advertisements that are excluded from control are those set out in Schedule 1 to the 2007 Regulations, subject to conditions and limitations in that schedule:

- Class A – An advertisement displayed on enclosed land.
- Class B – An advertisement displayed on or in a vehicle normally employed as a moving vehicle. When a vehicle containing an advertisement was not parked in the ordinary use of the vehicle, but was placed in a position for the purpose of displaying the advertisement, the exemption under Class B did not apply: see *Tile Wise Ltd* v *South Somerset District Council* [2010].
- Class C – An advertisement incorporated in the fabric of a building.
- Class D – An advertisement displayed on an article for sale or on the container in, or from which, an article is sold.
- Class E – An advertisement relating specifically to a pending Parliamentary, European Parliamentary or local government election or a referendum under the Political Parties, Elections and Referendums Act 2000.
- Class F – An advertisement required to be displayed by standing orders of either House of Parliament or by any enactment or any condition imposed by any enactment on the exercise of any power or function.

- Class G – A traffic sign.
- Class H – The national flag of any country, a flag of the Commonwealth, the European Union, the United Nations, of any English county, or any saint.
- Class I – An advertisement displayed inside a building.

11.6. Advertisements with deemed consent

Although all the advertisements in this group are within control, consent is deemed to be granted by the regulations (see regulation 6, and Part I of Schedule 3), and such deemed consent is subject to standard conditions set out in Schedule 2, such as requirements that all advertisements are displayed with the consent of the site owner and shall not cause any dangers. Save for Classes 12 or 13, the Secretary of State can, on a proposal by a local planning authority, make a direction that deemed consent to display any advertisements shall not apply to specified classes in a specified area for a specified period or indefinitely (regulation 7). The deemed consent is more restricted in areas of special control (regulation 20). The following is a list of classes of advertisements for which there is deemed consent. In most cases, there are limitations and conditions.

- Class 1 – Functional advertisements of local authorities, statutory undertakers and public transport undertakers, and Transport for London.
- Class 2 – Miscellaneous advertisements relating to the premises on which they are displayed.
- Class 3 – Miscellaneous temporary advertisements.
- Class 4 – Illuminated advertisements on business premises.
- Class 5 – Other advertisements on business premises.
- Class 6 – An advertisement on a forecourt of business premises.
- Class 7 – Flag advertisements.
- Class 8 – Advertisement on hoardings.
- Class 9 – Advertisements on highway structures.
- Class 10 – Advertisements for neighbourhood watch and similar schemes.
- Class 11 – Directional advertisements.
- Class 12 – Advertisements inside buildings.
- Class 13 – Advertisements on sites used for the preceding 10 years for the display of advertisements without express consent. However, in R (*J C Decaux UK Ltd*) v *Wandsworth London Borough Council* [2009], the alteration of the means of support for an existing advertisement hoarding, from a wall-mounted to a self-supported structure, took it outside this class.
- Class 14 – Advertisements displayed after expiry of express consent.
- Class 15 – Advertisements on balloons.
- Class 16 – Advertisements on telephone kiosks.

A 'Let by' board showed that a tenancy had been granted, and the board had to be removed within 14 days: see *Barbara Rees Ltd* v *Cardiff City Council* [2006] 1617. In R (*Clear Channel Ltd*) v *Hammersmith & Fulham LBC* [2010], a static advertisement was replaced by a digital display advertising hoarding; this allowed images to be displayed at intervals, but it did not enjoy deemed consent under Class 13. Class 13 of Schedule 3, in its form provided for by the Regulations (1992) in force prior to the 2007 Regulations, permitted the display of advertisements on sites used on 1 April 1974 without express consent which had been so used continually since that date. *Barking and Dagenham Borough Council* v *Mills & Allen Ltd* [1997] was a case involving the display of advertisements since 1970 by a company on two wooden hoardings, one being attached to the flank wall of an adjacent building. By 1994, the wooden hoardings had been replaced by a double-sided freestanding unipole hoarding within the company's site away from the flank wall. The council contended that since the rearranged hoardings were no longer within the site previously having deemed consent, the company was in breach of the 1992 Regulations. It was held, the word 'site' was a parcel of land upon which the advertisement was displayed, rather than the specific position within that part of the land at which the display was located: *Scotts Restaurant plc* v *City of Westminster* [1993] was applied.

The deemed consent, under Class 14, to continue to display an advertisement after the expiration of an express consent, only applies where there has been a continuity of display: see *Clear Channel Ltd* v *Southwark LBC* [2007].

11.7. Discontinuance of deemed consent

Under regulation 8 of the 2007 regulations, a discontinuance notice may be served by the local planning authority requiring the discontinuance of the display of an advertisement, or the use of a site for the display of an advertisement, for which deemed consent is granted. The local planning authority may serve a discontinuance notice if they are satisfied that it is necessary to do so to remedy a substantial injury to the amenity of the locality or a danger to members of the public (regulation 8(1)). In *Cheque Point UK Ltd* v *Secretary of State for the Environment* [1995], a discontinuance notice served on the grounds 'unduly detrimental to the appearance of the building and the street' within a conservation area, was held to be a valid notice. In that case, the court held that the Secretary of State had used the appropriate test when concluding that the continued display of a sign was 'substantially injurious to the interests of amenity and incompatible with the conservation status of the locality'.

Where an advertisement falls within Class 12 in Schedule 3 (a display inside a building), the authority may not serve a discontinuance notice if that advertisement also falls within Class E or F in Schedule 2 (Parliamentary advertisements).

Any failure to comply with the obligations under a discontinuous notice will be a breach of the 2007 Regulations (see the following text, under Offences).

11.8. Advertisements requiring express consent

All other advertisements that are not either excluded from control (see part **11.5** above), or for which consent is deemed granted (see part **11.6** above), need the express consent of the local planning authority. Applications for consent are required to be made in accordance with regulation 9, namely in writing on a form provided by the local planning authority and giving the particulars required by that form.

The authority is required to consult with certain bodies (regulation 13) and it may then grant consent, in whole or in part, subject to standard conditions and to such additional conditions as they think fit (regulation 14(1)(a)), or refuse it (regulation 14(1)(b)), decline to determine the application in accordance with section 70A of the 1990 Act (which applies in relation to the application subject to the modifications specified in Part I of Schedule 4). Every consent is normally for a limited period of five years (regulation 14(7)(b)), but may, if appropriate, be for a longer or shorter period (regulation 14(7)(a)).

In exercising their powers, the authority is required to consult such bodies as neighbouring local planning authorities, the Secretary of State (where a trunk road may be affected) and the operators of such facilities as railways where safety may be relevant. The authority should also consider the general characteristics of the locality and the presence of any feature of historic or architectural interest; have regard to the safety of persons who may use roads, etc., and the effect of an advertisement on any traffic sign, railway signal, etc. (regulation 13). There can be no control over the content of an advertisement other than in the interests of amenity and public safety.

There is a right of appeal to the Secretary of State against an adverse decision under sections 78 and 79 of the 1990 Act, subject to the modifications specified in Part 3 of Schedule 4 of the 2007 Regulations (see regulation 17).

11.9. Advertisements: areas of special control

These have already been referred to (see part **11.4** above). Section 221 additionally contains powers for the Secretary of State to make special regulations for advertisement control in conservation and other areas. He may make or approve an order for such an area. The area must then be kept under review (regulation 20). The section states that different provision in respect to different areas may be made. Such areas can be either conservation areas, or other areas that appear to require special protection on grounds of amenity or public safety.

In an area of special control, there is some restriction on the advertisements that have deemed consent under regulation 21. Subject to the provisions in regulation 21, no advertisements may be displayed in an area of special control unless they fall within any class within Schedule 1 (advertisements otherwise excepted from control), and Classes 1–3, 5–7 and 9–14 in Schedule 3 (advertisements having deemed consent).

In respect of other advertisements, the power of the local planning authority is limited in an area of special control. Express consent may not be given for certain illuminated advertisements (regulation 21(3)).

11.10. Offences

Section 224 of the 1990 Act makes it an offence if any person displays an advertisement in breach of the regulations. The offence can be committed either by the owner or occupier of the land, or by the person whose goods, trade or business or other concerns are advertised (section 224(4)). In the case of advertisements displayed under deemed consent, the deemed consent can be terminated by the service of a discontinuance notice (see part **11.7** above). It is then an offence if the advertisement is not removed by the specified time.

The Localism Act 2011 introduced time limits within which a prosecution must be brought. Proceedings may be brought within six months beginning with the date on which evidence sufficient to justify proceedings came to the prosecutor's knowledge, but proceedings cannot be commenced more than three years after the date of the offence: see section 224 of the 1990 Act as amended.

A person is liable for a fine on a certain scale, and for each day during which the offence continues (section 224(3)).

If there has been a decision of a court as to whether or not an offence has been committed, that disposes of the criminality up to that point. In *O'Brien* v *London Borough of Croydon* [1998], a court determined that an accused had deemed consent; that did not prevent the service of a discontinuance notice thereafter from founding a fresh offence. The deemed consent is determined by a valid discontinuance notice.

Delay in serving a discontinuance notice is not a defence to a discontinuance action. In *Swishbrook Ltd* v *Secretary of State for the Environment* [1990], where there had been a delay of three-and-a-half years between the council's decision to take discontinuance action and the service of the notices, the court held that the advertiser had not suffered substantial prejudice by reason of the council's failure to serve the notice as the regulations required. The defence of limitation provided under section 171B(3) of the 1990 Act is not available to advertisers, because the offence of unlawfully displaying advertisements is governed only by the regime for the control of advertisements (1992 Regulations), and not that for breaches of planning control: *Wyatt* v *Jarrad* [1998]. However, see now the time limits within which a prosecution must be brought.

Irregularities in the service of a discontinuance notice are unlikely to amount to a defence unless there is substantial prejudice: see *Nahlis* v *Secretary of State for the Environment* [1995]. Nor is a discontinuance notice rendered invalid in its entirety by failure to serve it on the 'advertiser'. The requirement to serve the advertiser is directory only: see *O'Brien* v *London Borough of Croydon* [1998].

11.10.1. An abuse of process

An abuse of process is a defence to a discontinuance action. Where proceedings are instituted against the advertiser, in circumstances where he has been led to

believe by officers of the local planning authority that displaying an advertisement would not be unlawful, the advertiser is entitled to rely on the advice given
to him by the council's officials: see *Postermobile plc* v *Brent London Borough
Council* [1997].

11.10.2. Burden of proof

The burden of proof in proceedings for discontinuance actions rests with the
prosecution to prove the display, it is then for the defence to prove authorisation.
It is not for the prosecution to prove absence of lawful authority, which principles
equally apply to cases involving express and deemed consent: see *R* v *O'Brien
and Hertsmere District Council* [1997].

11.10.3. Knowledge or consent

There is a defence if the person charged can show that the advertisement was
displayed without his knowledge or consent. This defence succeeded in *John* v
Reveille Newspapers Ltd [1955] where the newspaper company were able to
prove that an advertisement of theirs was displayed on a public urinal without
their consent. However, the defence failed in *Preston* v *British Union for the
Abolition of Vivisection* [1985], because the offence was regarded as a continuing
one, and the advertiser had been informed of the display. In *Wycombe District
Council* v *Michael Shanley Group Ltd* [1994], the Divisional Court followed the
ruling in *Merton London Borough Council* v *Edmonds* [1993], and held that
the words 'knowledge or consent' in section 224(5) were to be read disjunctively.
Accordingly, where the owner knew that advertisements were being displayed
on his land by another, he was still permitted to prove that he did not consent to
their display. In *Porter* v *Honey* [1988], where an estate agent erected a second
display of 'for sale' signs without the knowledge or consent of the first estate
agent, the House of Lords held that where deemed consent for the display of a
second 'for sale' sign is given pursuant to Class III(a) of regulation 14 of the 1984
Regulations (now Class 3 to Schedule 3 of the 2007 Regulations), the first display
is to be taken to be the one permitted.

Display of advertisements without consent of different advertisements at a
site constitutes a series of different offences and not a single offence: see
Kingston-upon-Thames Royal London Borough Council v *National Solus Sites
Ltd* [1994].

In the case of *Wadham Stringer (Fareham) Ltd* v *Fareham Borough Council*
[1987], the defendant was convicted for the unlawful display of an advertisement
on a tethered balloon attached to a motor vehicle on his garage premises. The
court held that the balloon was attached to the 'site' on which it was displayed for
the purposes of the 1984 Regulations (now the 2007 Regulations; Schedule 2
Class 15).

11.11. Power to remove structures used for unauthorised display

The Localism Act 2011 introduced new provisions into the Town and Country Planning Act 1990 to give powers to a local planning authority to remove and then dispose of any display structure which has been used for the display of advertisements in contravention of the advertisement regulations: see section 225A. The power may not be used unless a removal notice has first been served on the person who appears to be responsible for the erection or maintenance of the display structure. If no such person can be identified, then a removal notice must be affixed to the structure and served on the occupier of the land giving a period of time for the removal of the display structure. Where the removal notice is not complied with, the authority may remove the structure and recover its expenses. There are detailed provisions for the contents of a removal notice, and the definition of a display structure, which includes a moveable structure. Presumably, this would include the odd lorry or tractor-trailer occasionally seen from a motorway.

Section 225B of the 1990 Act contains provisions for appealing a removal notice to the magistrate's court.

11.12. Remedying persistent problems with unauthorised advertisements

Where a local planning authority has reason to believe that there is a persistent problem with the display of unauthorised advertisements on the surface of any building, wall, fence, or other structure or erection, or on any apparatus or plant, they may serve an action notice on the owners or occupier: see section 225C of the 1990 Act introduced by the Localism Act 2011. The action will specify the measures that are to be taken by a specified time. The measures must be reasonable measures to prevent or reduce the frequency of the display of unauthorised advertisements. One has in mind cases where trespassers or unknown parties continuously use, say, a particular wall for unauthorised advertisements and where the application of some suitable material on the surface of the wall might reduce the problem.

Section 225C contains provisions for enforcement of an action notice by the authority carrying out the measures itself with a right to recover its expenses. There are rights of appeal to the magistrate's court. In the case of statutory undertakers' operational land, the statutory undertaker may serve a counter-notice within 28 days specifying alternative measures: see section 225E.

11.13. Remedying defacement of premises

The Localism Act 2011 introduced additional provisions into the Town and Country Planning Act 1990 to deal with a sign on a surface, readily visible from a public place, that a local planning authority considers to be detrimental to the amenity of the area or offensive: see section 225F of the 1990 Act. A sign may include any writing, letter, picture, device or representation, but may not include an advertisement. A notice may be served on the owner or occupier requiring the sign to be removed or obliterated by a specified time. If the requirement of the notice is not complied with, the authority may take the action and recover its expenses. The powers can be used in respect of post boxes, bus shelters and other street furniture. There are rights of appeal to the magistrate's court.

The powers to deal with the defacement of premises can also be used in respect of any operational land of a statutory undertaker, with appropriate modifications.

11.14. Enterprise zones

Part XVIII of the Local Government, Planning and Land Act 1980 made provision for the designation of enterprise zones. Substantial relaxation of planning control, and exemption from rates for non-domestic hereditaments, was seen as the tonic for the areas selected, together with other fiscal incentives.

The government then saw the purpose of these zones as an experiment to test how far industrial and commercial activity can be encouraged on a few selected sites. Areas with problems of economic and physical decay were chosen. A number of zones were designated between 1981 and 1986, each to last for a period of 10 years. During 2011, the government announced the creation of 21 new enterprise zones where, among other advantages, there will be generous capital allowances to encourage investment.

12 Trees, minerals and caravans

12.1. Trees: general

There are two principal controls that protect trees: one is found in planning legislation, and concerned with the protection of trees because of the interests of amenity; and another is found in the Forestry Act 1967 and controls felling, through the issue of felling licences, with a view to the proper conservation and management of timber resources.

There is a general duty, in section 197 of the 1990 Act, on a local planning authority to impose conditions, where appropriate, in a grant of planning permission for the preservation or planting of trees, or to make tree preservation orders to give effect to such conditions or otherwise. The protection afforded to trees by the 1990 Act does not extend to hedgerows. Special controls to prevent the removal and destruction of hedgerow were introduced by section 97 of the Environment Act 1995. These special controls were given effect by the Hedgerows Regulations 1997 (SI 1997/1160) which introduced a new prohibition on the removal of certain hedgerows, unless proceeded by certain procedural steps.

12.2. Tree preservation orders

A local planning authority may make a tree preservation order (TPO) in the interests of amenity for the preservation of such trees, groups of trees or woodlands as may be specified in the order (section 198(1)). Most TPOs are made by district planning authorities; county planning authorities can make TPOs only in relation to a grant by them of a planning permission, in relation to land covering more than one district planning authority area, where the county hold an interest in the land, or in relation to land in a national park (Schedule 1, para 13). For guidance, see the *Tree Preservation Orders: Guide to the Law and Good Practice*, 2000, published by the Department of Communities and Local Government.

The Town and Country Planning (Trees) Regulations 2012 specify a model form of TPO in accordance with section 198(3). The model TPO and the Regulations contain provisions:

(a) For prohibiting the cutting down, topping, lopping, uprooting, wilful damage or wilful destruction of trees without consent, or causing such acts;

(b) For securing the replanting of any part of a woodland area felled in the course of felling operations permitted by or under the order, in compliance with a direction;

(c) For the making of written applications for consent in relation to prohibited acts, to which the provisions of the 1990 Act apply as if the application were one for planning permission;

(d) For the payment of compensation for certain losses and damage.

A map detailing the trees protected by a TPO is annexed to and is part of the order: see regulation 3.

12.2.1. Criminal offences

It is a criminal offence to contravene a TPO by the cutting down, uprooting, wilful destruction, or wilful damage, topping or lopping of a tree without consent (section 210(1)). The offence is one of strict liability; the tree feller who fells a tree in ignorance of the TPO is committing an offence, and is liable to the stiffer penalty: *Maidstone Borough Council* v *Mortimer* [1980]. The landowner who orders the tree felling is only liable to the lighter penalty as his offence of causing or permitting the felling appears to fall under section 210(2): *R* v *Bournemouth Justices, ex p Bournemouth Corporation* [1970]. Liability under section 210 is absolute: see *R* v *Alath Construction Ltd* [1991], where the ingredients of the offence were set out by the Court of Appeal as follows: that (1) there is a TPO in place; (2) the defendant has cut down the tree or carried out activities contrary to the appropriate section; and (3) the felling or other activity was without consent.

Where a TPO covers an area of woodland, it may be important to determine what is a 'tree', as it is only in relation to trees that an offence may be committed. In *Kent County Council* v *Batchelor* [1976], Lord Denning MR opined that a tree should be something of seven or eight inches in diameter. This case was not followed in *Bullock* v *Secretary of State for the Environment* [1980], where the landowner had argued that a coppice does not contain trees, as what grew in a coppice were too small and anyway were cut in rotation which was inconsistent with a TPO. This argument did not prevail; the word tree should be given its ordinary meaning, and trees in a coppice can be the subject of a TPO; provision can also be made for the proper management of a coppice.

The Localism Act 2011 introduced time limits within which a prosecution must be brought. Proceedings may be brought within six months beginning with the date when on which evidence sufficient to justify proceedings came to the prosecutor's knowledge, but proceedings cannot be commenced more than three years after the date of the offence: see section 210 of the 1990 Act as amended.

The penalty does not include imprisonment, but in case of deliberate and threatened contravention, the court may grant an injunction, which, if disobeyed amounts to contempt; the court may then imprison; see *Attorney-General* v *Melville Construction Co Ltd* [1968].

12.2.2. Exemptions

It is not an offence to do work to a tree if it is dying, dead, or has become danger-ous, the work is required by some other statute, or is necessary to abate a nuisance (section 198(6)). Where a person relies on any exception or exemption in section 198(6), the burden of proof rests on him to show that he falls within that provision: see *R* v *Alath Construction Ltd* [1991]. Whether trees have become dangerous for the purposes of the statutory exception is a question of fact for the justices where information is laid against the defendant. The test in deter-mining whether a tree is dangerous is that of a sensible approach taken by a prudent citizen in considering its state of danger: see *Smith* v *Oliver* [1989]. Nevertheless, a replacement tree must be planted (section 206(1)). There is provi-sion for enforcement, and a right of appeal (section 207(1)).

The 2012 Regulations set out a number of exemptions.

12.2.3. Procedure for making a TPO

Section 199(3) provides for the making of regulations: the Town and Country Planning (Trees) Regulations 2012. Further guidance was found in Circular 36/78 – Trees and Forestry, however, this circular has been replaced: see now the appropriate *Guides* published by the Department of Communities and Local Government.

The regulations specify the form of the TPO. It is the TPO itself, it will be recalled, that not only carefully identifies the trees, groups of trees, or woodland to be protected, but also sets out the prohibited acts in relation to the trees (regu-lation 3).

The local planning authority, after making the TPO, must deposit details for public inspection, and serve a copy of the TPO on the owners and occupiers of the affected land. The owners and occupiers are also entitled to a notice containing certain particulars. These include the reasons for the order and that objections and representations may be made to the authority (regulation 5).

If no objections or representations are received within 28 days, the TPO may be confirmed by the local planning authority (regulation 7). If objections or representations are made, and they must include the grounds upon which they are based, the local planning authority must take these into consideration. They may then decide to confirm the TPO with or without modifications, or not to confirm it.

When the TPO is confirmed, the local planning authority must notify the owners and occupiers of the affected land (regulation 8). Although reasons must be given, it is not generally necessary to give separate reasons in respect of each of the individual trees and a group of trees: see *R (Brennon)* v *Bromsgrove DC* [2003].

The TPO is also registered: see regulation 12: this is good notice to any purchaser. Regulation 10 enables a TPO to be varied.

12.2.4. Provisional tree preservation order

A TPO does not take effect until the local planning authority has confirmed it (section 199(1)). Because this may involve several months' delay, and the trees concerned may meanwhile be felled or otherwise damaged, the local planning authority may, in making a TPO, include a direction that section 201 of the 1990 Act shall apply: see article 4 of the model TPO. The effect of this is that the TPO takes effect on a specified date, and before it is confirmed. It is then a provisional TPO, lasting for six months, or until confirmed, and protects the trees for the time being as if the TPO were confirmed (section 201(2)).

12.2.5. Consents under a tree preservation order

A TPO must be made substantially in the form of the model TPO contained in the 2012 Regulations. The 2012 Regulations set out the procedure for obtaining consents, making appeals and making claims for compensation.

If a person wishes to do any of the prohibited acts to trees protected by a TPO, he must do so by applying for consent from the district planning authority. The application is treated as an application for planning permission. If permission is granted, it may be subject to such conditions as the authority thinks fit (such as to require replacement trees). The TPO contains a power for the authority to issue a direction as to replanting following forestry operations.

The 2012 regulations contain a number of exemptions.

There is a right of appeal to the Secretary of State against an adverse decision of the local planning authority: see regulation 19 of the 2012 Regulations. There is also a right to claim compensation if consent is refused for certain matters prohibited by the TPO (section 203 of the 1990 Act). (See Chapter 19.)

12.2.6. Replacement of trees

Section 206 of the 1990 Act imposes a duty on the landowner to replace any protected trees removed, uprooted or which are destroyed or die unless the local planning authority dispenses with the requirement. The duty is fulfilled by planting another tree of an appropriate size and species at the same place as soon as reasonable. In the case of trees in a woodland, the same number of trees should be planted on or near the land where the original trees stood, or as may be agreed with the local planning authority.

12.3. Trees and felling licences

Unless the developer is felling a large number of trees, he is unlikely to need a felling licence under the Forestry Act 1967. In any event, no licence is required if the felling is necessary for the purpose of carrying out development for which planning permission has been granted (section 9(4) of the Forestry Act 1967).

If an application is made to the Forestry Commission for a felling licence in respect of trees which are the subject of a TPO, the commission may, and must if they propose granting the licence, refer the application to the local planning authority (section 15(1) of the Forestry Act 1967). If the authority objects to the granting of a licence, the matter is referred to the Secretary of State.

Where a person has been convicted of felling trees without a felling licence, the Forestry Act 1967 (as amended by the Forestry Act 1986) contains powers for the Forestry Commission to require the restocking of the felled land with trees. Where a person is charged with the offence of felling without a licence, the presumption of *mens rea* is displaced as section 17 of the Forestry Act 1967 imposes an absolute offence: see *R (Grundy & Co Excavations Ltd)* v *Halton Division Magistrates' Court* [2003]. However, there is a defence under section 9(2)(b) of the Forestry Act 1967 where the land is a private garden: see *Rockall* v *Department of Environment Food and Rural Affairs* [2008] and *McInerney* v *Portland Port Ltd* [2001].

12.4. Trees in conservation areas

Section 211 of the 1990 Act provides protection for trees in conservation areas. The protection is equivalent to the protection afforded to trees which are subject to a TPO. The prohibited acts are those found in section 198 (see part **12.2** above). It is a defence to any criminal proceedings that the person charged either obtained consent for the prohibited acts, or he gave notice to the local planning authority and six weeks have elapsed since that notice and the act is done not more than two years from the date of the notice (this is to give time, if appropriate, for the making of a draft TPO) (section 211(3)). In *R* v *North Hertfordshire District Council, ex p Hyde* [1989], it was held that these provisions do not prevent a local planning authority from making a TPO within or outside the six-week period.

The protection under section 211 does not apply to dead or dying trees, to the activities exempted under article 5 of the model TPO, trees within a forestry dedication or approved forestry plan made with the Forestry Commission, trees for which a felling licence is granted, works to trees by or on behalf of the local planning authority, to trees less than 75 mm in diameter (or 100 mm if the work is done to improve the growth of trees), and to topping and lopping of a tree whose diameter does not exceed 75 mm: see the Town and Country Planning (Trees) Regulations 2012.

12.5. Control over mineral workings

Although somewhat outside the scope of this book, there follows a very brief outline of the control over mineral workings that may interest the general developer. The Town and Country Planning (Minerals) Act 1981, which enacted many of the recommendations of the Stevens Committee 'Planning Control over Mineral Workings' (1976), amended the provisions for the control of mineral workings. These amendments are now found in the Town and Country Planning Act 1990. There are further provisions in the Environment Act 1995 that affected what were referred to as 'old mineral permissions'.

Policy guidance is now found in the National Planning Policy Framework (March 2012), which also repealed most of the series of Minerals Policy Statements and Guidance Notes (MPSs and MPGs). The NPPF now advises that minerals are essential to support sustainable economic growth and, when determining planning applications, local planning authorities should, inter alia, give great weight to the benefits of mineral extraction, among other considerations. The NPPF also contains guidance to mineral planning authorities on minerals in the preparation of local development documents.

12.5.1. Wider definition of mining operations

It will be recalled that planning control arises in respect of development, and development is defined to include mining operations (section 55 of the 1990 Act). Mining operations are now defined to include the removal of material of any description from a mineral working deposit, a deposit of pulverised fuel ash, furnace ash or clinker, or a deposit of iron, steel or other metallic slags; and the extraction of minerals from a disused railway embankment (section 55(4)). A mineral working deposit is that left after the extraction of minerals from land (see section 336(1) of the Town and Country Planning Act 1990). Minerals include all minerals and substances in or under land of a kind ordinarily worked for removal by underground or surface working (section 336).

The purpose of this wider definition is to bring the working of slag and other waste heaps clearly within control.

12.5.2. Minerals planning authority

Outside Greater London and the metropolitan areas, the county councils and national park authorities are the minerals planning authorities. The minerals planning authorities in Greater London in the metropolitan areas and in the unitary areas are the London borough councils, the metropolitan district councils and the unitary authorities, respectively.

There is a duty on every county council minerals planning authority to prepare and maintain a minerals and waste development scheme: see section 16 of the Planning and Compulsory Purchase Act 2004. The resulting policy documents will be called the Minerals and Waste Development Framework – in some areas, these may be separated into two frameworks a Minerals Development Framework and a Waste Development Framework. These will form part of the development plan.

12.5.3. Environmental impact assessment

By their nature, mining operations frequently have an impact on the environment. Council Directive 85/337/EEC requires an assessment of the effects of certain public and private projects on the environment. The Town and Country Planning (Environmental Impact Assessment) (England and Wales) Regulations 1999 give effect to the directive (see Chapter 8). An environmental impact assessment is required for quarries and opencast mining where the surface of the site exceeds 25 hectares and other operations in the extractive industries involving quarries, opencast mining, peat extraction and underground mining. There is a screening procedure to determine whether the development requires an EIA.

12.5.4. Permitted development

The Town and Country Planning (General Permitted Development) Order 1995 contains several classes of development relating to mining operations that is permitted development (see Chapter 4). These classes include operations for the erection of plant, machinery buildings and structures on land used as a mine; certain waste tipping at a mine; mineral exploration on any land during a period not exceeding 28 consecutive days; the removal of material from a stockpile or deposit; and certain coal mining by licensees of the Coal Authority.

12.5.5. Notification and publicity for planning applications

In connection with an application in respect of underground mining operations, notice of the application must be given to all owners of land (including those with an interest in a mineral in the land except oil, gas, coal, gold or silver) (section 67(1)).

12.5.6. Aftercare conditions

Conditions may be imposed upon the grant of planning permission to ensure the restoration and aftercare of land after the completion of mining operations. A restoration condition requires the site to be restored by the use of soils; and an aftercare condition requires steps to be taken to bring the land to a required standard for agriculture, forestry or amenity uses (Schedule 5, para 2).

Aftercare may include planting, cultivating, fertilising, watering, draining or otherwise treating the land: a period of time may be specified for any of these steps; the aftercare period itself being five years after compliance with the restoration condition (Schedule 5, paras 5–7).

The required standard for agriculture is that which prevailed before the mining operations, where the original use was agriculture, provided the Minister of Agriculture has stated what the original physical characteristics of the land were, and it is practicable to achieve these characteristics again (Schedule 5, para 3(1)).

There are also provisions defining the required standard for agriculture (in other circumstances), forestry or amenity uses (Schedule 5, paras 3 and 4).

In *R* v *Derbyshire County Council, ex p North East Derbyshire District Council* [1980], a question arose as to whether a condition imposed in a planning permission to extract minerals, that the excavated land should be filled with the remaining overburden and other fill that may be necessary, to restore the area, authorised the use of the land for tipping waste. The Divisional Court decided that the condition did authorise the tipping of waste, because if the condition required infilling of the site, it must authorise the necessary activities, and these could include infilling with waste.

Where planning permission for mineral extraction has been granted subject to conditions requiring the restoration of the quarry, and an application for tipping waste materials finished with a topsoil covering is approved, that approval may be treated as approval of the restoration conditions which had originally been reserved: see *R* v *Surrey County Council, ex p Monk* [1986].

12.5.7. Duration of planning permission

Unless some other period is specified, planning permission for the winning and working of minerals lasts for 60 years. The permitted operations must cease within the period of time (Schedule 5, para 1).

12.5.8. Positive planning powers

These have already been mentioned. The 1990 Act gives powers to minerals planning authorities to deal with sites where the winning and working of minerals becomes inconsistent with the proper planning of their areas. It may be a site that has not yet been worked, is in the process of being worked, or work has been abandoned. The use of a site may become inconsistent with adjoining uses or development, become an eyesore; or, in some cases, prevent the land from being used for some other purpose.

The ordinary powers of making revocation or modification orders to an existing planning permission, or a discontinuance order requiring a use to cease and buildings or works removed, are available under the 1990 Act subject to some amendments (sections 97 and 102(1)). For the purpose of a discontinuance order, use includes the winning and working of minerals, and plant or machinery can be ordered to be removed as well as buildings or works (section 102(8)). In respect of any of the foregoing orders, aftercare conditions may be added. The basis of compensation is considered in the following text.

Schedule 9 to the 1990 Act contains provisions for the discontinuance of mineral working. These provisions are in place of the power to make a discontinuance order under section 102. An order may be made that the use of land for mineral working or the deposit of refuse or waste minerals should be discontinued. There are also powers to make an order prohibiting the resumption of mineral working or to make an order after the temporary suspension of mineral working.

A prohibition order may be made where work has not been carried out to any substantial extent for two years, and it appears that it is unlikely to be resumed. The order may specify the steps to be taken to remove plant or machinery and to safeguard the amenities; it may include a restoration condition and an aftercare condition. The order must be confirmed by the Secretary of State, to whom there is a right of appeal (para 4 of Schedule 9 to the 1990 Act).

A suspension order may be made where any work has been suspended for 12 months, but may be resumed. The purpose of the order is to require that steps be taken for the protection of the environment. The steps mean those necessary for the preservation of the amenities of the area during suspension, the protection of the area from damage and the prevention of any deterioration in the condition of land (Schedule 9, para 5). A suspension order must be confirmed by the Secretary of State, to whom there is a right of appeal (Schedule 9, para 7). The winning and working of minerals, suspended by an order, may be recommenced: a notice must first be served on the minerals planning authority. If they fail to revoke the order within two months, application may be made to the Secretary of State, for revocation (Schedule 9, para 10(5)). It is an offence to contravene a prohibition or suspension order without planning permission.

Under section 96 of, and Schedule 14 to, the Environment Act 1995, there can be a periodic review of mining planning permissions. If a restriction on working rights is imposed, this is treated as if a modification order had been made under section 97 of the 1990 Act, as in the preceding text. Chapter 19 deals with claims for compensation for an inability to work minerals.

12.5.9. *Compensation in connection with positive planning powers*

Compensation is ordinarily available to a person with an interest in land following an order revoking or modifying planning permission, or a discontinuance order (see Chapter 19). In either case, a person entitled only to an interest in the mineral may now be entitled to compensation (section 107(1) and 115(2)). Other provisions provide a right to claim compensation in connection with prohibition and suspension orders. The right to compensation is set out in the Town and Country Planning (Compensation for Restrictions on Mineral Working and Mineral Waste Depositing) Regulations 1997. It is also explained in MPG 4, a mineral planning guidance note.

The making of revocation and modification orders, discontinuance orders, prohibition orders and suspension orders may attract compensation from the minerals planning authority. The 1997 Regulations define the circumstances in which compensation is not payable. The circumstances vary as between the different types of order (see regulations 3–6).

12.6. Caravans

The ordinary developer is unlikely to be concerned with the problems of caravans; what follows is therefore a very brief outline of the appropriate controls.

Caravans are the subject of general planning control as well as the special controls found in the Caravan Sites and Control of Development Act 1960. The 1960 Act was passed because of the enforcement difficulties experienced by planning authorities in applying the development controls of the planning Acts.

12.6.1. Control under the 1990 Act

Under the 1990 Act, the use of land for caravans may involve a material change of use requiring planning permission: see *Guildford Rural District Council* v *Fortescue* [1959]. It is unlikely that the stationing of a caravan in the drive or garden of a private dwellinghouse during such times as it was not used for holiday purposes would be development; therefore no planning permission would be required.

If a caravan becomes a permanent fixture to land, perhaps because its wheels are removed or it is connected to services, it may be regarded as a building or structure within Lord Parker CJ's tests in *Cheshire County Council* v *Woodward* [1962] (see Chapter 2).

However, where a caravan is stationed on agricultural land and used for the storage of feed for agricultural purposes, this involves no material change of use: *Wealden District Council* v *Secretary of State for the Environment* [1986].

Planning permission is deemed granted by the General Permitted Development Order 1995 in respect of the matters described in paragraphs 2–10 of Schedule 1 to the Caravan Sites and Control of Development Act 1960. These purposes are set out in the following text. In the case of paragraph 10, the purpose does not include use for winter quarters.

12.6.2. Control under the 1960 Act

Any person who wishes to use land as a caravan site must first obtain planning permission in the ordinary way under the 1990 Act (see the preceding text). Appropriate conditions may be attached to any such permission.

Having obtained planning permission, the occupier of the land still cannot use it as a caravan site unless he obtains a site licence under the 1960 Act. Failure to do so is a criminal offence (section 1). A caravan site 'means land on which a caravan is stationed for the purposes of human habitation and land which is used in conjunction with land on which a caravan is stationed' (section 1(4)).

Schedule 1 to the Act contains a list of exemptions to this requirement of a site licence (section 2). The exemptions include: land within the curtilage of a dwellinghouse if the use is incidental to the enjoyment of that dwellinghouse (para 1); land used for a person travelling with a caravan, staying not more than two nights, and which is otherwise not so used for more than 28 days in the preceding 12 months, and only by one caravan at a time (para 2); land that exceeds 5 acres and which is not used by a caravan on more than 28 days in the preceding 12 months, and with no more than three caravans at a time (caravans on farms during holiday seasons) (para 3); land used by an organisation holding a certificate of exemption (para 4); land approved by exempted organisations (maximum five caravans) (para 5); land used by meetings of exempted organisations (maximum five days) (para 6); seasonal accommodation of agricultural or forestry workers on agricultural or forestry land (paras 7–8); use of land, adjoining a site where building or engineering operations are taking place, for accommodation of building or engineering workers employed on the site (para 9); use of land by a certified travelling showman (para 10); and, the use of land by the local authority (para 11).

It must be emphasised that these are exemptions from the need for a site licence; planning permission may still be required in some of these cases.

Application for a site licence is made to the district or borough council. Planning permission must first be obtained; if granted, otherwise than by a development order (see Chapter 4), the site licence must be issued within two months (section 3). A site licence cannot be issued for a limited period only, unless the planning permission is limited in duration, in which case the site licence expires when the period limited in the planning permission ends (section 4).

Conditions may be attached to the site licence: they may restrict the numbers of caravans, their type and state of repair and distribution on the site; they may concern the preservation and enhancement of the amenities; the provisions for fire prevention; and the proper arrangements for sanitary facilities (section 5). Apart from these particular matters, conditions are otherwise imposed in the interests of the caravan dwellers, other persons, or of the public at large (section 5(1)).

There is a right of appeal, against any condition that is unduly burdensome, to the magistrate's court within 28 days of the issue of the site licence. The court may vary or cancel any condition having regard to the 'Model Standards' issued by the Department of the Environment (section 7).

Conditions may be changed, deleted or added to a site licence at any time by the local authority. The licence holder may first make representations. He then has a right of appeal to the magistrate's court within 28 days of the written notification of the altered site licence (section 8).

It is an offence if the occupier fails to comply with a condition of a site licence. Persistent offenders risk the revocation of the licence (section 9).

13   Environmental controls

13.1. Introduction

This chapter outlines some of the environmental controls. Two policy ideas form the background to environmental controls; the adoption by Parliament of an integrated pollution control in the Environmental Protection Act 1990 and the Environment Act 1995; second, the development of the legislative programme of the European Union on environmental protection. The latter programme included directives relating to a requirement for environmental impact assessments for major development schemes; environmental impact assessments are considered in Chapter 7. A number of legislative provisions give powers to deal with contamination of land, noise, pollution of water supplies and the atmosphere, and the storage and use of hazardous materials. Although these controls are largely additional to planning control, proposals involving environmental risks are carefully considered at a planning application stage. The Environmental Protection Act 1990 attempted to integrate various statutes such as the Control of Pollution Act 1974 and the Health and Safety at Work, etc., Act 1974, with the aim of providing an integrated system of environmental protection of nature and the conservation of scare resources.

13.2. Principal legislation

The Environmental Protection Act 1990 was enacted to 'make provision for the improved control of pollution arising from certain industrial and other processes; to re-enact the provisions of the Control of Pollution Act 1974 relating to waste on land with modifications as respects the functions of the regulatory and other authorities concerned in the collection and disposal of waste … to restate the law defining statutory nuisances … to provide for the extension of the Clean Air Acts to prescribed gases and … to make provision for the control of genetically modified organism', among many other matters.

It is important that the 1990 Act be read together with the Control of Pollution Act 1974, the Water Resources Act 1991, the Water Act 1991 and the Clean Air Act 1993 as some of the provisions of those Acts are also in force. However, the 1990 Act has provided an integrated system of environmental control, the object being to prevent pollution at source.

The Environment Act 1995 was enacted specifically to provide for the establishment of the Environment Agency and to make 'provision for the control of pollution, the conservation of natural resources and the conservation or enhancement of the environment'. The principal aim of the agency is to discharge its functions so to protect or enhance the environment, taken as a whole, to make the contribution towards attaining the objective of achieving sustainable development (see section 4(1)).

The functions for environment and pollution controls of a number of bodies were transferred to the Environment Agency. The bodies whose functions have been transferred include the Chief Inspector under the Radioactive Substances Act 1993, the Inspectorate under the Alkali, etc., Works Regulation Act 1906, the Health and Safety Executive, in relation to the service and improvement and enforcement notices under the Health and Safety at Work Act 1974, the National Rivers Authority, the waste regulation authorities, the waste disposal authorities in relation to the Control of Pollution Act 1974 and certain functions under the Water Industry Act 1991. The Environment Agency has pollution control powers for the purposes of preventing or minimising, or remedying or mitigating the effects of, pollution of the environment under section 5. Many of these functions will be regulation through the comprehensive Environmental Permitting (England and Wales) Regulations 2007.

The 1995 Act inserted a series of new provisions into the Environmental Protection Act 1990 dealing with contaminated land as from July 1999 (see part **13.4** below). The Act also sets out provisions relating to air quality, waste and drainage.

This chapter does not address any liability at common law for any matters of an environmentally damaging nature, such as noise or pollution. Where such events take place, an adjoining landowner may have a remedy at common law in the law of nuisance or negligence. Indeed, the case of *Rylands* v *Fletcher* [1868] is an early example of strict liability for the escape of non-natural matters brought onto land.

13.3. Health and safety at work

The Health and Safety at Work Act 1974 provides for securing the health and safety of persons in the work place. This is administered by the Health and Safety Executive. In appropriate cases, a local planning authority will consult the HSE where there are particular concerns relating to such matters. Thus, in *R* v *Tandridge District Council, ex parte Al Fayed* [1999], the local planning authority sought the view of the HSE about the safety of radio emissions from a radio mast.

13.4. Waste management and contamination control

13.4.1. Waste management licences (environmental permits)

Section 33 of the Environmental Protection Act 1990 provides for a system of waste licensing. The deposit, treatment or disposal of controlled waste in or on land can only be carried out with a waste management licence. 'Waste' means any substance or object in the categories set out in Schedule 2B to the Act. 'Controlled waste' is household, industrial and commercial waste.

Section 34 of the Act imposes a duty of care on persons to prevent the contravention of the provisions relating to licensing and to prevent the escape of waste and secure its proper transfer.

Waste management licences may relate to a specific site (site licence). The Act provides for conditions to be imposed on a site licence. There are also provisions for the variation, transfer and surrender of licences. The Environmental Permitting (England and Wales) Regulations 2007 now provides a new regime for waste and hazard licensing. From 1 April 2008, all pollution and Prevention and Control permits and waste management licences became environmental permits.

Where the owner of a site that is the subject of a site licence is a company, and goes into liquidation, the liquidator can disclaim the licence under the Insolvency Act 1986: see *Official Receiver* v *Environment Agency* [1999].

13.4.2. Contaminated land

Part IIA of the Environmental Protection Act 1990 contains powers for the Environment Agency to deal with contaminated land. Under section 78A, contaminated land is any land which appears to the local authority to be in such a condition, by reason of substances, in or under that land, that a significant harm is being caused or there is a significant possibility of such harm being caused, or there is, or is likely to be pollution to certain waters such as rivers and streams.

Certain sites may be designated as 'special sites' by the local authority under section 78C. Such sites are those where the land has or may have a serious environmental impact or where the Environment Agency is likely to have the requisite expertise (see section 78C).

Where a site has been designated as a special site, or a local authority has identified any contaminated land, then the appropriate authority shall take steps to serve a remediation notice requiring remediation: see section 78E. The appropriate authority will be the Environment Agency in respect of special sites and the local authority for other sites; in either case, the Act uses the expression 'the enforcing authority'. Section 78F contains provisions for determining the appropriate person to bear the responsibility of remediation. Generally, the appropriate person is the person who caused the contamination. The principles for deciding whether a person has 'caused' pollution were set out by Lord Hoffmann in *Empress Car Co (Abertillery) Ltd* v *National Rivers Authority* [1998].

13.5. Noise control

Part III of the Control of Pollution Act 1974 contains provisions to control noise. Local authorities shall inspect their area from time to time to decide whether to designate any noise abatement zones. These are zones where noise levels can be monitored and steps taken to reduce noise levels for the improvement of the local environment. The 1974 Act also contains provisions to control noise on construction sites, in streets and from new buildings.

13.6. Statutory nuisances

The control of statutory nuisances was largely contained in the Public Health Act 1936. It is now found in sections 79–82 of the Environmental Protection Act 1990. A number of matters constitute statutory nuisances under section 79. Essentially, they are matters that are prejudicial to health or a nuisance.

Where a local authority is satisfied that a statutory nuisance exists, or is likely to occur or reoccur, they shall serve an abatement notice. An abatement notice may require the abatement of the nuisance and or the execution of works (see section 80(1)). Where a person on whom an abatement notice is served contravenes or fails to comply with any requirement or prohibition imposed by the notice, he shall be guilty of an offence.

The grant of planning permission does not confer immunity against an action for nuisance. In *Wheeler* v *JJ Saunders Ltd* [1995], the planning consents granted did not prevent the plaintiffs from succeeding in their claim in nuisance.

In *R* v *Kennet District Council, ex parte Somerfield Property Co* [1999], the relationship between environmental protection legislation and pure planning legislation was considered. The local planning authority imposed a condition in a planning permission more onerous then a condition agreed following the service of an abatement notice under the Environmental Protection Act 1990. The court declined to interfere with what it regarded as a factual issue for the local planning authority to resolve.

Noise can also be controlled by the use of an abatement notice served under section 79(1)(g) of the Environmental Protection Act 1990; this applies to statutory nuisances. *Manly* v *New Forest DC* [1999] concerned an abatement notice alleging that noise from dogs, kept at premises used for the business of breeding dogs, amounted to a statutory nuisance. It was held that a defence could be advanced that the 'best practicable means' does not necessarily involve moving the kennels to a more appropriate location. The intention of Parliament was to provide the business operator with a defence to the nuisance his business created so long as the best practicable means had been employed.

13.7. Miscellaneous controls

13.7.1. Radioactive materials and other controls

Where premises were to be used for the keeping and use of radioactive material, they had to be registered with the Chief Inspector for England and Wales under the Radioactive Substances Act 1993. No premises were to be used for the purposes of installing or operating a nuclear installation without a Nuclear Site Licence granted by the Health and Safety Executive under the Nuclear Installations Act 1965. Under the Environment Act 1995, the functions of the chief inspector and the functions of the inspectors relating to the improvement and prohibition notices under the Health and Safety at Work Act 1974, have now been transferred to the Environment Agency among many others. The functions of the Drinking Water Inspector have not been transferred to the agency as those are more related to public health then the environment. The general functions of the agency are exercisable for the purposes of 'preventing or minimising, or remedying or mitigating the effects of, pollution of the environment'. Where a minister thinks appropriate he may require the agency to carry out an environmental assessment report of the effect, or likely effect, on the environment of existing or potential levels of pollution of the environment.

The Environment Agency has power to bring these controls within the scope of the comprehensive Environmental Permitting (England and Wales) Regulations 2007, or something similar, after 1 April 2010.

13.7.2.   *Entertainment, theatre, cinema and sex establishments*

Under the Theatres Act 1968, no premises may be used for the public performance of any play without a licence granted by the local authority. The licence may contain terms and conditions. Cinemas must also be licensed by a local authority under the Cinemas Act 1985. The basis of licensing theatres and cinemas is to protect the public against unsafe premises.

The Local Government (Miscellaneous Provisions) Act 1982 contains provisions to license sex establishments. A sex establishment means a sex cinema or sex shop. A sex cinema is one primarily devoted to the stimulation of or encouragement of sexual activity, force or restraint, or which primarily deals with genital organs, urinary or excretory functions: see Schedule 3. A sex shop includes premises for the sale, hire, loan, display or demonstration of sex articles or other things related to sexual activity. A local authority may resolve under section 2 of the Act that the provisions shall apply to their area. A licence is then required for the use of premises as a sex establishment on such terms, conditions and restrictions as may be specified.

Part IV

Positive planning and enforcement

14 Revocation, modification and discontinuance orders

14.1. Revocation or modification of a planning permission

A local planning authority may, by order, revoke or modify any planning permission that has been granted upon an application. They may do so having had regard to the provisions of the development plan and to any other material considerations (section 97 of the 1990 Act). This power is used sparingly; first, because the authority is liable to pay compensation; and, second, because a planning permission that is not implemented expires after three years; outdated planning permissions will hopefully just lapse.

If planning permission has been granted in an improper or invalid manner, the High Court might quash the permission on an application for judicial review, and no compensation would then become payable: see *Corbett* v *Restormel BC* [2001].

14.1.1. General

The power to revoke or modify a planning permission may only be exercised before the carrying out of building or other operations have been completed, or before a change of use has taken place. A revocation or modification order cannot affect so much of any building or other operations already carried out (section 97(4)). Once the building or other operations have been completed, or the change of use has taken place, the local planning authority is limited to the use of a discontinuance order (see the following text) if they wish to take positive action against development that is unacceptable in planning terms, but is otherwise perfectly lawful. A revocation or modification order is registrable as a local land change. An order cannot be made to withdraw any permission granted by the Town and Country Planning (General Permitted Development) Order 1995; permitted development rights can only be withdrawn by the making of a direction under article 4 of that order (see Chapter 4).

There are two procedures for the making of these orders by the local planning authority; the second of these will be appropriate where the landowner is not going to oppose the order; this may be the case if the local planning authority is prepared to grant an alternative planning permission.

14.1.2. Procedure where order is likely to be opposed

Notice of the making of an order, and its submission to the Secretary of State for his approval, must be served on the owner and occupier of the land affected, and upon any other person who in the opinion of the local planning authority will be affected by the order. Any person upon whom the notice has been served may require that he be heard before a local inquiry or hearing. The Secretary of State, having considered, if the order is opposed, any report made by his inspector, may confirm the order, with or without modifications; the order only then takes effect (section 98). The costs of a successful objector at an inquiry may be awarded against the local planning authority (Circular 8/93 – *Award of Costs incurred in Planning and Compulsory Purchase Order Proceedings*).

14.1.3. Unopposed procedure

Orders made under this procedure do not need the confirmation of the Secretary of State (see section 99). The procedure may be used where the owner and occupier of the affected land, and any other person affected by the order, have notified the local planning authority in writing that they do not object to the order. The local planning authority must advertise the fact that they have made an order under section 97, and also notify the parties who have previously said in writing that they will not object.

Any person who may be affected by the order may give notice to the Secretary of State that they wish to be heard before a local inquiry or hearing (section 98).

If no notice of objection is made, and the Secretary of State, who must have been sent a copy of the advertisement, does not call in the order for his consideration, the order then takes effect without the Secretary of State's confirmation at the end of a period of time specified in the advertisement (section 99(6) and (7)).

14.1.4. Alternative methods of terminating a planning permission

Rather than using the procedure outlined in the preceding text, it is always possible for the landowner to agree not to implement a planning permission through the use of an agreement or planning obligation under section 106 of the 1990 Act (see Chapter 9). Further, if a planning permission has been wrongly issued, an application could be made to quash the permission on the ground that it was unlawfully issued (see *R* v *Bassetlaw District Council, ex p Oxby* [1997]). There would then be no liability to pay compensation. However, an application was dismissed in *R* v *Restormel BC, ex p Parkyn* [2001], where the only purpose of the application to judicially review and quash the permission was to deny the landowner compensation.

14.1.5. Compensation and purchase

An order may cause a depreciation in land values or give rise to other costs; the land may be rendered incapable of reasonably beneficial use. The remedies of compensation (see Chapter 19) or a purchase notice (see Chapter 18) may be available to the landowner.

14.1.6. Mining operations

There are provisions in Schedule 5 of the 1990 Act that allow minerals planning authorities to modify planning permissions by the imposition of restoration and aftercare conditions. There is a modified compensation basis. The Town and Country Planning (Compensation for Restrictions on Mineral Working and Mineral Waste Disposing) Regulations 1997 provide that no compensation is payable where an order is made modifying planning permission for development consisting of the winning and working of minerals or the deposit of mineral waste if certain conditions are satisfied.

14.2. Discontinuance orders

If it is considered by a local planning authority that it would be in the interests of the proper planning of their area (including the interests of amenity) that: (a) any use of land should be discontinued, or that any condition should be imposed on the continuance of a use of land, or (b) any buildings or works should be altered or removed, they may then achieve such objectives by the making of a discontinuance order, due regard being paid to the provisions of the development plan and to any other material consideration (section 102(1) of the 1990 Act). The importance of a discontinuance order is that it may be directed against any existing lawful use of land, or require the removal or alteration of any building or works that are lawful. Its limitation is that compensation becomes payable.

The storage and sorting of scrap, an activity that might well be the subject of a discontinuance order, was held in *Parkes* v *Secretary of State for the Environment* [1979] CA to be a use of land within (a) above, and therefore properly within the scope of these orders.

The demolition of a disused coastguard lookout station was held in *Re Lamplugh* [1967] to be within the scope of (b) above: (b) was a separate case for the use of a discontinuance order, it was not necessarily ancillary to (a), which refers to a discontinuance of a use.

A discontinuance order will specify the steps to discontinue the use, or to alter or remove the buildings or works, or it will specify the conditions being imposed. It may grant planning permission for any development of land to which the order relates; or grant planning permission for the retention of buildings or works or the continuance of a use of the land which existed or commenced, as the case may be, before the order was submitted for confirmation (section 102(1) to (4)).

A discontinuance order does not take effect until confirmed by the Secretary of State (section 103(1)). He may vary the order to include a grant of planning permission (section 103(2)). The reason for including a power, either for the local planning authority, or the Secretary of State, to grant a planning permission for the affected land, is to ensure that the owner is able, following the discontinuance order, to use his land for some lawful purpose.

Notice that the discontinuance order has been submitted to the Secretary of State for confirmation must be served on the owner and occupier of the affected land, and on any other person who the local planning authority believe will be affected by the order. Such persons may then require that their objections be heard before a local inquiry or hearing (section 103(4) and (5)).

Once the Secretary of State has confirmed the order, the local planning authority must serve a notice to that effect on the owner and occupier of the affected land (section 103(7)). These orders are registrable as land charges.

Unless planning permission has been obtained, it is a criminal offence to use the land for the purpose that is required to be discontinued by an order; or to use land contrary to any conditions imposed by an order. It is also an offence if a person causes or permits the affected land to be so used in contravention of an order (section 189). If the discontinuance order required the alteration or removal of any buildings or works, and this has not been achieved within the specified time, the local planning authority may do the work (section 189).

For the payment of compensation in respect of discontinuance orders, see Chapter 19. For purchase notices, see Chapter 18.

Schedule 9 to the 1990 Act sets out a special procedure relating to the discontinuance of mineral working. The Schedule also contains powers to make other types of orders prohibiting or suspending mineral working (see Chapter 12).

15 Enforcement

15.1. Introduction

It is not a criminal offence to carry out development without planning permission or to fail to comply with any planning condition. The scheme of the Town and Country Planning Act 1990 is to first define what is meant by a breach of planning control and then to provide that a local planning authority may take steps to enforce a breach of planning control within specified time limits. Enforcement involves the issue of notices specifying the alleged breaches and requiring the breaches to be remedied. Subject to any right of appeal, it is only at that stage that a criminal offence may be committed if the requirements of such a notice are not complied with.

In 1989, the Carnwath Review, *Enforcing Planning Control*, was published. The report made a number of recommendations for the improvement of the law. The Planning and Compensation Act 1991 introduced amendments to the Town and Country Planning Act 1990 to give effect to the Report's recommendations.

15.2. Breach of planning control

Under the Town and Country Planning Act 1990, enforcement action may only be taken in respect of 'a breach of planning control' (see section 172). Under section 171A, the following constitute a breach of planning control:

(a) *Carrying out development without the required planning permission, or*
(b) *Failing to comply with any condition or limitation subject to which planning permission has been granted.*

This definition is relevant not only for the purposes of considering whether enforcement action can be taken, but also in those cases where there has been a breach of planning control, no enforcement action has been taken within the time limits (see the following text), and the owner is entitled to a certificate of lawful use and development under section 191 (see part **15.8** below).

15.2.1. Development without planning permission

The meaning of development was dealt with in Chapters 2 and 3. Development consists of the carrying out of certain operations or the making of a material change of use. A planning permission will include a permission granted upon an application to the local planning authority, a permission granted following a successful appeal to the Secretary of State or a permission deemed granted by the General Permitted Development Order 1995.

In *Wyatt* v *Jarrad* [1998], it was held that, because the Town and Country Planning (Control of Advertisements) Regulations 1992 provide a self-contained code for control of advertisements, the general enforcement provisions in this part of the 1990 Act relating to enforcement notices, and the time limits within which they must be issued, do not apply.

15.2.2. Non-compliance with conditions or limitations

A planning permission may be granted subject to a condition (see Chapter 9). However, the word 'limitation' in (*b*), at page 159 has not always been wholly clear: in *Cynon Valley Borough Council* v *Secretary of State for Wales* [1986], it was said that where planning permission was granted by a general development order for shop use, except certain excluded uses, the permission was subject to a limitation. In *I'm Your Man Ltd* v *Secretary of State for the Environment* [1998], where the court was considering a planning permission granted for a 'temporary period of seven years', it was said that that was not a limitation and was ineffective because the purported time limit ought to have been imposed by way of a planning condition. It is generally considered that the word 'limitation' probably refers to the time limitations set out in various parts of the Town and Country Planning (General Permitted Development) Order 1995. For example the permitted development consisting of the right to use any land for any purpose for not more than 28 days in any calendar year.

15.3. Time limits

Until amendments were made in 1992 by the Planning and Compensation Act 1991, there were two time limits. In the case of the making of a material change of use (other than the use of a building as a single dwellinghouse), enforcement action could be taken in respect of a breach of planning control that took place after 1963. Any breach of planning control consisting of a material change of use (other than the use of a building as a single dwellinghouse) that took place before 1964 was immune from enforcement action. In the case of all other breaches of planning control, there was a four-year limitation period.

Following the amendments introduced by the 1991 Act, as amended by the Localism Act 2011, the position is now as follows.

15.3.1. The four-year limitation period

Under section 171B(1) of the 1990 Act, where there has been a breach of planning control consisting in the carrying out without planning permission of building, engineering, mining or other operations in, on, over or under land, no enforcement action may be taken after the end of the period of four years beginning with the date on which the operations were substantially completed. In *Sage* v *Secretary of State for the Environment, Transport and the Regions* [2003], where a building, having the external appearance of a dwelling, was substantially completed, save for internal works and some external works that would not have constituted development requiring planning permission, the House of Lords concluded that the time limit of four years only began when the whole intended operation of creating the dwellinghouse was substantially completed. Accordingly, the four-year period would not begin until the additional internal and the remaining external works were completed as a holistic approach should be adopted.

Under section 171B(2), where there has been a breach of planning control consisting in the change of use of any building to use as a single dwellinghouse, no enforcement action may be taken after the end of the period of four years beginning with the date of the breach.

One difficulty that the courts have had to address is whether a particular activity is one operation or a series of operations. For example, if engineering operations are carried out over a long period of time, could the owner contend that those works carried out more than four years prior to any enforcement action are immune? In *Ewen Developments Ltd* v *Secretary of State for the Environment* [1980], the court accepted that the construction of earth embankments amounted to one operation, so that even if some of the work was carried out more than four years before the enforcement action, the whole operation was a breach if carried out without planning permission, and the whole of the embankments, not just the part made within the four years, could be required to be removed.

Difficulties have arisen where a building or buildings are subdivided and a change of use to *more* than one single dwellinghouse occurs. Does the four-year limitation period apply or, because more than one single dwellinghouse use arises, does the 10-year limitation period apply? In *Van Dyck* v *Secretary of State for the Environment* [1993], the Court of Appeal held that the four-year rule applied to residential subdivisions where a single dwellinghouse was converted into two flats. A similar approach was adopted by the Court of Appeal in *Moore* v *Secretary of State for the Environment* [1999] in relation to the change of use of ten self-contained units of residential occupation in the outbuildings of a country house. The concept of the planning unit was rejected and each residential unit was treated as if it were a single dwellinghouse. Where a permission allowed use of a building solely for recreational purposes during the summer months and not for use at other times as permanent residential accommodation, it was held in *Bloomfield* v *Secretary of State for the Environment, Transport and the Regions* [1999] that continuous permanent residential use would not involve a change of use to a dwellinghouse, but a breach of condition to which the 10-year time limit (see below) applied. Although in *First Secretary of State* v *Arun DC* [2006] the Court of Appeal decided that residential use of a unit in breach of a planning condition was subject to the four-year limit.

In the case of mining operations, it was held in *High Peak Spar Ltd* v *Secretary of State for Communities and Local Government* [2009] that an enforcement notice directed to the winning and working not permitted by a planning permission was not time-barred.

15.3.2. The 10-year limitation period

In the case of any other breach of planning control, no enforcement action may be taken after the end of the period of 10 years beginning with the date of the breach: see section 171B(3). Any other breach will therefore consist of the making of a material change of use without planning permission, or the non-compliance with a condition or limitation subject to which planning permission was granted. There is just one exception to this. By virtue of section 171B(2), where there is a condition preventing the change of use of any building to use as a single dwellinghouse, a breach of such a condition is a breach of planning control to which the four-year limitation period applies (see earlier).

There must be continuity of the use to the date when enforcement action is taken.

Where a building has been erected without planning permission, the four-year limitation period, mentioned at part **15.3.1** above applies. However, on the expiration of that period, the use of a building immune from enforcement action in relation to the building operations is not immune from enforcement action in respect of the use: see *Sumner* v *Secretary of State for Communities and Local Government* [2010].

In material change of use cases, difficulties can arise in deciding when a material change of use took place where changes of use are gradual over a number of years. In *Thurrock BC* v *Secretary of State for the Environment, Transport and the Regions* [2001], the court approved the test of comparing the present use with the use in the base year, normally 10 years earlier. Breaks in periods of use could mean that there has been no continuity for the requisite 10 years: see *Miles* v *National Assembly for Wales* [2007].

15.3.3. Triggering the conditions

There can be no breach of a planning condition if the planning permission it is attached to is not implemented: *Newbury District Council* v *Secretary of State for the Environment* [1981].

In *Handoll* v *Warner Goodman & Streat* [1995], the Court of Appeal considered a preliminary matter in negligence proceedings against a firm of solicitors and the local planning authority. Although planning permission had been granted for the erection of an agricultural dwelling, subject to a condition restricting the occupation of the dwelling to persons employed in agriculture, the dwelling was erected some 90 feet west of the approved location. The court held that the decision of *Kerrier District Council* v *Secretary of State for the Environment* [1980] was wrongly decided; where development is carried out in breach of planning control, and not by way of implementation, the development is not subject to any conditions imposed by a planning permission.

It follows that in the factual circumstances that arose in the *Handoll* case, the erection of the unauthorised dwelling must be enforced within the four-year time limit, whereas if the building had been erected in the authorised location, the 10-year time limit would apply to any breach of condition.

15.3.4. *Extending the time limits*

Under section 171B(4), enforcement action can be taken outside the time limits in two circumstances. A breach of condition notice may be served in respect of any breach of planning control if an enforcement notice in respect of the breach is in effect (see section 171B(4)(a)). Second, a further enforcement notice may be issued within four years of any earlier enforcement action that has been taken or purported to have been taken (see section 171B(4)(b)). The purpose of this further provision would appear to authorise the service of additional enforcement action or to rectify a fault in some earlier enforcement action. The first enforcement action must be taken within the appropriate time limits specified in the earlier provisions of section 171B: see *William Boyer (Transport) Ltd* v *Secretary of State for the Environment* [1996]. The second enforcement action can be used to cover the same actual breach of development control, even though it may be described in a different way; what it cannot be used for is to cover two different physical developments or two different changes of use: see *Jarmain* v *Secretary of State for the Environment, Transport and the Regions* [2000].

15.3.5. *Time limits for enforcing concealed breaches of planning control*

In *Welwyn Hatfield Council* v *Secretary of State for Communities and Local Government* [2011], the landowner obtained planning permission to erect a hay barn. In fact, he built it as a dwellinghouse, and after four years applied for a certificate of lawful use relying on the time limit in section 171B(2) of the 1990 Act. Although the Court of Appeal decided he was so entitled, and that it was for Parliament to consider whether to legislate where a deception might be involved, the Supreme Court held that as the building was in every respect designed and built as a house, there had been no change of use that triggered the time limit, and that a person could not, in any event, rely on the expiration of the time limit in section 171B(2) as a result of a deceit. The Localism Act 2011 inserted a new section 171BA into the Town and Country Planning Act 1990 to deal with this. This provides that where it appears to a local planning authority that there may have been a breach of planning control, it may apply to a magistrate's court for a 'planning enforcement order'. If such an order is made, the local planning authority may take enforcement action within an 'enforcement year', being a year beginning at the end of 22 days from the day of the court order. However, there is a detailed procedure for applying for a planning enforcement order. An application must be made within six months of the date when sufficient evidence of the apparent breach of planning control came to the local planning authority's notice as would justify an application. It must produce to the magistrate's court a certificate with its application, and notice of the application must be served on the affected owner and occupier, and other persons with an interest in the land. The magistrates court may make an order if it is satisfied, on a balance of probabilities, that the apparent breach, or any of the matters constituting the apparent breach, has been deliberately concealed by any person. Any order made must identify the apparent breach and state the date of the order. A planning enforcement order will be entered in the register of enforcement and stop notices.

15.4. Enforcement notices

15.4.1. *Planning contravention notices*

Where it appears to the local planning authority that there may have been a breach of planning control in respect of any land, they may serve a planning contravention notice (see section 171C). A planning contravention notice may be served on any person who is the owner or occupier of the land or has any other interest in it or who is carrying out operations on the land or using it for any purpose.

A planning contravention notice may require the person on whom it is served to give information about any operations being carried out on the land, any use of the land and any other activities being carried out on the land. A notice may also require information about any matter relating to planning conditions or limitations subject to which any planning permission in respect of the land has been granted. A planning contravention notice may require the person on whom it is served to provide certain information, such as whether or not the land is being used for any purpose specified in the notice or any operations or activities specified in the notice are being or have been carried out on the land. The notice may also request dates when uses, operations or activities began, the names and addresses of any persons who have used or carried out operations or activities on the land, information about any planning permissions, and the nature of the interest in the land and the names and addresses of any other person known to have an interest.

In *R* v *Teignbridge District Council, ex p Teignmouth Quay Co Ltd* [1995], it was held that a local planning authority was not entitled to serve a planning contravention notice merely at the request of local residents objecting to a particular development activity; the local planning authority must first form the view that it appears to them that a breach of planning control has taken place.

The purpose of planning contravention notices is to provide a warning that enforcement action may be taken, to encourage compliance with planning control without the necessity of an enforcement notice, and to provide the local planning authority with information enabling them to decide whether to take enforcement action or not.

Under section 171D, it is a criminal offence to fail to comply with the requirements of a planning contravention notice within 21 days.

15.4.2. Issue of enforcement notice

Under section 172 of the 1990 Act, a local planning authority may issue an enforcement notice where it appears to them that there has been a breach of planning control and that it is expedient to issue the notice, having regard to the provisions of the development plan and to any other material considerations. Although it is not necessary for the authority to be satisfied of the actual existence of a breach of planning control, they must show that it appears to them that a breach has occurred: see *R* v *Rochester City Council, ex p Hobday* [1989].

It is clear that enforcement action is discretionary. It would obviously not be necessary to issue an enforcement notice in circumstances where the activity is one for which planning permission would have been granted without any conditions. However, guidance on the use of enforcement notices is contained in the Government Policy Guidance Note – PPG18 – *Enforcing Planning Control*.

Under section 172, a local planning authority issues an enforcement notice and are then required to serve copies of the enforcement notice on the owner and on the occupier of the land to which it relates and on any other person having an interest in the land, being an interest which, in the opinion of the authority, is materially affected by the notice.

The enforcement notice must be served not more than 28 days after the date of issue and not less than 28 days before the date specified in the notice as the date on which it is to take effect.

15.4.3. Contents of an enforcement notice

Under section 173(1) of the 1990 Act, the enforcement notice must state the matters which appear to the local planning authority to constitute the breach of planning control and whether the breach falls within paragraph (*a*) or (*b*) of section 171A. These two paragraphs set out, respectively, the carrying out of development without the required planning permission and the failure to comply with any condition or limitation subject to which planning permission has been granted.

The enforcement notice must specify the steps which the authority requires to be taken, or the activities which the authority requires to cease, in order to achieve, wholly or partly, certain purposes: see section 173(3). Those purposes are remedying the breach by making any development comply with the terms (including conditions and limitations) of any planning permission, by discontinuing any use of the land or by restoring the land to its condition before the breach took place or remedying any injury to amenity which has been caused by the breach. Section 173(5) gives examples of the steps that may be specified in an enforcement notice. An enforcement notice may require the alteration or removal of any buildings or works, the carrying out of any building or other operations, any activity on the land not to be carried out except to the extent specified in the notice, or the contour of a deposit of refuse or waste materials on land to be modified by altering the gradient or gradients of its sides.

In the case of an enforcement notice issued in respect of the demolition of a building, in breach of planning control, the enforcement notice may require the construction of a replacement building.

The enforcement notice must specify the date on which it is to take effect (see section 173(8)). Subject to its suspension while there is any appeal, the enforcement notice will take effect on that specified date.

An enforcement notice must specify the period at the end of which any steps are required to have been taken or any activities are required to have ceased. This is the period for compliance.

The Town and Country Planning (Enforcement Notices and Appeals) (England) Regulations 2002, require that the notice shall specify the precise boundaries of the land to which it relates by reference to a plan or otherwise. The same regulations also require that an enforcement notice must specify the reasons why the local planning authority considers it expedient to issue it, all policies and proposals in the development plan which are relevant to the decision to issue the enforcement notice, and that the enforcement notice should be accompanied by an explanatory note. The explanatory note sets out the relevant provisions of the 1990 Act, or a summary of those provisions, and explains that there is a right of appeal, how that right can be exercised, and the grounds upon which it may be brought.

15.4.4. *Effect of under-enforcement*

There may be circumstances where an authority, in specifying the steps to be taken to remedy any breach of planning control, does not require all matters to be remedied. For example, a building may have been erected in breach of planning control and the enforcement notice specifies only the partial alteration of the building or the remedying of an injury to amenity. By section 173(11), where an enforcement notice could have required any buildings or works to be removed or any activity to cease, but does not do so, and all the requirements of the notice having been complied with, then, so far as the notice did not so require, planning permission is treated as having been granted in respect of the development not required to be remedied by the enforcement notice: see *Tandridge District Council* v *Verrechia* [1999], where the limits of this power are explained.

15.4.5. Variation and withdrawal of enforcement notices

Under section 173A, a local planning authority may withdraw an enforcement notice issued by them or waive or relax any requirement of such a notice and, in particular, may extend any period specified as the period for compliance. An enforcement notice might be withdrawn if negotiations with the landowner have produced a change in the circumstances such that it would no longer be appropriate to continue with the enforcement notice, or possibly new facts come to light such that a new notice ought to be served.

15.4.6. Service of copies of the enforcement notice

It is a defence to criminal liability under section 179 for a defendant to show that he was not served with a copy of the enforcement notice, and the notice is not contained in the enforcement register. It follows that the persons who should be served with copies of the enforcement notice are the owner and occupier of the land and any other person having an interest materially affected by the notice (see section 172(2) of the 1990 Act).

15.4.7. Effect of an enforcement notice

Section 285(1) of the 1990 Act precludes any challenge to the validity of an enforcement notice in any proceedings on any grounds on which an appeal may be brought. That provision is far reaching, as the Court of Appeal explained in *Staffordshire CC* v *Challinor* [2007], where it held that an enforcement notice has the effect of taking away any lawful use, even one with the benefit of a CLEUD (see part **15.8.1** below).

15.4.8. Power to decline determination of retrospective planning applications

Where development has taken place in breach of planning control, and an enforcement notice has been served, a local planning authority now has power to decline to determine a retrospective application for planning permission for development that constitutes the breach of planning control: see the new section 70C of the 1990 Act introduced by the Localism Act 2011.

15.5. Appeals against enforcement notices

A person having an interest in the land to which an enforcement notice relates, or a relevant occupier, may appeal to the Secretary of State against an enforcement notice, whether or not a copy of it has been served on him (see section 174(1)). A relevant occupier means a person who on the date on which the enforcement notice was issued occupies the land to which the notice relates by virtue of a licence and continues so to occupy the land when the appeal is brought.

Where the person having an interest in the land is not the same party making the appeal, difficult issues can arise. In *R* v *Secretary of State for the Environment, Transport and the Regions, ex p Eauville Ltd* [2000], the inspector erred in failing to consider whether the person making the appeal was the owner's agent. In *Buckinghamshire CC* v *Secretary of State for the Environment, Transport and the Regions* [2000], it was held that the inspector failed to consider whether an exception could be made to the rule that the corporate veil could be pierced in circumstances where the owner was a company and the appellant its sole director.

The statutory appeal process, by which appeals are considered by the Secretary of State, together with the procedural safeguards and supervisory role of the High Court (see Chapters 17 and 20), means that Article 6 of the Convention of Human Rights (fair trial) is satisfied: see *R* v *Secretary of State for the Environment, Transport and the Regions, ex p Holding & Barnes plc* [2001] and *Bryan* v *United Kingdom* [1996].

15.5.1. Grounds of appeal

Under section 174(2), an appeal may be brought on any of the following grounds:

(a) That, in respect of any breach of planning control which may be constituted by the matters stated in the notice, planning permission ought to be granted or, as the case may be, the condition or limitation concerned ought to be discharged.

(b) That the matters alleged to be breaches of planning control have not occurred.

(c) That the matters alleged to be breaches of planning control (if they occurred) do not constitute a breach of planning control.

(d) That, at the date when the notice was issued, no enforcement action could be taken in respect of any breach of planning control which may be constituted by those matters, in other words the time limits for the taking of enforcement action have expired.

(e) That copies of the enforcement notice were not served as required by section 172.

(f) That the steps required by the notice to be taken, or the activities required by the notice to cease, exceed what is necessary to remedy any breach of planning control which may be constituted by those matters or, as the case may be, to remedy any injury to amenity which has been caused by any such breach.

(g) That any period specified in the notice as the period for compliance falls short of what should reasonably be allowed.

An appeal may not be brought on ground (a) in the preceding text where an enforcement notice was issued at a time after the making of a related planning application, but before the end of the period specified for the purposes of section

78(2) of the 1990 Act: see the new subsections (2A) and (2B) introduced into section 174 of the 1990 Act by the Localism Act 2011. That period, which is two months from the acceptance of an application, is relevant where an authority does not make a determination. On the expiration of that period, an applicant for planning permission can bring an appeal against non-determination.

15.5.2. Making an appeal

An appeal must be made before the date specified in the enforcement notice as the date on which it is to take effect. It will be remembered that the enforcement notice must specify the date on which it is to take effect, and that date is not less than 28 days following service. The time limit is absolute and cannot be extended: see *Howard* v *Secretary of State for the Environment* [1975] and *R* v *Secretary of State for the Environment, ex p JBI Financial Consultants* [1989].

Section 174(3) provides two ways of serving a notice of appeal. The first is by giving written notice to the Secretary of State before the specified date. The second is by posting notice to him before that date at such time as that, in the ordinary course of post, it would be delivered to him before that date.

The person making an appeal must submit with that appeal, or within 14 days of being requested by the Secretary of State so to do, a statement in writing specifying the grounds on which the person is appealing against the enforcement notice, and the facts in support of those grounds: see the Town and Country Planning (Enforcement Notices and Appeals) (England) Regulations 2002, regulation 6. The same regulations require the local planning authority to provide a statement summarising submissions that they will make in the appeal.

15.5.3. Form of appeal

An appeal will be by way of a public local inquiry or, if both parties agree, it may be determined by way of written representations. The general nature of public local inquiries is considered in Chapter 17. The ordinary inquiries procedure rules do not apply to enforcement notice inquiries. These are governed by the Town and Country Planning (Enforcement) (Inquiries Procedure) (England) Rules 2002. Under rule 8 of these rules, an appellant is required to submit a written statement of his case. The local planning authority is also required by the rules to prepare and serve a statement. Where the appeal is to be heard by way of a hearing, which is a more simplified and informal manner of proceeding, the Town and Country Planning (Enforcement) (Hearings Procedure) (England) Rules 2002 will apply. Under the Town and Country Planning (Enforcement) (Determination by Inspectors) (Inquiries Procedure) Rules 2002, and the Town and Country Planning (Enforcement) (Inquiries Procedure) (England) Rules 2002, the appellant is required to provide a statement of case. In simple cases, an appellant may opt to have the appeal heard by written representations under the

Town and Country Planning (Enforcement) (Written Representations Procedure) (England) Rules 2002.

Save for a small category of classes of appeal, the jurisdiction to determine all appeals has been given to planning inspectors: see the Town and Country Planning (Determination of Appeals by Appointed Persons) (Prescribed Classes) Regulations 1997, as amended.

The burden of proof in enforcement notice appeals is on the appellant: see *Nelsovil* v *Minister of Housing and Local Government* [1962]. The burden of proof is the civil standard of proof on a balance of probabilities: see *Thrasyvoulou* v *Secretary of State for the Environment* [1984]. Under sections 175 and 176 of the 1990 Act, the Secretary of State has power to quash, correct or vary the enforcement notice and to grant planning permission for all or part of the unauthorised development to which the notice relates. Under section 176, the Secretary of State may allow an appeal and quash the enforcement notice.

Following the decision of the Secretary of State, by his inspector, section 289 of the 1990 Act provides for an appeal to the High Court. However, the leave of the High Court is first required. The object of obtaining leave is to weed out cases that are bound to fail. In the past, it had often been the practice of appellants to appeal, not with a view of achieving success at an appeal, but to prolong the effect of the suspension of an enforcement notice until an appeal was determined. A lucrative use of land in breach of planning control might be continued for some considerable time until all the appeal proceedings had been concluded.

15.5.4. *The Mansi principle*

Section 176 of the 1990 Act enables the Secretary of State to correct any defect, error or description in an enforcement notice or to vary its terms. The use of this power frequently arises where the terms of an enforcement notice fail to allow for such activities, whether operations or uses, that are within the scope of any existing use or planning permission. The point is illustrated by the case of *Mansi* v *Elstree Rural District Council* [1964]:

> A greenhouse had been used for horticulture since 1922, and there had been some incidental retail sales. The sales use became more significant, and eventually constituted the main use of the building. An enforcement notice required the use to cease. The Divisional Court held that, as some incidental or ancillary retail sales in association with the horticultural activity would not have been a breach of planning control, the enforcement notice went too far in prohibiting all sales.

Existing use rights can be lost by abandonment, substantive discontinuance, or the implementation of a planning permission inconsistent with the existing use rights. However, subject to the loss of such rights, the local planning authority, or an inspector on an appeal, must consider what existing use rights remain and ensure that they are not curtailed by the terms of the enforcement notice.

The *Mansi* principle protects existing rights from an excessive enforcement notice. However, in cases where a new building is erected under a planning permission, any previous rights are extinguished (*Petticoat Lane Rentals Ltd* v *Secretary of State for the Environment* [1971]), and the only rights flow from the permission itself.

15.5.5. The Secretary of State's decision

In deciding an appeal, the Secretary of State (by his inspector where appropriate) may:

(a) Grant planning permission in respect of the matters stated in the enforcement notice as constituting a breach of planning control, whether in relation to the whole or any part of those matters or in relation to the whole or any part of the land to which the notice relates;

(b) Discharge any condition or limitation subject to which planning permission was granted;

(c) Determine whether, on the date on which the appeal was made, any existing use of the land was lawful, any operations which had been carried out in, on, over or under the land were lawful or any matter constituting a failure to comply with any condition or limitation subject to which planning permission was granted was lawful and, if so, issue a certificate of lawful use or development under section 191.

In considering whether to grant planning permission, the Secretary of State is required to have regard to the provisions of the development plan, so far as material to the subject-matter of the enforcement notice, and to any other material considerations (see Chapter 8). In that regard, the determination must be made in accordance with the development plan unless material considerations indicate otherwise: see section 38(6) of the Planning and Compulsory Purchase Act 2004. When the Secretary of State discharges a condition or limitation, he may substitute another condition or limitation for it whether more or less onerous (see section 177(4)).

15.6. Prosecution and enforcement

15.6.1. Prosecution

The enforcement notice comes into effect on the date it specifies. This date may be suspended if an appeal is made to the Secretary of State, or there is a further appeal to the High Court: see sections 175(4) and 289. Once the notice is effective, and the periods specified for compliance have expired, a criminal offence is committed if the steps required to comply with the notice have not been taken (section 179(1) to (2)). At any time after an enforcement notice has been served on a person, the local planning authority can give that person a letter of assurance under section 172A of the 1990 Act. The assurance can be that that person is not at risk of prosecution, but that any time the assurance can withdrawn from a specified time sufficient to enable the person to avoid the risk of prosecution; presumably that is to enable a person to remedy the breach of planning control. This offence is one for which the owner of the land from time to time is liable. The owner means the freeholder owner or the person entitled to the rack rents. Where the owner transfers the land to a company without officers, as an attempt to avoid the criminal liability, the transfer might be disregarded as a sham: see *Buckinghamshire CC* v *Briar* [2002]. There is a second offence under section 179(4) to (5). This concerns a person who has control of or an interest in the land to which an enforcement notice relates (other than the owner); that person must not carry on any activity which is required by the notice to cease or cause or permit such an activity to be carried on.

There is a defence available to an owner under subsection (3). He may show that he did everything he could be expected to do to secure compliance with the notice. That defence would be available where the premises are let or the breach is being caused by some other person not within the owner's control. Impecuniosity would not normally be a good defence: see *Kent County Council* v *Brockman* [1996]. However, *Wycombe DC* v *Wells* [2005] shows that the owner of a site cannot rely on the defence that he could not find a new site to relocate to, and therefore he had done all that could be reasonably expected to close the subject caravan site. It is also a defence to show that the person charged with the offence was not aware of the notice where he has not been served with a copy of the enforcement notice and the enforcement notice was not contained in the appropriate register kept by the local planning authority.

Further offences may be committed even if an enforcement notice is initially complied with. Thus, in *Prosser* v *Sharp* [1985], the court held that where a landowner removed a caravan in compliance with an enforcement notice, but then replaced it with another caravan, he had committed an offence.

In criminal proceedings, it is not possible to challenge the validity of the original enforcement notice, save by way of one of the statutory rights of appeal to the Secretary of State (see section 285). However, it may be possible to challenge an enforcement notice by judicial review proceedings (see Chapter 20).

R v *Luigi and another* [2011] shows that a confiscation order can be made against offenders for failure to comply with an enforcement notice, even where the profit motive was not personal gain.

15.6.2. *Execution of work in default*

Where any steps required by an enforcement notice to be taken are not taken within the period for compliance with a notice, under section 178 of the 1990 Act, the local planning authority may enter the land and take the steps and recover from the person who is then the owner of the land any expenses reasonably incurred by them in doing so. Under regulation 14(2) of the Town and

Country Planning General Regulations 1992, the expenses of a local planning authority are a charge binding successive owners of the land.

15.6.3. Stop notices

It may be desirable that the activities believed by the local planning authority to be a breach of planning control should cease at the earliest opportunity rather than await the date specified in the enforcement notice, a date which may be suspended and put off by an appeal. Accordingly, where an enforcement notice has been served, an authority may serve a stop notice to prevent the carrying out of any activity included within the matters believed to be a breach of planning control (see section 183(1)). A stop notice may not be served where the enforcement notice has taken effect.

A stop notice cannot prevent the use of a building as a dwellinghouse. A stop notice cannot prevent the carrying out of any activity if the activity has been carried out (whether continuously or not) for a period of more than four years ending with the service of the notice (see section 183(5)).

A stop notice must refer to the enforcement notice to which it relates and have a copy of that notice annexed to it (see section 184(1)). A stop notice must specify the date on which it will take effect (and it cannot be contravened until that date). The date a stop notice takes effect cannot be earlier than three days after the date when the notice is served, unless the local planning authority considers that there are special reasons for specifying an earlier date and a statement of those reasons is served with the stop notice. The date specified in the stop notice as the date when the notice takes effect must not be later than 28 days from the date when the notice is first served on any person. A stop notice ceases to have effect when the enforcement notice to which it relates is withdrawn or quashed or the period for compliance with the enforcement notice expires (see section 184(4)). It is an offence to fail to comply with a stop notice (section 187(1)). If any defence is sought to be raised, such as that under the four-year rule for operational development (see the preceding text), then this should be raised at the magistrates court rather than by a judicial review application: *R (JRP Holdings Ltd)* v *Spelthorne BC* [2007].

Under section 186, compensation may be payable for loss due to a stop notice in certain circumstances. These arise where the enforcement notice is quashed on any of the grounds of appeal, save ground (a) of section 174(2) – that planning permission ought to be granted. Compensation is also payable if the enforcement notice is varied so that any activity, the carrying out of which is prohibited by the stop notice, ceases to be an activity which is the subject matter of the stop notice. If the enforcement notice is withdrawn by the local planning authority otherwise than in consequence of the grant by them of planning permission, or the stop notice is withdrawn, again compensation is payable.

A person who, when the stop notice is first served, has an interest in or occupies the land to which the notice relates shall be entitled to be compensated by the local planning authority in respect of any loss or damage directly attributable to the prohibition contained in the notice. The loss or damage includes any sum payable in respect of a breach of contract caused by the taking of action necessary to comply. Compensation can be claimed by a licensee who has a significant business presence, but not where the occupation lacks substance: see *International Traders Ferry Ltd* v *Adur DC* [2004]. Compensation is not payable in respect of the prohibition in a stop notice of any activity which at the time when the notice was in force constituted or contributed to a breach of planning control (see section 186(2) to (5)). In *Graysmark* v *South Hams District Council* [1989], the Court of Appeal upheld a decision of the Lands Tribunal awarding £1,976 as against a claim for loss due to a stop notice of £426,000. The claimant had sought compensation for a number of matters including his loss of credibility with his bankers, general embarrassment and loss due to worry and stress. The Tribunal assessed compensation for interest on expenditure incurred on the development which had been stopped and for deferment of the profit which would have been made if the development had not been stopped. All other items of claim were found to have arisen out of the claimant's own impecuniosity or were too remote.

15.6.4. Injunction

Where a local planning authority considers it necessary or expedient for any actual or apprehended breach of planning control to be restrained by injunction, they may apply to the court for an injunction, whether or not they have exercised or are proposing to exercise any of their powers under the 1990 Act (see section 187B). The court then has discretion to grant an injunction for the purpose of restraining the breach. In *Runnymede Borough Council* v *Harwood* [1994], the Court of Appeal said that this provision should be given a broad interpretation. In *Hambleton District Council* v *Bird* [1995], it was said that it was not a legitimate reason for refusing an injunction that there was a possibility of planning permission being granted in the future. In *Wealdon District Council* v *Kruschandal* [1998], the Court of Appeal accepted that, in appropriate circumstances, an injunction could involve a district-wide ban, such as a ban on siting a caravan.

In *South Buckinghamshire DC* v *Porter* [2003], the House of Lords approved the approach of Simon Brown LJ in the Court of Appeal that the judge was not required, or entitled, to form any view of the planning merits of the case, but in exercising his discretion the judge should not grant injunctive relief unless he would be prepared if necessary to contemplate committing the defendant to prison for breach of the order, and that he would not be of this mind unless he had considered for himself all questions of hardship for the defendant and his family if, in a domestic case, they are required to move. Simon Brown LJ also added that the following factors were also capable of being relevant in the exercise of the discretion to grant an injunction: the decision of a democratically elected authority to seek an injunction, and the degree of environmental damage resulting from the breach.

The court has a discretion as to whether it will require the local planning authority to give an undertaking in damages: see *Kirklees Borough Council* v *Wickes Building Supplies Ltd* [1992]. Section 187B confers jurisdiction to grant injunctions on the High Court and the County Court. It also provides that rules of court may provide for the granting of an injunction against a person unknown.

15.6.5. Continuing effect of an enforcement notice

Under section 180 of the 1990 Act, where after the service of a copy of an enforcement notice or a breach of condition notice (see the following text) planning permission is granted for any development carried out before the grant of that permission the enforcement notice (or breach of condition notice) shall cease to have effect so far as inconsistent with that permission.

Compliance with the terms of an enforcement notice do not discharge the notice (see section 181(1)). An enforcement notice is therefore a continuing obligation binding the land.

Under section 188 of the 1990 Act, every district planning authority or London borough is required to keep a register of enforcement notices, stop notices and breach of condition notices (see the following text). These registers provide the appropriate public notice of the existence of enforcement notices.

15.7. Breach of condition notices

The provisions of section 187A of the 1990 Act were introduced in 1991 following a recommendation in the Carnwath Report. The provisions provide a summary means of enforcing a breach of a planning condition. Where planning permission for carrying out any development of land has been granted subject to conditions, and any of those conditions are not complied with, a breach of condition notice may be served on any person who is carrying out or has carried out the development or any person having control of the land, requiring him to secure compliance with such of the conditions as are specified in the notice (see section 188A(1) to (2)).

The breach of condition notice must specify a period for compliance of not less than 28 days beginning with the date of service of the notice as may be specified in the notice. If at the end of the period of compliance the steps required to be taken by the breach of condition notice have not been complied with, then the person responsible is in breach of the notice and guilty of an offence (see section 187A(8) to (9)).

There are two statutory defences to an offence under section 187A(11). These are that the person charged took all reasonable measures to secure compliance with the conditions specified in the notice or, where the notice was served on him as the person having control of the land, he no longer has control of the land. Under the first defence, there must be a genuine lack of capacity to remedy the breach, such as illness or a failure of a contractor contracted to remedy the breach: see *Buckland* v *Secretary of State for the Environment, Transport and the Regions* [2001].

It can be seen that a local planning authority has a choice between enforcing a breach of a planning condition by the issue of an enforcement notice or the service of a breach of condition notice. In respect of the former procedure, there is a right of appeal on certain grounds (see part **15.5.1** above). In the case of a breach of condition notice, the planning Act provides no right of appeal as such. The recipient of a breach of condition notice might be able to challenge such a notice by way of judicial review proceedings (see Chapter 20) or by way of a defence in a prosecution: see *Dilieto* v *Ealing London Borough Council* [1998], but an application for judicial review must be made within strict time limits: see *Trim* v *North Dorset District Council* [2011].

15.8. Certificates of lawful use and development

Two provisions were introduced in 1991 that replace, and considerably expand on, certificates of established use under the old legislation. Under section 191 of the 1990 Act, a certificate of lawful use or development may be obtained in circumstances where there may have been a breach of planning control, but by reason of the expiration of the time limits, enforcement action can no longer be taken. Under section 192, a certificate may be obtained of the lawfulness of a proposed use or development.

15.8.1. Certificates of lawfulness of existing use or development (CLEUD)

If any person wishes to ascertain whether (a) any existing use of the buildings or other land is lawful, (b) any operations which have been carried out in, on, over or under land are lawful, or (c) any other matter constituting a failure to comply with a condition or limitation subject to which planning permission has been granted is lawful, he may make an application for the purpose to the local planning authority. In making that application, he must specify the land and describe the use, the operations or any other matter (see section 191(1)). The detailed requirements for any application are set out in Article 35 of the Town and Country Planning (Development Management Procedure) Order 2010.

Uses and operations are lawful at any time if two conditions can be satisfied. First, that no enforcement action may then be taken in respect of them; this might arise where the uses and operations did not involve development or require planning permission or because the time limits for the taking of enforcement action have expired. Second, they do not constitute a contravention of any of the requirements of any enforcement notice then in force (see section 191(2)). A CLEUD granted under section 191 must specify the land to which it relates, describe the use, operations or other matter in question, and give the reasons for determining the use, operations or other matters to be lawful. In addition, it must specify the date of the application for the certificate. There is then a presumption of the lawfulness of such matters (see section 191(5) to (6)). However, as *Ellis* v *Secretary of State for Communities and Local Government* [2010] shows, the relevant use or breach of planning control, such as non-compliance with a planning condition, must continue to the date of the application for the CLEUD. However, a CLEUD can be granted where seasonal use is established for the requisite 10 years; there is no fresh breach, with time running again, each time a use recommences at the beginning of each season: see *North Devon DC* v *First Secretary of State* [2004].

The principal purpose of CLEUDs is that where development has taken place in breach of planning control, such as by the erection of a building or part of a building, or a material change of use, and such matters are immune from enforcement action, there is now a procedure for proving conclusively these matters. This will make land more marketable.

In considering an application for a CLEUD, the local planning authority is only concerned with the necessary factual background and proof of the same. They are not concerned with any planning considerations; in particular, they are not concerned with whether or not they would have granted planning permission. Where there have been changes of use from time to time the local planning authority should follow the guidance in the following case:

Panton v *Secretary of State for the Environment, Transport and the Regions* [1999]

It was held that immunity against enforcement for material changes of use occurring before 1 July 1948 or 31 December 1963 is not lost by the provisions relating to certificates of lawful use introduced under the Planning and Compensation Act 1991. However, uses commenced prior to either of those two dates, or before the commencement of the 10-year period for acquiring immunity from enforcement action may be lost by operation of law in one of three ways: by abandonment, by the formation of a new planning unit or by way of material change of use. In considering an application for a certificate of lawful use, the decision-maker should adopt the following approach. First, ask when the breach of planning control occurred (this would be before 1 July 1948, by 31 December 1963, or at a date 10 years prior to the application for the certificate). Second, if the material change of use took place prior to one of those dates, has that use been lost by operation of law, in one of the three possible ways. Third, if it is satisfied that the description of the use specified in the application does not describe the nature of the use which resulted from the material change of use, then the decision-maker must modify/substitute such description so as properly to describe the nature of the material change of use which has taken place.

If an authority issues a certificate for the siting of two caravans on a farm with reference to a plan, and then a caravan site licence for the whole farm, it was not estopped from contending that the certificate was intended to apply only to the part of the farm on which the two caravans were sited: see *Keevil* v *Secretary of State for Communities and Local Government* [2012].

15.8.2.   *Certificates of lawfulness of proposed use or development*

Under section 192(1) of the 1990 Act, a person may wish to ascertain whether any proposed use of buildings or other land or any operations proposed to be carried out in, on, over or under land, would be lawful. He may make an application for this purpose to the local planning authority specifying the land and describing the use or operations in question. The detailed requirements for any application are set out in Article 35 of the Town and Country Planning (Development Management Procedure) Order 2010.

The local planning authority may then issue a certificate specifying the land and describing the use or operations in question, giving their reasons for determining the use or operations to be lawful.

The purpose of this provision is well illustrated by the decision in *Church Commissioners for England* v *Secretary of State for the Environment* [1995]. The landowners of a large shopping centre contended that the whole shopping centre was the appropriate planning unit so that a change of use of any part of the centre was not a change of use of the planning unit, but an adjustment of the ancillary uses. The local planning authority contended that each of the some 300 units was a separate planning unit so that a change of use of each unit constituted a material change of use. The court held that for the purposes of planning control, each individual shop unit was the planning unit. As *James Hay Pension Trustees Ltd* v *First Secretary of State* [2007] illustrates, a strict approach is adopted to the interpretation of the application, and therefore of the certificate purported to be granted.

15.9.   Enforcement and listed buildings

There are extensive provisions for the enforcement of controls, similar to the general enforcement controls explained in the preceding text, but with technical differences, in the Planning (Listed Buildings and Conservation Areas) Act 1990. For the rest of this chapter this Act will be referred to as the Listed Buildings Act 1990.

15.9.1.   *Listed building enforcement notice*

There are parallel provisions for the issue of a listed building enforcement notice where it appears to the local planning authority that there has been a contravention of the requirement, under section 9 of the Listed Buildings Act 1990, to obtain listed building consent for certain work (section 38). There is no limitation period, and an enforcement notice may address breaches going back to the date of listing and/or in respect of breaches by a predecessor in title: see *Braun* v *Secretary of State* [2003]. The notice may require steps to be taken for restoring the building to its former state (section 38(2)(*a*)); or, where this is not reasonably practicable, or would be undesirable, the notice may be limited to works 'necessary to alleviate the effect of the works carried out without listed building consent' (section 38(2)(*b*)); or bringing the building to the state in which it would have been if the terms and conditions of any listed building consent which has been granted for the works had been complied with (section 38(2)(*c*)).

A listed building enforcement notice cannot be used to compel an owner to repair a listed building that has fallen into a state of disrepair; its purpose is to address some positive act by which the owner has carried out works without listed building consent or in breach of conditions of such a consent: see *Robbins* v *Secretary of State for the Environment* [1989].

It will not always be the case that a building, which has been totally demolished, cannot be restored to its former condition. Thus, in *R* v *Leominster District Council, ex p Antique Country Buildings Ltd* [1988], the local planning authority was entitled to enforce the restoration of the building where some 70–80 per cent of the structural timbers were in existence.

The grounds of appeal are as follows (section 39(1)):

(a) The building is not of special architectural or historic interest.
(b) The matters alleged to constitute a contravention of section 9(1) or (2) have not occurred.
(c) The matters (if they occurred) do not constitute a contravention of section 9(1) or (2).
(d) The works were urgently necessary in the interest of safety or health or for the preservation of the building and that it was not practicable to secure safety, health or preservation by temporary works of repair or support, and the work done was the minimum.
(e) Listed building consent ought to be granted, or any condition discharged or substituted.
(f) Copies of the notice were not properly served.
(g) Except where authorised by section 38(2)(*b*) or (*c*) of the Planning (Listed Buildings and Conservation Areas) Act 1990 (see the preceding text), the steps required for the restoration of the building go too far.
(h) The period specified for the steps required to be taken are less than should be reasonably allowed.
(i) The steps specified for restoring the character of the building to its former state would not serve that purpose.
(j) Alleviate the steps required by virtue of section 38(2)(*b*) exceed what is necessary to the effect of the unauthorised works.
(k) The steps required by virtue of section 38(2)(*c*) exceed what is necessary to make the building conform to the terms of a listed building consent.

The enforcement provisions are otherwise, with technical differences, similar to those described for general development control (sections 39–46).

15.9.2. *Urgent works for preservation of listed buildings*

Under section 54 of the Listed Buildings Act 1990, provision is made for the carrying out of works urgently necessary for the preservation of a listed building.

Powers are given to a local planning authority to execute any works which appear to them to be urgently necessary (see section 54(1)). The Secretary of State may authorise English Heritage to carry out such works (see section 54(2)). The works may include works for affording temporary support or shelter for the building. If the building is occupied, works may be carried out only to those parts which are not in use (see section 54(4)). The owner of the building must be given not less than seven days notice in writing of the intention to carry out the works, and the notice must describe those works.

In *R v Secretary of State for the Environment, ex p Hampshire County Council* [1980], a notice under this section was considered inadequate in not specifying with some degree of particularity the works to be carried out. In *R v Camden London Borough Council, ex p Comyn Ching & Co (London) Ltd* [1983], a requirement to take all such steps as may be necessary to preserve the building was also too imprecise and held to be invalid.

Under section 55 of the Listed Buildings Act 1990, the authority that carried out works of urgent repair may give notice to the owner of the building requiring him to pay the expenses of the works. The owner has 28 days within which to make a representation to the Secretary of State that, for example, some or all of the works were unnecessary for the preservation of the building or that the amount required is unreasonable. In *R v Secretary of State for Wales, ex p Swansea City and County Council* [1999], it was held that the Secretary of State was not entitled to review the question of urgency; he could only consider whether the works were necessary for the preservation of the building.

15.9.3. Compulsory purchase

The final measure available to a local planning authority where a listed building is allowed to fall into disrepair is compulsory purchase of the building (section 47). Minimum compensation, which excludes any development value, may be directed by the Secretary of State if he is satisfied that the building has been deliberately allowed to fall into disrepair for the purpose of justifying its demolition and the development or redevelopment of the site, or adjoining site (section 50).

However, compulsory purchase may not commence unless at least two months previously a repairs notice was served on the owner of the building (section 48). The repairs notice must specify works necessary for the proper preservation of the building, and explain the effect of the compulsory purchase powers if the notice is not complied with.

16 Development for nationally significant infrastructure projects and by public authorities

16.1. Introduction

In the case of nationally significant infrastructure projects, the Infrastructure Planning Commission had power to issue a development consent; such a consent obviates the need for planning permission, and authorises the project and the compulsory acquisition of any land that may be required. Although some public authorities may use the powers in this Act, they are far more likely to be used by developers and large utility companies: see the relevant provisions of the Planning Act 2008 considered in the following text.

There is plenty of precedent for a development role for public authorities. The consequences of the Great Fire of London were such that wide powers of street improvement were conferred upon the Corporation. In the nineteenth century, the new local authorities exercised a wide range of powers to provide and improve highways, sewers and water supplies; and to deal with unsatisfactory houses. In this century, the erection of houses, and the building of the new towns have been significant elements in the development process. Although it has always been a matter of great debate as to whether public authorities should have wider development powers, in practice compulsory powers have been widely used over the last 15 years for town centre schemes, often financed and promoted by a private developer.

This chapter also deals with some of the present development powers conferred upon certain public authorities. However, by way of introduction, some history of past attempts to secure a wider development role for public authorities needs to be considered.

Present development control has its genesis in the Town and Country Planning Act 1947; from 1 July 1948, all development required planning permission. A necessary corollary of development control was the existence of development plans to provide the basis for the implementation of development control decisions. In effect, the Act nationalised development rights; they were thereafter no longer part of a landowner's property rights. Landowners who lost development value as a consequence of this Act in 1948 made a claim for compensation. On the other hand, landowners securing a planning permission for development could not, logically, keep any development value that arose: they were required to pay a charge to the state. The 1947 Act did not contain wide powers for public authorities to purchase and develop land; the powers were limited to postwar problems for existing urban areas.

A change of government in 1951 left the main provisions of the 1947 Act on the statute book, but eventually removed the landowners' obligation to pay the development charge upon obtaining planning permission, and limited his rights to claim compensation. Rather illogically, development rights remained national-ised, but development value could be retained whenever it arose. During the course of the Conservative Government's years in office, others, including the Labour Party, were arguing for more public control over the ownership of land in order to secure its development in accordance with perceived social policies.

Undoubtedly, the existence of public control over the planning and develop-ment of land is not easily reconciled with the continuance of private ownership of suitable development land. There are two basic problems. First, the decision to develop or not remains with the private landowner, and his development ideas may be influenced by economic considerations rather than social policies as expressed in development plans. For example, Blackacre is considered as appropriate for residential development in a development plan, but because of its location, it would enjoy a higher 'rent' if available for industrial development.

The second problem lies in the complex tapestry of landownership: a deve-lopment plan may consider a certain form of development as appropriate for a particular area, but the complexity of landownership in the area may mitigate against the realisation of that development; however willing some landowners may be, development may not be possible without the co-operation of all; and, there may be tenants with unexpired leases and security of tenure effectively preventing the release of such an area.

Some argue that these two problems can easily be met by a better co-operation between planners and landowners (see *Prospects of Co-operative Planning* – DR Denman 1974); others that the public ownership of development land is the only solution, thus bringing together the planner and the landowner in one entity. The Labour Party has made two attempts when in office to partly realise this solution; the Land Commission Act 1967 and the Community Land Act 1975. Both were repealed by the succeeding Conservative Governments, and nothing comparable has since been enacted. However, the powers described in this chapter are now certainly more extensive than those of the 1947 Act. In recent years, we have seen the urban development corporations, with wide powers, dealing with dereliction in specific urban areas, and the regional development agencies supporting a wide range of projects.

16.2. Development for nationally significant infrastructure projects

The Planning Act 2008 was the then government's answer to the problems and delays in achieving large infrastructure projects. The planning process involved in obtaining planning permission and other orders required for Terminal 5 at

London Heathrow took several years, not merely months. With the possibility that the government may wish to encourage the provision of new nuclear power stations, and an additional runway at one or more of London's airports, and other large projects, the 2008 Act provides a more streamlined set of procedures.

An Infrastructure Planning Commission ('the IPC') was originally created under the 2008 Act. As from 1 April 2012, the IPC was abolished and its functions were transferred to a new national infrastructure directorate in the Planning Inspectorate, with the Secretary of State as the decision-maker on all national infrastructure applications for development consent. National policy statements are issued setting out a national policy in relation to one or more specified descriptions of development. Nationally significant infrastructure projects are defined (see part **16.2.3** below), and provision is made for the making of an application for, and the grant by the Secretary of State, of a 'Development Consent' for a nationally significant infrastructure project. Such a development consent will be granted by a development consent order, and such an order may include provision for authorising the compulsory acquisition of land. The effect of a development consent order will be such that the order will have the same effect as if planning permission had been granted (as well as any other necessary consents such as listed building consent), and a compulsory purchase order had been made and confirmed. Therefore, no separate or additional compulsory purchase order will be required.

16.2.1. Infrastructure planning commission

This was a corporate body with the functions conferred on it by the 2008 Act. The IPC processed applications for development consent orders, and was, until its abolition, the decision-maker. Since 1 April 2012, the application processing is done by the Planning Inspectorate, and the Secretary of State is the decision-maker; the Secretary of State grants or refuses to grant a development consent. A raft of Rules and Orders came into effect in 2010 to deal with the processing of applications for development consent orders and the model provisions. It is likely that, with the abolition of the IPC, some of these orders with be revoked or amended.

16.2.2. National policy statements

The Secretary of State may designate a statement as a national policy statement for the purposes of the 2008 Act if the statement is issued by the Secretary of State and sets out national policy in relation to one or more specified descriptions of development. Before designating such a statement, the Secretary of State must carry out an appraisal of the sustainability of the policy (section 5). The Act contains consultation and publicity requirements in relation to national policy statements (sections 7, 8 and 9). A national policy statement may be designated in respect of, for example, the construction or extension of electricity generating power stations. Having regard to that example, the policy set out in the national policy statement may in particular (section 5(5)):

(a) Set out, in relation to a specified description of development, the amount, type or size of development of that description which is appropriate nationally or for a specific area.

(b) Set out criteria to applied in deciding whether a location is suitable (or potentially suitable) for a specified description of development.

(c) Set out the relative weight to be given to specified criteria.

(d) Identify one or more locations as suitable (or potentially suitable) or unsuitable for a specified description of development.

(e) Identify one or more statutory undertakers as appropriate persons to carry out a specified description of development.

(f) Set out circumstances in which it is appropriate for a specified type of action to be taken to mitigate the impact of a specified description of development.

A national policy statement will also set out the criteria to be taken into account in the design of the particular description of development, and give reasons for the policy, and how it takes account of government policy relating to the mitigation of, and adaptation to, climate change. National policy statements are subject to publication (section 5(9)), and review (section 6). In relation to consultation and publicity, the Secretary of State will prescribe by regulations persons, and descriptions of persons, that shall be consulted (section 7). The Secretary of State must consult each local authority for the location of a proposed project (section 8). The parliamentary requirements in section 9 are somewhat curious. Where the Secretary of State proposes to designate a national policy statement (or an amendment to an existing one) the proposal must be laid before Parliament. If either House of Parliament makes a resolution with regard to the proposal (or a committee of either house), then the Secretary of State may lay before Parliament a statement setting out the Secretary of State's response to the resolution or recommendations. The Secretary of State himself prescribes any relevant periods.

Any legal challenge relating to a national policy statement must be by way of a claim for judicial review, and the claim form must be filed during the period of six weeks beginning with the day on which the statement is designated as a national policy statement or, if later, the day on which the statement is published (section 13). A number of other matters are expressly made subject to applications for judicial review. For judicial review, see Chapter 4.

By March 2012, some 12 National Policy Statements had been made or designated. Six statements covered various aspects of energy; three dealt with transport issues, such as ports, rail, roads and aviation; and the others concerned waste and water treatment.

16.2.3. Nationally significant infrastructure projects

A nationally significant infrastructure project will consist of any of the following:

(a) The construction or extension of a generating station;
(b) Development relating to underground gas storage facilities;
(c) The construction or alteration of a liquid natural gas (LNG) facility;
(d) The construction or alteration of a gas reception facility;
(e) The construction of a pipe-line by a gas transporter;
(f) The construction of a pipe-line other than by a gas transporter;
(g) Highway-related development;
(h) Airport-related development;
(i) The construction or alteration of harbour facilities;
(j) The construction or alteration of a railway;
(k) The construction or alteration of a rail freight interchange;
(l) The construction or alteration of a dam or reservoir.

Under section 14, the Secretary of State may add new types of project or vary or remove any existing type. New types of projects may only be added if they are within the fields of energy, transport, water, wastewater or waste (section 14(5)–(6)). A number of projects are the subject of additional qualification and description.

Thus, in the case of generating stations in England or Wales, they must have capacity more than 50 megawatts or, if an offshore generating station, then a capacity of more than 100 megawatts (section 15). In relation to electric lines, the nominal voltage must not be less than 132 kilovolts if above ground (section 16). A number of conditions relate to underground gas storage facilities. The principal condition must be that the working capacity is expected to be at least 43 million standard cubic metres, or the maximum flow rate of the facilities is expected to be at least 4.5 million standard cubic metres per day (section 17). The same storage and flow rates apply to LNG facilities (liquid natural gas) (section 18). Such facilities are for the reception of liquid natural gas from outside England, storage and the regassification of such gas. A gas reception facility is one where the maximum flow rate will be at least 4.5 million standard cubic metres a day, and is a facility for the reception of natural gas from outside England, and the handling of natural gas (other than its storage) (section 19). Gas transporter pipelines must be more than 800 mm in diameter and more than 40 km in length or the construction of a pipeline is likely to have a significant effect on the environment. In either case, the operation pressure must be more than seven bar gauge, convey gas for supply to at least 50,000 customers or potential customers (section 20). In the case of other pipelines, these must be cross-country pipelines that require authorisation under section 1(1) of the Pipelines Act 1962 (section 21).

Highways will be nationally significant infrastructure projects where their construction, improvement or alteration will be carried out by or for the Secretary of State, or the Secretary of State is or will be the highway authority for the highway and, in the case of an improvement to a highway, the improvement is likely to have a significant effect on the environment (section 22).

Airport-related development consists of the construction or alteration of an airport, or will permit an increase in the permitted use of the airport, where in either case this will result in air passenger transport services for at least 10 million passengers a year, or air cargo transport services for at least 10,000 air transport movements of cargo aircraft per year. An alteration includes the construction, extension or alteration of a runway or a building at the airport, or a radar or radio mast, antennae or other apparatus at the airport (section 23).

The construction or alteration of harbour facilities will only be nationally significant infrastructure projects if there is the creation of, or an increase in, the quantity of material exceeding 'the relevant quantity' (section 24). This quantity is, in the case of container ships, 500,000 TEU, for ro-ro ships, 250,000 units, and for other cargo ships, 5 million tonnes. There are complicated fractions where the facilities will be mixed.

In a case of rail freight interchanges, the land on which the interchange is situated must be at least 60 hectares in area capable of handling goods from more than one consignor to more than one consignee and at least four goods trains per day (section 26).

In the case of the construction of a dam or a reservoir, the volume of water, or the additional volume of water must exceed 10 million cubic metres (section 27). The transfer of water resources must be expected to exceed 100 million cubic metres per year and enable transfer between river basins, water undertakers' areas or between a river basin and a water undertaker's area (section 28). Water waste treatment plants must be expected to have a capacity exceeding a population equivalent of 500,000 (section 29). The construction or alteration of hazardous waste facilities must result in more than 100,000 tonnes per year, in the case of landfill or deep storage, or 30,000 tonnes per year, in any other case (section 30).

16.2.4. Requirement for development consent

'Development consent' is required for development to the extent that it forms part of a nationally significant infrastructure project (section 31). The Secretary of State has power to make directions as to development that requires a development consent (section 35).

Development has the same meaning as in section 55 of the Town and Country Planning Act 1990 (building mining and engineering operations and material change of use): see section 32 of the 2008 Act. Certain additional matters are treated as development, and therefore will require a development consent. Thus, the following are treated as material changes in the use of the relevant generating station, cavity or strata, or airport, as the case may be, namely the conversion of a generating station with a view to its being fuelled by crude liquid petroleum, a petroleum product or natural gas, the use of a cavity or strata for the underground storage of gas, and an increase in the permitted use of an airport (section 32(2)). Certain works are treated as development 'to the extent that they would not be otherwise'. These include the demolition, alteration or extension of a listed building, the demolition of a building in a conservation area, and certain works that will affect a scheduled monument (section 32(3)).

Where development consent is required, then there is no separate requirement to seek the following consents or permissions: planning permission, Consent under the Green Belt (London and Homes Counties) Act 1938, authorisation under the Pipelines Act 1962, authorisation under the Gas Act 1965 (storage of gas underground), notice under the Energy Act 1976 (conversion of generating station from one fuel to another), consent or notice under the Ancient Monuments and Archaeological Areas Act 1979, consents under section 36 or 37 of the Electricity Act 1989 (construction of generating stations and overhead lines), listed building consents (see section 33). Further, where development consent is required for the construction, improvement or alteration of a highway, a number of orders and schemes specified in section 33(4) are not required.

16.2.5. Applications for development consent orders

Where a person is contemplating the provision of a nationally significant infrastructure project, then application must be made for a development consent order to the Infrastructure Directorate of the Planning Inspectorate: see section 37 of the 2008 Act. Model provisions have been prescribed, although since repealed, are a useful guide: see the Infrastructure Planning (Model Provisions) Order 2009. Prior to any application, the applicant must consult a number of persons about the proposed application: see section 42. These include any local authority if the land is within the authority's area, or on its boundary (section 43), persons who may be prescribed, and other persons such as affected owners, lessees, tenants or occupiers of the land: see section 44. There is a deadline of not less than 28 days, from the receipt of the consultation documents, within which a person required to be consulted must provide a response, if any: see section 45. The applicant must also prepare a statement setting out how the applicant proposes to consult people living in the vicinity of the affected land: see section 47. There is an obligation to publicise the application in accordance with regulations, which will be prescribed: see section 48. Section 49 imposes an obligation on an applicant, when deciding whether the application, which the applicant is actually to make, should be in the same terms as the proposed application, having regard to any relevant responses from the persons consulted, or in consequence of publicity.

A number of provisions in the 2008 Act, as amended by the Localism Act 2011, permit the Secretary of State to make regulations for the giving of advice to the applicants or potential applicants (section 51), enable an applicant or proposed applicant to obtain information about interests in land (section 52), rights of entry for the purposes of surveying and taking levels and related matters (section 53).

16.2.6. Handling of application by the Secretary of State

The Secretary of State must, within a 28-day period, decide whether or not to accept an application for a development consent order: see section 55(2). In making that decision, the Secretary of State must conclude that the various procedural requirements have been satisfied as regards form and consultations. The Secretary of State must then give notice of the application to a number of persons. These include such persons as may be prescribed, each local authority affected and persons who are owners, lessees, tenants, or occupiers of the land, or have some interest in the land, or might be entitled to make a claim for compensation: see section 57. Persons in the last category would include those entitled to make claims for compensation under section 10 of the Compulsory Purchase Act 1965 or Part I of the Land Compensation Act 1973.

The Secretary of State will then decide whether the application is to be considered by a panel or a single appointed person: see section 61. The procedure, whether by way of a panel or a single appointed person, is outside the scope of this book. However, in broad terms, a panel or a single appointed person, examine and make a report to the Secretary of State, setting out findings, conclusions and recommendations. The 2008 Act uses the expression 'examining authority', to describe either a panel or a single appointed person when the process of the examination of an application takes place. That process may be by way of written representations: see section 90. Alternatively, as the examining authority may decide, by way of hearings necessary to ensure the adequate examination of a specific issue or issues, or that an interested party should be given a fair chance to put their case: see section 91.

In the case of an application for a development consent order which includes powers to authorise the compulsory acquisition of land or of an interest or a right over land, known as a 'compulsory acquisition request', the examining authority must fix and notify the deadline by which an affected person must notify the Secretary of State that that person wishes a compulsory acquisition hearing to be held: see section 92(2). If the Secretary of State receives notification from at least one affected person before the deadline, then the examining authority must cause a compulsory acquisition hearing to be held: see section 92(3). At a compulsory acquisition hearing, the applicant and each affected person is entitled to make oral representations: see section 92(4). An affected person is a person with an interest in the land to which the compulsory acquisition request relates: see section 59. A number of the provisions of the 2008 Act make provision for hearings (sections 93 and 94) and procedural rules (section 97). The Localism Act 2011 introduced additional provisions by which certain categories of persons can become 'interested parties' with consequential rights to be heard (sections 102A and 102B). Importantly, the examining authority is under a duty to complete the examination of an application by the end of the period of six months from a preliminary meeting: see sections 88 and 98. The examining authority may appoint an assessor to assist in the examination of the application: see section 100. In addition, the authority may also appoint a lawyer to provide legal advice and assistance: see section 101.

A decision on an application for development order consent must either be an order granting that consent or a decision refusing the same: see section 114. The Secretary of State must give reasons for deciding to make an order granting development consent, or refusing one: see section 116.

16.2.7. Legal challenges

A court may only entertain proceedings for questioning a development consent order if the proceedings are brought by way of judicial review and the claim form is filed during the period of six weeks beginning with the date on which the order is published, or, if later, the day on which the statement of reasons making the order is published: see section 118(1). The same applies where the decision is a refusal. The principles for legal challenge, considered earlier in this chapter, will have application here.

16.2.8. Compulsory acquisition

A development consent order may include provision authorising the compulsory acquisition of land if the decision-maker is satisfied that the land:

(a) Is required for the development for which the development consent relates;
(b) Is required to facilitate or is incidental to that development; or
(c) Is replacement land which is given in exchange for the order land (commons and open spaces), and that there is a compelling case in the public interest for the land to be acquired compulsorily: see section 122.

Further, a development consent order may only include a provision authorising the compulsory acquisition of land if certain conditions are met: see section 123. These are that the application for the order included a request for compulsory acquisition, or all persons with an interest in the land consent to the inclusion of the provision, or that the prescribed procedure has been followed in relation to the land: see section 123.

Where a development consent order includes a provision authorising the compulsory acquisition of land, then Part I of the Compulsory Purchase Act 1965 (procedure for compulsory purchase) applies to the compulsory acquisition of the land as if there was a compulsory purchase under Part II of the Acquisition of Land Act 1981, and as if the order was a compulsory purchase order under that Act: see section 125(2). However, Part I of the 1965 Act has effect with the omission of certain provisions; these are section 4 (time limits for exercise of compulsory purchase powers), section 10 (compensation for injurious affection) (considered further in the following text), and paragraph 3(3) of Schedule 3 (provision as to giving of bonds): see section 125(3). However, section 125(6) provides that these requirements are subject to any contrary provision made in the development consent order. Section 126 then provides that a development consent order may not include a provision the effect of which is to modify or exclude the application of a compensation provision, except, in relation to the modification of such a provision, to the extent necessary to apply the provision to the compulsory acquisition of land authorised by the order: see section 126. A compensation provision is defined as a provision under an Act which relates to compensation for the compulsory acquisition of land: see section 126(4). It is therefore somewhat curious as to why section 125(3)(*b*) excludes compensation for injurious affection under section 10 of the 1965 Act if section 126 actually provides that no order may generally exclude the application of such a provision. Presumably, any model provisions for development consent orders in relation to the compulsory acquisition of land will include compensation provisions, such as that in section 10 of the 1965 Act.

Sections 127–132 contain provisions dealing with the acquisition of land from statutory undertakers, the National Trust and land which is common land or an open space. These provisions are broadly similar to those considered in Chapter 21. Where a development consent order includes rights in connection with underground gas storage facilities, then the provisions in section 131 for the protection of commons and open spaces are excluded: see section 133(3). However, if the order authorises the compulsory acquisition of a new right, such as a right to store gas in underground gas storage facilities, to stop up a well, borehole or shaft, or prevent its use by another person, or is a right of way over land, then the special provisions relating to statutory undertakers, local authority, National Trust and commons and open spaces, are omitted.

16.2.9. *Notice of authorisation of compulsory acquisition*

Section 134 of the 2008 Act makes provision for notice of authorisation of compulsory acquisition. In this section, the order land means any land over which a right is authorised by a development consent order to be exercisable, and any land authorised to be compulsorily acquired by such an order. The prospective purchaser is the person for whose benefit the order authorises the creation of the right, or the person authorised by the development consent order to compulsorily acquire the land: see section 134(2). After the development consent order has been made, the prospective purchaser must serve a compulsory acquisition notice and a copy of the order on certain persons, and affix a compulsory acquisition notice to a conspicuous object or objects on or near the order land, and make a copy of the order available for public inspection at reasonable hours: see section 134(3). The persons who must be given a compulsory acquisition notice are qualifying persons for the purposes of section 12(1) of the Acquisition of Land Act 1981 (owners, lessees and occupiers). The compulsory acquisition notice, must be addressed to persons occupying or having an interest in the order land, and, where affixed to the land, must be kept in the place for a period of six weeks. The compulsory acquisition notice must also be published in local newspapers: see section 134(5) and (6). The compulsory acquisition notice must describe the order land, in the case of an order authorising the compulsory acquisition of a right, describe the right, state that the development consent order includes provision authorising the compulsory acquisition of a right over the land or, as the case may be, the compulsory acquisition of the land, and state that a person aggrieved by the order may challenge the order in accordance with section 118 (by judicial review): see section 134(7).

16.2.10. *Compensation where no right to claim in nuisance*

Presumably, section 10 of the 1965 Act is omitted on the basis that development authorised by a development consent order may be carried out by a person who is not necessarily exercising statutory powers. Section 152 of the 2008 Act does, however, provide that if, by virtue of section 158 or a development consent order, there is a defence of statutory authority in civil or criminal proceedings for nuisance in respect of any authorised works, then compensation is payable to any person whose land is injuriously affected by the works, and section 10(2) of the 1965 Act, and Part I of the Land Compensation Act 1973, have effect. Section 158 confers statutory authority for the carrying out of development for which consent is granted by an order granting development consent, and doing anything else authorised by an order granting development consent. That authority is conferred only for the purpose of providing a defence in civil or criminal proceedings, and, in any event, is subject to any contrary provision made in any particular case by a development consent order.

It follows that where a development consent order has been made, and the order authorises the compulsory acquisition of land, then that order is regarded as a compulsory purchase order. It also follows that the procedures for acquiring the land, either by notice to treat or general vesting declarations, then have application: see Chapter 7 of *Compulsory Purchase and Compensation* by B Denyer-Green [2013].

16.3. Land acquisition and development powers

16.3.1. *Compulsory purchase*

The principal powers of compulsory purchase of land for planning purposes are now found in sections 226 and 228 of the 1990 Act, as amended by the Planning and Compulsory Purchase Act 2004.

A local authority may be authorised by the Secretary of State to acquire compulsorily any land (section 226(1)) in their area (a) if the authority thinks that the acquisition will facilitate the carrying out of development, re-development or improvement on or in relation to the land, or (b) which is required for a purpose which it is necessary to achieve in the interests of the proper planning of an area in which the land is situated.

However, subsection (1A) provides that a local authority must not exercise the power under paragraph (*a*) of subsection (1), in the preceding text, unless they think that the development, re-development or improvement is likely to contribute to the achievement of any one or more of the following objects:

(a) The promotion or improvement of the economic well-being of their area;
(b) The promotion or improvement of the social well-being of their area;
(c) The promotion or improvement of the environmental well-being of their area.

Land may facilitate development, redevelopment or improvement either because it is land required for such purposes, or because it is needed to facilitate such purposes. In *R (Sainsbury's Supermarkets Ltd)* v *Wolverhampton City Council* [2010], it was held that the acquiring authority must first be satisfied of the well-being benefits that will be achieved from the subject site itself but, if so satisfied, it is precluded from considering off-site benefits that could be made available.

Where land is being compulsorily purchased under the preceding powers, any adjoining land required for executing works for facilitating the development or land use of the principal land, and any exchange land needed to replace any common land, open space or fuel or field garden allotment being acquired, may also be compulsorily purchased (section 226(3)).

Land may still be compulsorily purchased for the purposes set out in the preceding text even if the local authority acquiring the land does not propose to carry out the development or improvement themselves (section 226(4)). This enables the land to be developed by private developers, or another authority with the necessary powers, following a sale, lease arrangement, or a partnership agreement. Local authorities have commonly adopted a certain development model, over the last few years, to develop town centre and similar schemes under these powers. They appoint a preferred developer who may design, fund and promote a scheme carrying out all the necessary steps to obtain a compulsory purchase order, and even indemnifying the local authority in respect of all compensation claims. In return, the authority makes the CPO and agrees to grant appropriate interests in the acquired land to enable the development to proceed. This is obviously useful where there are land assembly problems, and other difficulties between rival developers: see *Standard Commercial Property Securities Ltd* v *City of Glasgow* [2006].

The Secretary of State also enjoys wide powers of compulsory purchase of land necessary for the public service under section 228.

16.3.2. *Purchase by agreement*

Under section 227, land may also be acquired with the agreement of the owner for any of the purposes for which it may be compulsorily acquired under section 226. In addition, a local authority may acquire any building appearing to them to be of special architectural or historic interest, and other surrounding land required for the preservation, amenity, access or management of such a building.

16.3.3. Disposal and development

Land that has been acquired, or appropriated to, planning purposes, may be sold or leased by the local authority in whatever manner and subject to whatever conditions that will secure the best use of that or other land, or that will secure the development of that land needed for the proper planning of the area (section 233(1)). This section may enable a partnership scheme between a local planning authority and a developer to be agreed (see *Jones* v *Secretary of State for Wales* [1974]).

Many local authorities use powers in local Acts to secure partnership scheme agreements, although adequate power would seem to be available under section 233(1) of the 1990 Act. The power in section 106 to make a planning agreement or obligation is limited to circumstances where the land is not owned by the local authority and the need is to restrict or regulate the development or use of that land.

Local authorities are authorised to carry out any development themselves by the erection or construction of buildings or works (section 235(1)). This power is hedged about with a number of restrictions, not least being those of a financial nature exercised by the Secretary of State.

Where persons are displaced from land following local authority acquisition, and the subsequent development, the authority should offer accommodation or business or other premises to such persons on completion of the scheme, 'so far as may be practicable' (section 233(5)). Whether it is so practicable is a question for the local authority to decide, not for a court: *Crabtree & Co Ltd* v *Minister of Housing and Local Government* [1965].

Town centre redevelopment schemes – the provision and improvement of land for industry, and facilities for recreation and leisure, may constitute aspects of a development plan or a policy accepted by a local planning authority. These matters can then be achieved with the statutory powers so far discussed.

The procedure for the acquisition of land by a public authority, and for assessment of compensation, is dealt with in the author's companion volume *Law of Compulsory Purchase and Compensation* (Tenth Edition, Estates Gazette, 2013).

16.4. Homes and Communities Agency

The Housing and Regeneration Act 2008 established the Homes and Communities Agency (HCA). The objects of the HCA are (section 2):

(a) To improve the supply and quality of housing in England;
(b) To secure the regeneration or development of land or infrastructure in England;
(c) To support in other ways the creation, regeneration or development of communities in England or their continued well-being;
(d) To contribute to the achievement of sustainable development and good design in England with a view to meeting the needs of people living in England.

Startling, for a body created by statute, the principal power of the HCA is to do anything it considers appropriate for the purposes of its objects or for purposes incidental to those purposes (section 3). More specifically it has powers to provide housing or other land (section 5), powers for regeneration, development or effective use of land (section 6), and powers in relation to the provision of necessary infrastructure (section 7). The HCA may carry out, or facilitate the carrying out of, any of the following activities in relation to land by (section 8):

(a) Acquiring, holding, improving, managing, reclaiming, repairing or disposing of housing;
(b) Acquiring, holding, improving, managing, reclaiming, repairing or disposing of other land, plant, machinery, equipment or other property;
(c) Carrying out building and other operations (including converting or demolishing buildings).

It has wide powers of compulsory acquisition of land (section 9), and extensive powers to override rights and extinguish public rights of way (Schedule 3). The HCA also has wide powers in relation to the extinguishment of rights of statutory undertakers and for the removal of their apparatus, powers for undertakers to carry out works, for the extension or modification of functions of undertakers, and for relieving undertakers of obligations (Schedule 4). Under section 14, the HCA can, by an order, be designated the local planning authority for an area or a designated area with such functions as may be provided for. The HCA will have borrowing powers and the ability to operate through subsidiaries. Section 31 provides for the provision of low cost rental accommodation and that a relevant provider of such accommodation shall be the landlord.

16.5. Public authorities and planning permission

16.5.1. The Crown

The planning Acts bind the Crown: see section 292A of the 1990 Act, as inserted by the Planning and Compulsory Purchase Act 2004, and equivalent provisions in the Listed Buildings Act 1990. However, there is provision for urgent planning applications relating to Crown land to be made to the Secretary of State: see section 293A, and the consent of the appropriate authority, in most cases the Secretary of State, is necessary to take any enforcement action. However, no act or omission by the Crown can constitute an offence: see section 296A. There are special provisions where issues of national security arise.

Crown land includes interests held in land by Her Majesty in right of the Crown, and those held by government departments held in trust by the Crown (section 293).

Because, until 2004, permission could not be granted for Crown land, difficulties began to arise over the disposal of surplus Crown land. To enable such land to be sold with the benefit of planning permission, listed building or conservation area consents, section 299 of the 1990 Act contained provisions authorising the grant of planning permission in anticipation of the disposal of Crown land. Obviously, those provisions are no longer necessary.

16.5.2. *Local planning authorities*

Local planning authorities are subject to development control, and require planning permission for any development of their own land or of any land of which they are the local planning authority as well as an applicant for consent. Permission may be granted in the Town and Country Planning (General Permitted Development) Order 1995 for certain minor forms of development (see Chapter 4). For all other development, the procedure for dealing with an application for planning permission by the local planning authority is governed by the Town and Country Planning General Regulations 1992 (section 316 of the 1990 Act). The details of the procedure are beyond the scope of this work, but because the procedure may result in the local planning authority giving planning permission to itself, the courts are often strict about the proper and full compliance with it. In *Steeples* v *Derbyshire County Council* [1981], planning permission granted to the county council was declared null and void because of a failure to comply with all steps of the procedures involved; the council were also bound by the rules of natural justice and had failed to show lack of bias in relation to an agreement with a developer to take all reasonable steps to obtain planning permission for a partnership scheme.

More recent cases involving a possible conflict of interest between an authority's role as planner, and as developer, have not followed the test of bias, based on the 'reasonable man' in the *Steeples* case (see further part 8.8 above).

Where a local government officer makes an application for permission for development that will affect a conservation area, there must be strict compliance with the regulations: *R* v *Lambeth London Borough Council, ex parte Sharp* [1986].

16.5.3. *Statutory undertakers*

Since the privatisation of several previously publicly owned industries, the number of bodies that are statutory undertakers has been reduced. Under section 262 of the 1990 Act, statutory undertakers are now principally railway, road and water transport undertakers. Special provisions are found in Part XI of the 1990 Act where development is to be carried out by these bodies in respect of their operational land. Apart from development authorised by the Town and Country Planning (General Permitted Development) Order 1995 (see Chapter 4), planning permission is required for the development of operational land. However, where the matter comes before the Secretary of State following a call-in or an appeal, the Secretary of State makes his decision in conjunction with the minister appropriate to the undertaking concerned.

Where development proposed by a local authority or a statutory under-taker requires the express consent of a government department, that consent may state that planning permission is deemed granted (section 90 of the 1990 Act).

The privatised industries, such as the telecommunications, electricity, water and gas, are subject to planning control in the ordinary way. Nevertheless, the installation and replacement of apparatus are largely permitted by the General Permitted Development Order (see Chapter 4 and Appendix B). Doubtless, for nationally significant infrastructure projects, they will be using the powers under the Planning Act 2008.

16.6. New towns and development corporations

The legislative powers in respect of new towns originated in the New Towns Act 1946. The essential features of these powers was that the Secretary of State designated an area of land as a new town; a development corporation was established to secure the laying out and development of the new town; and the development corporation was granted the necessary powers to achieve those objects.

Urban development corporations, modelled on the new town corporation, were set up in the early 1980s to regenerate designated urban development areas. The first two designated areas were the docklands in London and the south docks in Merseyside.

The new town development corporations have now been wound up and their assets transferred to the new towns residuary body. Urban development corpora-tions were set up for a limited time period only that has now expired.

16.7. Register of land held by public bodies

Part X of the Local Government, Planning and Land Act 1980 provides for a register of land held by public bodies and contains powers for the Secretary of State to direct a sale of any land entered on the register. It is said that much derelict or unused land in urban areas is owned by public bodies. The Secretary of State wants information and powers to deal with the situation.

16.7.1. The public bodies

The bodies affected are set out in Schedule 16. They include not only local authorities but also many of the nationalised industries and all statutory undertak-ers. Private companies authorised to run docks or harbours are therefore presently included in the list. Further bodies can be added to or deleted from the list by statutory instrument (section 93).

16.7.2. Areas

The Secretary of State may make orders specifying the registration areas for the operation of this Part of the Act (section 94).

16.7.3. The register

The Secretary of State may compile the register 'in such form as he thinks fit'. Land may be entered on the register if it satisfies certain conditions: either a freehold or a leasehold interest is held by one of the affected public bodies or a subsidiary; it is in a registration area or adjoins land which is in such an area; and 'in the opinion of the Secretary of State the land is not being used or not being sufficiently used for the purposes of the performance of the body's functions or of carrying on their undertaking' (section 95).

Although government departments are not in the list of affected public bodies (see the preceding text), the Secretary of State may enter any Crown land on to the register (section 95(4)).

16.7.4. Public access

A copy of the register for any area (including any amendments) shall be sent to the district council concerned. This shall be made available for public inspection. Copies of any information in the copy of the register shall be supplied to any member of the public upon payment of a reasonable charge (section 96).

16.7.5. Powers of the Secretary of State

The Secretary of State will have power to require information from a public body (section 99). He may also direct that a public body take steps to dispose of land entered on the register (section 98(1)). Before giving a direction, the Secretary of State will serve a notice of his intention on the public body; it may then make representations within a period of 42 days. He may not issue his direction to a public body which has made representations unless he is satisfied that the land can be disposed of without causing serious detriment to the performance of that body or the carrying on of its undertaking (section 99).

Any direction issued to a local authority to sell surplus land operates as consent to sell under section 123 of the Local Government Act 1972: the sale can therefore be by auction and without a reserve (*R* v *Secretary of State for the Environment, ex parte Manchester City Council* [1987]).

16.8. Regional development agencies and local enterprise partnerships

The Regional Development Agencies Act 1998 provides for the establishment of development agencies for each of the nine regions into which England is divided.

Section 4 of the Act sets out the purposes of an agency. These included the furtherance of the economic development and regeneration of an area and the promotion of employment.

An agency had power to acquire land by agreement or compulsorily. The regional development agencies were abolished in March 2012. Local enterprise partnerships have been set up in a number of areas to drive local economic growth and create conditions for private sector job growth.

Part V
Rights and remedies

17 Appeals and inquiries

17.1 Introduction

Local planning authorities possess wide discretionary powers to implement planning policies and make development plans; that discretion is also present in their development control powers and duties. Probably because planning and development control were regarded in 1948 as interferences to private interests in property rights, the planning Acts still contain many opportunities for the merits of the decisions of local planning authorities to be appealed to the Secretary of State. The right of appeal is granted to the applicant for planning permission or the landowner. The merits of development control decisions cannot be appealed by disappointed neighbours or pressure groups. There is an exception to this in relation to the making of development plans, any person may make representations or objections, and these were considered at a subsequent examination in public of a structure plan, or now, by the independent examination of development plan documents. There is also an exception where a decision of a local planning authority is being challenged, because it is wrong in law, or the proper procedure has not been followed; a person with sufficient interest (a group of local residents: *Covent Garden Community Association Ltd* v *Greater London Council* [1981]) may make a claim for judicial review seeking a quashing order in the High Court in respect of a planning decision (see *R* v *North Herts District Council, ex p Sullivan* [1981] where wrong procedure was used). The judicial and administrative supervision of local planning authorities and the Secretary of State are considered in Chapter 20.

This chapter deals with the more usual appeal to the Secretary of State against a refusal of planning permission by the local planning authority, and the manner in which such an appeal is heard. Some 85 per cent of such appeals are determined by written representations, with the occasional private hearing, and the balance by public local inquiry. Appeal also arises in connection with revocation or modification orders, discontinuance orders, refusal of listed building consent, and many other matters. The appeal against an enforcement notice was considered in Chapter 15, but the form of the public inquiry is similar to that dealt with here.

17.2. Appeal to the Secretary of State

Section 78(1) of the 1990 Act makes provision for an appeal where the following applications have been made to a local planning authority:

(a) For planning permission to develop land;
(b) For any consent, agreement or approval of that authority required by a condition imposed on a grant of a planning permission; or
(c) For any approval of that authority required under a development order.

and that permission, consent, agreement or approval is refused or granted subject to conditions. The right to appeal also arises following the issue of an enforcement notice, the refusal of certificates under sections 191, 192 of the 1990 Act (see Chapter 15), the refusal of listed building consent or conservation area consents (see Chapter 10) and listed building enforcement notices, refusal of consent under a tree preservation order and in relation to advertisements (see Chapter 11). Appeal is also allowed where the local planning authority fails to reach a decision within eight weeks, save in the case of 'major development', when the period is 13 weeks, or notification that the application is referred to the Secretary of State under section 78(2): see Article 29 of the Town and Country Planning (Development Management Procedure) Order 2010 (DMPO): the periods of time run from the date when the local planning authority accepts that a valid application has been received and the necessary fee paid. Where the local planning authority denies that an application has been validly made, for example they say that they have received insufficient details, the applicant can probably appeal under this subsection and the Secretary of State decides whether a valid application has been made: see *R* v *Secretary of State for the Environment, Transport and the Regions, ex p Bath and North East Somerset District Council* [1999].

In the case of an appeal against the refusal of planning permission, or the grant of permission subject to conditions, notice of appeal must normally be made within six months of the decision or, where no decision is made within the time limit, within six months of the expiration of the time limits, as set out in the preceding text (DMPO, article 33). For householder appeals, the time limit is 12 weeks. In the case of an application for development which is the same or substantially the same as development the subject of an enforcement notice served no earlier than two years beforehand, the time limit is 28 days running from different specified dates in Article 33 of the DMPO. There is a special form, obtainable from the Department of Communities and Local Government (DCLG), that must be completed and returned with copies of all relevant documents. The appellant is now required to send a duplicate copy of the appeal form to the local planning authority. Appeals may also be made online through the Planning Inspectorate's website.

In a case where a decision appears bad in law, and is not simply disputed on its planning merits, application may be made direct to the High Court for an order quashing that decision (*R* v *Hillingdon London Borough Council, ex p Royco Homes Ltd* [1974]). The legal validity of planning conditions may also be tested by seeking a declaration in the High Court (*Pyx Granite Co* v *Ministry of Housing and Local Government* [1958]). These remedies should only be considered in place of the statutory appeal where the issues are solely legal and do not involve planning merits; these remedies are considered in Chapter 20.

If either the applicant or the local planning authority requests it, the Secretary of State must afford to each an opportunity of being heard by an inspector, before the appeal is determined (section 79(2)). This is either by a public local inquiry, private hearing or by written representations. Appeals in respect of applications for planning permission, and appeals against enforcement notices are now determined by an inspector appointed by the Secretary of State. The only appeals now decided by the Secretary of State himself are those by statutory undertakers and which concern their operational land (Town and Country Planning (Determination of Appeals by Appointed Persons) (Prescribed Classes) Regulations 1997, although he may direct that he will hear any appeal if he so decides (Schedule 6 to the 1990 Act)). Inspectors have the power to determine appeals relating to listed building consents (except applications relating to Grade I and Grade II buildings and grant aided buildings (see the 1997 Regulations)).

17.3. Public local inquiry

The public local inquiry is a familiar process in many fields of public administration, apart from planning and development control. It is a quasi-judicial process in that the inspector who conducts it is impartial and also an expert with knowledge and experience of the area of administration being considered. The public local inquiry tends to mirror court procedure in being adversarial rather than inquisitorial. In other words, it depends on the parties present to put their 'case' and not upon the inspector to take the major initiative in examining the disputed decision; although he may well ask questions to clarify matters necessary for him to reach his own conclusions. In any adversarial process, parties are likely to be represented by persons skilled and qualified in advocacy whose task is to present their respective clients' evidence, and to test, by cross-examination, the evidence of opposing parties. It is not unusual for a non-lawyer to present a 'case' at a public local inquiry, indeed he deserves every encouragement, but the advantages of a legal presence should be that the matters of dispute can be identified from the common ground, and the procedure can be conducted fairly to ensure that the irrelevant and hearsay are excluded.

Where the Secretary of State is to decide the appeal himself, the conduct of an inquiry is subject to the Town and Country Planning (Inquiries Procedure) Rules 2000; where the inspector is to determine the appeal, the Town and Country Planning Appeals (Determination by Inspectors) (Inquiries Procedure) (England) Rules 2000, as amended by Town and Country Planning (Hearings and Inquiries Procedures) (England) (Amendment) Rules 2009, and in respect of appeals against enforcement notices, the Town and Country Planning (Enforcement)

(Determination by Inspectors) (Inquiries Procedure) (England) Rules 2002, apply.

It can be argued that the inquiry process has now become too judicial; that the trend of the decisions of the courts when reviewing post-inquiry decisions places such a constraint on what the inspector may or may not refer to in his written decision that his main task is to use the right 'incantation' in that decision and, that as a consequence, planning decision-making suffers because alternatives and solely planning issues seem subordinated to the requirements of a fair procedure. However, the courts have decided that the decision of an inspector is not to be read or construed as if it were a piece of legislation, neither is it essential that an inspector expresses his views in respect of every minor point provided that it is clear that he has come to his decision in a fair way.

The various inquiries procedure rules share a number of features. What is described here are the more important rules in relation to a planning appeal in the Town and Country Planning Appeals (Determination by Inspectors) (Inquiries Procedure) (England) Rules 2000.

The 2000 Rules set out a timetable for various steps to be taken.

17.3.1. Preliminary information and notification of inquiry

Under rule 3A, the Secretary of State gives notice that the appeal shall proceed to an inquiry; this is referred to as the 'starting date'. Under rule 4, on receipt of that notice, the local planning authority shall forthwith inform the Secretary of State and the appellant of the name and address of any statutory party who has made representations. Statutory parties are those who have made representation under section 71(2)(a)–(b) of the 1990 Act (i.e., owners and certain tenants); and of any directions or advice received from a government department.

17.3.2. Service of statements before inquiry

Under rule 6, the local authority shall, not later than six weeks after the starting date, serve a statement of case on the Secretary of State, the appellant and any statutory party. Where any directions have been issued or any advice received by the local planning authority, these must be included in the statement of case.

Not later than six weeks after the starting date, the appellant shall serve a statement of case on the Secretary of State, the local planning authority and any statutory party.

Either party may require the other to send them a copy of any document referred to in the list of documents in a parties' statement of case.

17.3.3. Other preliminary matters

Not later than 12 weeks after the starting date, the inspector may serve on the parties a statement of the matters about which he particularly wishes to be informed for the purposes of his consideration of the appeal (see rule 7(1)).

The inspector shall hold a pre-inquiry meeting where he expects the inquiry to last for eight days or more, or otherwise he considers it necessary (see rule 7(2)).

The inspector may, at any time, arrange a timetable for the proceedings (see rule 8(1)). The timetable may specify the dates for the service of proofs of evidence.

The Secretary of State will normally fix the date of the holding of the inquiry within 20 weeks of the relevant date (see rule 10(1)). The local planning authority will then be required to publish and serve notices of the inquiry, giving the date, time and place of the inquiry together with a description of the land and of the subject matter of the appeal (see rule 10(5) to (7)).

17.3.4. *Appearances at inquiry*

The persons entitled to appear at the inquiry include: the applicant; the local planning authority; the county or district council if not the local planning authority; the section 66 parties (owners and certain tenants of the land); the parish or community council where they have made representations; and other persons listed in rule 11 of the 2000 Inquiries Rules. Any other person has no right to appear, but may do so at the inspector's discretion.

A party may appear on his own behalf or may be represented by counsel, solicitor or any other person (such as a town planning consultant, surveyor or agent).

17.3.5. *Representatives of government departments*

Where a government department is involved because of a direction or has given advice, the applicant may make a request at least four weeks before the inquiry that a representative of that department be available at the inquiry (see rule 12). Such a person may give evidence and be cross-examined, although he cannot be required to answer any question which in the opinion of the inspector is directed to the merits of government policy.

17.3.6. *Proofs of evidence*

Where a person entitled to appear at an inquiry proposes to give or to call another person to give evidence by reading a proof of evidence, the proof and a written summary must be sent to the inspector at least three weeks before the date of the inquiry (see rule 14). Copies of proofs must be served on the other parties. At the inquiry only the written summary will normally be read.

17.3.7. *Procedure at the inquiry*

The procedure is at the discretion of the inspector to the extent that the rules do not otherwise apply. The applicant will ordinarily begin, and has the right of final reply. The practice is for the applicant to make an opening speech outlining his case and the evidence he is going to rely on; he calls his expert witness or witnesses who read their proofs of evidence. They may then be cross-examined by the other parties entitled to be present at the inquiry. The purpose of this cross-examination is to enable the evidence to be tested by direct questions. The local planning authority will present their case in a similar manner, and their expert witnesses may also be cross-examined.

Ordinary objectors who are present at the discretion of the inspector are generally limited to making a statement. They are not usually allowed to call evidence or to cross-examine other witnesses. This is very much at the discretion of the inspector.

The local planning authority closes their case with a speech urging the relevance and significance of their evidence; and the applicant then makes his own final reply in a similar vein.

The inspector will record all the evidence called before him, and make a note of the main points argued by the parties. He will be in breach of the rules of natural justice if he relies on any other evidence not before the inquiry, or not in issue at the inquiry, even if that evidence comes to him because he is an observant expert in his own right and sees something he thinks relevant at his site inspection: *Fairmount Investments Ltd* v *Secretary of State for the Environment* [1976].

17.3.8. Site inspection

The inspector may make an unaccompanied site visit before the inquiry. He may, and must if requested, make a site visit after the inquiry. On the second occasion, the parties entitled to be present at the inquiry may accompany him. The purpose of this is to identify any features referred to during the course of the inquiry. Because no further arguments can be put to the inspector, the lawyers are not present at these site visits. Where the landowner fails to appear at a pre-arranged site visit, it is an error of law by the inspector to speak to the planning officer, as justice must be seen to be done: see *R (Tait)* v *Secretary of State for Communities and Local Government* [2012].

17.3.9. Procedure after inquiry

Where the decision is that of the Secretary of State, the inspector will make a report in writing with his findings and recommendations. In other cases, the inspector will proceed to formulate his decision and his reasons.

If the Secretary of State is deciding the appeal under the Town and Country Planning Appeals (Inquiries Procedure) (England) Rules 2000, and differs from his inspector on a finding of fact, or wishes to take into consideration new evidence or any new issue of fact (not government policy) not considered at the inquiry, and as a consequence disagrees with a recommendation of his inspector, he must inform the parties (see rule 16). They may make further representations, or, where new evidence or a new issue of fact is raised, may ask within 21 days that the inquiry be reopened. A difference of opinion on the merits of siting a house in a walled garden is not a difference on a finding of fact: *Lord Luke of Pavenham* v *Ministry of Housing and Local Government* [1968]. The test is between a finding on an existing state of affairs as opposed to a subjective opinion as to the future (*Pyrford Properties Ltd* v *Secretary of State for the Environment* [1977]). Only if the Secretary of State disagrees with the former, need he reopen the inquiry. The Secretary of State cannot decide an appeal on a ground not considered at all at the inquiry (*Webb* v *Secretary of State for the Environment* [1972]).

The inspector is bound by the rules of natural justice to give the Secretary of State a fair account of the inquiry but he need not record every irrelevant submission: *North Surrey Water Co* v *Secretary of State for the Environment* [1977]. He must give adequate reasons as was explained in *Bolton Metropolitan Borough Council* v *Secretary of State for the Environment* [1995]. The inspector must address the principal issues: see *MJT Securities* v *Secretary of State for the Environment* [1997], and give reasons: see rule 19 of the Town and Country Planning Appeals (Determination by Inspectors) (Inquiries Procedure) (England) Rules 2000. The Secretary of State is required to give the reasons for his decision in writing (see rule 18 of the Town and Country Planning Appeals (Inquiries Procedure) (England) Rules 2000).

Where the inspector decides an appeal he is required to inform the parties if he proposes to take into consideration any new evidence, or any new issue of fact not raised at the inquiry. The parties may make representations, and request that the inquiry be reopened. Where evidence has been presented at the inquiry, or where a site visit has been undertaken, the inspector is entitled to make inferences, such as of architectural values; such inferences will not be new evidence: *Winchester City Council* v *Secretary of State for the Environment* [1978]. The inspector's obligations with regard to providing a reasoned decision letter are the same as the report to and decision of the Secretary of State considered in the preceding text.

17.3.10. Notification of the decision

The decision, and the reasons for it, are made in writing and notified to the applicant, the local planning authority and the other parties entitled to be at the inquiry, and to any other person, who has requested to receive it. The decision is final (section 79(5)), although an application may be made to the High Court on a point of law (section 288(1)). This is dealt with partly later in this chapter and partly in Chapter 20.

17.4. Hearings

This involves a more simple examination of the matters arising under the appeal. The procedure is appropriate where evidence does not need to be tested by formal cross-examination, the issues are straightforward, and the matters can be dealt with in one day. The Town and Country Planning (Hearings Procedure) (England) Rules 2000, as amended by the Town and Country Planning (Hearings and Inquiries Procedures) (England) (Amendment) Rules 2009, set out the procedures. The inspector leads the discussion, but must ensure that each party has a fair hearing: see Circular 05/2000 – *Planning Appeals: Procedures (Including Inquiries into Called-in Planning Applications)*.

17.5. Written representations

A large number of appeals are now decided on the basis of written representations in place of the public local inquiry or hearing; they are appropriate for simple cases. The Town and Country Planning (Appeals) (Written Representations Procedure) (England) Regulations 2009 prescribe time limits for the various stages of the procedure. General guidance is to be found in Circular 05/2000 – *Planning Appeals: Procedures (Including Inquiries into Called-in Planning Applications)*.

When the appellant gives notice of appeal to the Secretary of State, he may request that the appeal be disposed of by written representations, and should enclose his written statements of case. Copies of the notice of appeal and all supporting documents must be sent to the local planning authority. Where the Secretary of State accepts that written representations are appropriate, and so notifies the parties, that is the 'starting date' for the procedure laid down in the 2009 Regulations.

In the case of the simple 'householder' appeals, within five working days of the starting date, the local planning authority, if they accept the procedure, must notify the authorities and other parties entitled to be consulted over a planning application. The appeal notice and the documents accompanying it constitute the appellant's representations: see regulation 7.

In the case of appeals that are not 'householder' appeals, Part 2 of the regulations provides for rather longer timetables. The local planning authority must complete a questionnaire and submit this to the appellant and the Secretary of State within 14 days of the starting date. The local planning authority may submit their written representations at this stage, but in any event within six weeks of the starting date. The appeal and the documents with it may constitute the appellant's representations, but otherwise the appellant has six weeks of the starting date to make representations. The written representations put in by either party should deal concisely with the facts, the policies applied and the reasons why the appeal should, or as the case may be, should not be allowed.

After receipt of the representations, each party has nine weeks from the starting date to give comments on each other's representations, by way of reasoned replies.

The appointed inspector will usually make a site visit immediately following the closing date for representations. A decision letter will then be issued.

The inspector can only decide an appeal on the matters raised in the written representations, he cannot take into account matters which a party has not had an opportunity of considering (*Wontner-Smith* v *Secretary of State for the Environment* [1977]). If he receives any document or information that was before the appellant and the local planning authority, he will be acting unfairly if he fails to send a copy to an objector, who should be entitled to comment on it: see *Jory* v *Secretary of State for Transport, Local Government and the Regions* [2003]. In R (*Ashley*) v *Secretary of State for Communities and Local Government* [2012], a planning permission granted on appeal was quashed; the developer had submitted an expert report on the last date for making representations, and the inspector relied on this report. The objector was unaware of the report and had been unfairly prejudiced. The Court of Appeal recommended that the procedure should be reconsidered to avoid the prejudice of a party putting in evidence at the last moment that objectors would not see. He should provide a reasoned decision letter in accordance with the guidance in *North Wiltshire District Council* v *Secretary of State for the Environment* [1992].

17.6. The decision of the Secretary of State or inspector

The Secretary of State may allow or dismiss an appeal, or may reverse or vary any part of the decision of the local planning authority, and may deal with the application as if it had been made to him in the first instance (section 79(1)). An inspector determining an appeal enjoys the same powers.

Because of this wide discretion on the part of the Secretary of State, careful tactics are required in appealing against planning conditions. The Secretary of State may take the opportunity of re-deciding the whole or part of the original planning permission. It may be preferable to submit a further application to the local planning authority to have the condition varied or removed; and if refused, then appeal to the Secretary of State. In this way, the original planning permission remains.

The decision may be quashed by the High Court if the reasons given are obscure and leave in doubt the matters which the Secretary of State took into account, or did not take into account (*Givaudan & Co Ltd* v *Minister of Housing and Local Government* [1967]). Where the Secretary of State makes a decision following an inquiry and his inspector's report, he is not bound by the recommendations of his inspector; but if he relies on a policy document to which the inspector attached little evidential value, and reaches a contrary conclusion and decision to his inspector, his decision will be quashed if he fails to reopen the inquiry (*French Kier Developments Ltd* v *Secretary of State for the Environment* [1977]). These matters are more fully considered in Chapter 20 in relation to statutory appeals and judicial review applications to the High Court.

Policies set out in ministerial circulars are material considerations, and where relevant, must therefore be taken into account: *Pye (Oxford) Estates Ltd* v *West Oxfordshire County Council* [1982]. Nevertheless, it would be wrong for the Secretary of State to apply a policy that is inconsistent with the clear provisions of the development plan unless there are reasons for departing from the plan: *Reading Borough Council* v *Secretary of State for the Environment* [1986].

Indeed, decisions must be made in accordance with the development plan unless material considerations indicate otherwise: see section 38(6) of the Planning and Compulsory Purchase Act 2004.

The Secretary of State is entitled to have a policy, but in determining an appeal, he must give genuine consideration to the issues and merits raised (*Lavender* v *Minister of Housing and Local Government* [1970]). He is, however, entitled to make reference to a policy statement of another government department in his decision without having to reopen the inquiry on the basis that he is relying on new evidence (*Kent County Council* v *Secretary of State for the Environment* [1977]).

Further grounds for challenging the decision are considered in Chapter 20. Undoubtedly, the Secretary of State has a difficult task. On the one hand, he is deciding an appeal between two parties and is required to show impartiality by reason of the procedure rules and the rules of natural justice. Yet, on the other hand, he is entitled to regard an appeal as if an application were being made to him for planning permission; the appeal process gives him an opportunity to put into effect ministerial or government policy.

17.7. Costs

There is power to award costs under section 250 of the Local Government Act 1972. In practice, costs have usually only been awarded to a party in a case of unreasonable behaviour of another party (*R* v *Secretary of State for the Environment, ex p Reinisch* [1971]).

Circular 8/93 – *Award of Costs Incurred in Planning and other (including Compulsory Purchase Order) Proceedings* – explains the government's policy on the award of costs. Costs are not awarded on the basis of success or failure; they are to be awarded where the unreasonable behaviour of one party has caused another party unnecessary expense. Unreasonable behaviour includes a failure of a local planning authority to make a decision within the statutory time limits; or, the action of an appellant in pursuing an appeal over a refusal of planning permission against a history of such refusals.

In the various inquiries procedure rules, there is also a rule that deals with costs. If a person makes an application for an award of costs, the inspector must make a report on this and on any considerations which may be relevant for the Secretary of State to decide the question. In the case of transferred appeals, where the inspector makes the decision, the inspector now has power to award costs in accordance with the policy in Circular 8/93.

This discretionary power to award costs is no longer limited, in appeal cases against a planning refusal, to those where an inquiry has been held. There is now power to award costs where the appeal is heard by written representations (amendments introduced by the Housing and Planning Act 1986 to section 250 of the Local Government Act 1972). In enforcement notice appeals, the discretion to award costs arises once proceedings have commenced, and therefore costs can be awarded if the enforcement notice is withdrawn or the matter determined by written representations.

Section 42 of the Housing and Planning Act 1986 provides that where a minister is authorised to recover costs incurred by him in relation to an inquiry, the costs shall be the entire administrative costs of the inquiry including his departmental overheads. The minister may make regulations to provide for the recovery of these costs in relation to planning.

17.8. Independent examination of development plan documents

The local planning authority must submit every development plan document to the Secretary of State for independent examination: see section 20(1) of the 2004 Act. The examination is carried out by an examiner appointed by the Secretary of State: see section 20(4). Representations can be made, and the examiner must make recommendations, and give his reasons: see section 20(7). There is a right to be heard at the examination, although that must be considered in the context of an examination at which cross-examination might not be allowed, and topics for consideration might be kept to those of general and particular importance.

18 Purchase and blight notices

18.1. Introduction

In some situations, a decision of a local planning authority may render land 'incapable of reasonably beneficial use'. This chapter deals with certain rights of a landowner to compel the local planning authority to purchase from him such land. It may also happen that land may be 'blighted' by the proposals of a local authority or some other body, or by some indication on a development plan; the affected land may then be difficult to sell. In this case, certain owners have the right to compel the appropriate authority to purchase such land by serving a blight notice. Land may be blighted by a proposal that ultimately envisages the compulsory purchase of the land, such as a proposed road improvement. Although an outline of blight notices is given in this chapter, the service of a blight notice is a form of inverse compulsory purchase. That topic is more fully considered in *Compulsory Purchase and Compensation* by B Denyer-Green (Routledge, Tenth Edition, 2013).

18.2. Purchase notices

A purchase notice may be served where an owner has an interest in land affected by a planning decision, and must be considered as a remedy against an adverse planning authority decision. An owner for this purpose is a person entitled to receive the rack rent. An owner of a reversionary interest, subject to a lease at a rent less than the rack rent, is not entitled to serve a purchase notice (*London Corporation* v *Cusack-Smith* [1955]). A number of matters must be satisfied before a purchase notice compelling a local planning authority to purchase the land can be effectively served. The law and procedure is described in Circular 13/1983 – *Purchase Notices*.

18.2.1. The planning decisions

A purchase notice may only be served in respect of land affected by a planning decision; the decisions giving rise to the right to serve a purchase notice are found in the 1990 Act:

(a) *Decision in connection with a planning application*
Where an application has been made for planning permission to develop any land, a decision which is a refusal of permission or a grant subject to conditions (section 137(1)(a)).
(b) *Revocation or modification orders*
An order under section 97, revoking any planning permission or modifying any planning permission by the imposition of any conditions, may enable the owner to serve a purchase notice (section 137(1)(b)). There is an alternative right to claim compensation in those circumstances (see Chapter 19).
(c) *Discontinuance orders*
Where an order under section 102 requires the discontinuance of any use of land, or imposes a condition on the continuance of a use, or requires the alteration or removal of any buildings or works, the owner may be able to serve a purchase notice (section 137(1)(c)). There is also an alternative right to claim compensation.
(d) *Decisions in connection with a listed building consent application*
Where an application has been made for listed building consent, a decision which is a refusal of consent or a grant subject to conditions, or a revocation or modification of a listed building consent (section 32 of the Planning (Listed Buildings and Conservation Areas) Act 1990). Again, there may be an alternative right to claim compensation in some cases.
(e) *Trees and advertisements*
Decisions in connection with tree preservation orders and under the advertisement control regulations may enable the owner of the affected land to serve a purchase notice (section 198(4)).

In each case, the decision is usually that of the local planning authority, and there is no need to appeal to the Secretary of State before serving a purchase notice.

18.2.2. Incapable of reasonable beneficial use

After establishing one of the appropriate planning decisions, the owner must then show that:

(a) 'The land has become incapable of reasonably beneficial use in its existing state'; or where he has received a decision subject to conditions, 'the land cannot be rendered capable of reasonably beneficial use by the carrying out of the permitted development (or of the works, in the case of listed buildings) in accordance with those conditions', and
(b) The land cannot be rendered capable of reasonably beneficial use by the carrying out of any development for which planning permission has been granted or for which the local planning authority or the Secretary of State has undertaken to grant planning permission.

It was decided in *Wain* v *Secretary of State for the Environment* [1982] CA, that the owner can only serve a purchase notice in respect of that part of his land that is incapable of reasonably beneficial use in its existing state. He cannot include other land that is capable of reasonably beneficial use. 'Land' is defined as including a building (section 336). Where a purchase notice is served following the refusal of planning permission or the grant of planning permission subject to conditions, the notice must relate to the whole of the land that is the subject of that planning decision: see *Cook* v *Winchester City Council* [1994]. Further, the claimant must own the whole of the land: see *Smart & Courtney Dale Ltd* v *Dover Rural District Council* [1972].

'Incapable' means that the land cannot be put to a reasonably beneficial use because of some practical or legal reason. The words 'reasonably beneficial use' must be considered together. The plain meaning suggests that no use can be found for the land that is reasonably beneficial to the owner; thus, a reasonably beneficial use must be distinguished from a use that is not reasonably beneficial. The size, location and character of the land will be relevant; for example, a small piece of derelict land adjoining private gardens may have a reasonably beneficial use as an addition to one or more of those gardens, whereas a large area of derelict land in a similar position may have no reasonably beneficial use.

In *Purbeck District Council* v *Secretary of State for the Environment* [1982], Woolf J decided that it would be wrong to confirm a purchase notice where the land had become 'incapable of reasonably beneficial use' due to some activity, such as tipping, which was a breach of planning control. The Court of Appeal in *Balco Transport Services Ltd* v *Secretary of State for the Environment* [1985] narrowed the effect of the *Purbeck* case by deciding that a purchase notice could be confirmed where, although the land was incapable of reasonably beneficial use due to a breach of planning control, the breach had become immune from enforcement proceedings.

From the cases, two tests have emerged which can be applied to determine whether land has become incapable of beneficial use.

(1) Is the existing use reasonably beneficial?

The important requirement is that the land has become incapable of reasonable beneficial use: see section 137(3). Reasonable beneficial use does not necessarily mean profitable use (although profitability is often relevant), nor does it mean that the land is less useful to the owner in its present state than if developed: see *R* v *Minister of Housing and Local Government, ex parte Chichester Rural District Council* [1960] QBD. In *Colley* v *Secretary of State for the Environment* [1998], it was said that the word 'use' in the expression in section 137(3) 'incapable of reasonably beneficial use' was concerned with use rather than the value of land.

If only part of the land is incapable of reasonable beneficial use, and part is capable of such use, then the owner can serve a notice only in respect of the part that is incapable of reasonable beneficial use: *Wain* v *Secretary of State for the Environment* [1982] CA. 'Land' includes a building: section 336(1).

In deciding whether land has any reasonable beneficial use, one first considers if there is any existing use that is beneficial (waste land that was partly used as a sports ground and could, in part, be let for grazing, had a beneficial use: *General Estates Company Ltd* v *Minister of Housing and Local Government* [1965]). In that case, part of the land involved was let to a sports club at a rent of £52 pa. Although the rent was low in comparison with the potential value of the land with planning permission, had it been granted, this was not considered sufficient to render the land incapable of beneficial use. Similarly, it was decided in *R* v *Minister of Housing and Local Government, ex parte Chichester Rural District Council* [1960] that the fact that the land is of less value to the owner than it would have been had the planning decision been more favourable is again irrelevant. In that case, the existing use of the land was as a number of plots let as caravan sites during the summer under a temporary planning permission; this was held to be reasonably beneficial use.

Examples of existing uses that are not reasonably beneficial would include waste or derelict land, the site of a building that has been seriously damaged by fire or has fallen into dereliction, and amenity land or land designated as public open space.

If it is considered that the existing use is reasonably beneficial, no purchase notice can be served; but if the existing use is not reasonably beneficial, it will be necessary to consider the second test.

(2) Where the existing use is not reasonably beneficial, is some prospective use reasonably beneficial?

If there is no existing use that is beneficial, the next question is whether there are any other uses that could be reasonably beneficial; any use in Part I of Schedule 3 to the 1990 Act can be considered provided the development can be carried without an express grant of planning permission. A use which is 'new development' ('new development' is any development not in Schedule 3) must not be considered. Whether any use is or is not reasonably beneficial will be a question of fact and degree, and for the minister to decide: *Brookdene Investments* v *Minister of Housing and Local Government* [1970] QBD. Any unauthorised prospective use is to be disregarded: see section 138 of the 1990 Act.

In *Colley* v *Secretary of State for the Environment* [1998], it was held that some woodland could still be capable of reasonable beneficial use as commercial woodland even though it was protected by a tree preservation order.

The Secretary of State takes the view that any prospective use of the land must be considered even where it is not development and therefore does not need planning permission. In one decision (ref PLUP2/5319/176/1), a small site, in a back-land position in a residential area, was considered to be of reasonably beneficial use as garden ground despite its existing neglected condition. Again, in the *General Estate Co* case (see the preceding text), part of the land could have been let for grazing purposes and this was held to be a reasonably beneficial use. Prospective use for agriculture or forestry would not involve development and could, therefore, be considered in suitable cases.

Section 138 of the Act states that in considering any prospective use, no account shall be taken of any unauthorised prospective use of the land. Prospective use of land is regarded as development if it would involved the carrying out of any development other than that in paragraphs 1 or 2 of Schedule 3 to the 1990 Act, or, in the case of a purchase notice following the refusal or conditional grant of planning permission, if it would contravene the condition in Schedule 10. The effect of section 138 is that only rather minor forms of development can be considered, such as the rebuilding of a building or its alteration, within the Schedule 10 space limitations, or the use of a single dwellinghouse as two or more dwellinghouses.

If planning permission has been granted for development or the local planning authority or the Secretary of State has undertaken to grant planning permission, the owner serving a purchase notice must show that the land could still not be rendered capable of reasonable beneficial use. However, the limitation in section 138 (see the preceding text) would seem also to apply here.

The expression 'has become' suggests that its state of incapability of reasonable beneficial use is a new state and is attributable to the planning decision. This interpretation is not accepted by the minister (see para 13 of Circular 13/83), who is prepared to accept that the criteria include the physical state of the land. In *Purbeck District Council* v *Secretary of State for the Environment* [1982], the Secretary of State's confirmation of a purchase notice was quashed on the ground that the minister had failed to consider that the reason for the state of the land was an unauthorised activity. That case was distinguished in *Balco Transport Services Ltd* v *Secretary of State for the Environment* [1985], where the Court of Appeal decided that if an unauthorised activity could be the subject of an enforcement notice, and so made reasonably beneficial of use, the owner could not serve a valid purchase notice: if the unauthorised activity was immune from enforcement, then he could serve a valid purchase notice.

The final point, which applies generally to the problem of reasonably beneficial use, is that, in deciding what is beneficial, it is relevant to make a comparison between the existing use value and the value of the land with the benefit of any relevant Schedule 3 development (see *Brookdene Investments Ltd* v *Minister of Housing and Local Government* [1969]). Such a comparison may be one of the matters which the minister considers (see also Circular 13/1983 – *Purchase Notices*).

18.2.3. *Amenity conditions*

Where planning permission has been granted for an area of land and a condition imposed restricting the use of the small part, e.g., as open space or play area in a residential estate, then a purchase notice cannot be used if planning permission is subsequently refused to develop that area of restricted use (see section 142).

18.2.4. Existing state

One is required to consider whether land is capable of reasonably beneficial use in its existing state. This must mean that land without buildings must be considered as land without buildings, amenity land as amenity land, and so on. It is doubtful whether existing state refers to the state of repair of buildings, although there seems no reason why it should not refer to the age and suitability of a particular building as a description of its general characteristics. A building of the nineteenth century may be in good or bad repair, but its existing state is the present building without improvement. If the building is in a very bad state of repair, a state of dereliction, repair may then be relevant and existing state could in those circumstances refer to the building in a state of dereliction.

The following examples may clarify some of these points:

(1) A strip of land adjoining a highway is used by a nursery-man to grow roses. An application for planning permission to build a petrol-filling station is refused. Can a purchase notice be served? In applying the first test, it would probably be conceded that growing roses was a reasonably beneficial use of the land, and the fact that it was less profitable than a petrol-filling station is not relevant. It would be unnecessary to apply the second test; a purchase notice would be unsuccessful.

(2) A rambling old building in a conservation area and in a very bad state of repair has, until recently, been used for offices. The low rental value of this building for office purposes renders it uneconomic to carry out repairs and improvements to the present building. An application to demolish and redevelop the site is refused. In considering whether a successful purchase notice could be served, the first test requires us to decide whether the existing building, in its existing state, is incapable of reasonably beneficial use. Although unoccupied, the present use is offices, but in its existing state of age and dereliction, that use is not reasonably beneficial, because rental income would not cover outgoings. The second test is then applied. Does the building, in its existing state, have a reasonably beneficial use for a purpose other than offices which does not involve new development? The answer is probably no, as a change of use will almost certainly involve new development: a purchase notice is likely to succeed in this example.

18.2.5. The procedure

The purchase notice is served on the district council (or London borough) within 12 months of the decision concerned. Although a notice cannot be amended, an owner can serve further notices until he gets one right: see *Herefordshire CC* v *White* [2007].The procedural requirements, and time limit, are found in the Town and Country Planning General Regulations 1992. In the Scottish case of *Reside* v *North Ayreshire Council* [2000], it was decided that the start date of the 12-month time limit, where there is an appeal, was the date of an inspector's decision.

Within three months of the service of the purchase notice, the council must serve a response notice stating that:

(a) The council is willing to comply and to purchase the land; or
(b) Another specified authority is willing to purchase; or
(c) Neither the council nor any other authority is willing to accept the purchase notice, and that a copy has been sent to the Secretary of State with reasons for not complying with the notice.

If (c) applies, the Secretary of State invites representations from the owner, the council and the county planning authority. And if he has in mind substituting any other authority for the council as being required to comply with the purchase notice, he seeks their views as well. If required to, the Secretary of State holds a hearing or public local inquiry; he may then confirm the purchase notice, substitute another authority for the council, refuse to confirm the purchase notice and grant planning permission or vary any conditions; he may also in lieu of confirming the purchase notice, direct that planning permission shall be granted for some form of development other than that applied for.

The Secretary of State need not confirm a purchase notice covering land which includes some land the subject of restrictions against development imposed by a planning condition or was clearly intended as amenity land in a planning application (section 142 of the 1990 Act).

Unless there is an appeal against the planning decision before the Secretary of State, the purchase notice is deemed confirmed if the Secretary of State fails to make a decision before the end of the relevant period (see section 143(2) of the 1990 Act). The relevant period is nine months from the service of the purchase notice or, if earlier, six months from the date on which a copy of the notice was transmitted to the Secretary of State (section 143 of the 1990 Act).

18.2.6. Listed buildings

Section 32 of the Planning (Listed Buildings and Conservation Areas) Act 1990 makes stricter provision for the service of a purchase notice where listed building consent is refused, granted subject to conditions or is revoked or modified. The procedure and conditions for service are stricter than under the Town and Country Planning Act 1990.

18.2.7. Deemed compulsory purchase

If a purchase notice is accepted by the council or some other authority, or confirmed by the Secretary of State, the authority acquiring are deemed to have served a notice to treat: section 143(1) of the 1990 Act. A notice to treat is the first step in the use of the acquisition powers in the more usual compulsory purchase order situation. Notices to treat are therefore considered in Part II of this book. However, by section 67 of the Planning and Compensation Act 1991, a deemed notice to treat will cease to have effect after three years where the planning decision which founded the purchase notice was made after 25 September 1991. The point is that once a purchase notice is accepted, the authority is deemed to be compulsorily purchasing the land, and the rules and basis of compensation for a compulsory purchase should apply. Once a purchase notice is accepted or confirmed, as a deemed notice to treat it cannot be withdrawn under section 31 of the Land Compensation Act 1961 (see section 143(8) of the 1990 Act). However, there appears to be no prohibition on the withdrawal of a purchase notice before it is accepted or confirmed.

Where a purchase notice has been accepted or confirmed, a claim for damages may be brought against the authority if they fail to negotiate with the owner in good faith: *Bremer* v *Haringey London Borough Council* [1983].

18.2.8. Agricultural land

There are special provisions where part of an agricultural unit is subject to a deemed notice to treat following the service of a purchase notice and the remainder of the unit is not economically viable (see sections 145–147 of the 1990 Act).

18.3. Blight notices

Certain owners can serve a blight notice to compel an authority to acquire their 'blighted' properties. 'Blight' in the legal sense is not the generalised description adopted by lay people in their reference to a state of affairs where property values are affected by local authority planning decisions or other activities, although such a state of affairs may include circumstances in which a blight notice can be served. Blight has a specific meaning and the circumstances are specified in the Town and Country Planning Act 1990. The right to serve a blight notice is limited to certain categories of owner-occupiers. Broadly, the purpose of blight notices is to compel authorities to purchase land in advance of their compulsory purchase needs in order to mitigate hardship.

18.3.1. Classes of blighted land are specified in Schedule 13 to the 1990 Act

18.3.1.1. Land allocated for public authority functions on development plans, etc.

(1A) Land which is identified for the purposes of relevant public functions by a development plan document for the area in which the land is situated. A development plan document is part of the replacement provisions for development plans introduced by the 2004 Act; it will include an adopted or approved document, a revision, or one submitted to the Secretary of State for independent examination. Relevant public functions include the functions of a government department, local authority, national park authority or statutory undertakers. They also include the establishment of running of a public telecommunications operator of a telecommunications system.

(5) Land indicated in a plan (other than a development plan) approved by resolution of a local planning authority as land which may be required for the purposes of paragraph 1A.

(6) Land indicated in a plan, which is not a development plan, but which is approved by the local planning authority for the purpose of 'safeguarding' when exercising development control powers because the land is required for the purposes of a public authority. If land is required for a public authority purpose, that fact may not necessarily be indicated in the development plan (see 1 in the preceding text); the planning authority may refuse permission to develop that land to safeguard it for the intended purpose.

18.3.1.2. New towns and urban development areas

(7) Land within an area described as the site of a proposed new town in a draft order.

(8) Land designated as a site of a new town in an order in operation under section 1 of the New Towns Act 1981.

(9) Land which is intended or has been designated as an urban development area.

18.3.1.3. Clearance and renewal areas

(10) Land declared to be clearance area by resolution under the Housing Act 1985.

(11) Land surrounded by or adjoining an area declared to be a clearance area and land the authority has resolved to purchase under section 290 of the 1985 Act.

(12) Land indicated by information under section 92 of the Local Government and Housing Act 1989 as land to be acquired as a renewal area.

18.3.1.4. Highways

(13) Land indicated in a development plan (other than a planning document) as land proposed for a highway or a highway improvement or alteration, or

(14) Land on or adjacent to the line of a highway in an order or scheme for a new highway or a highway improvement made under the Highways Act 1980, including land in any proposed order or scheme which has been submitted to the Secretary of State for confirmation, and land required under the Land Compensation Act 1973 for works of mitigation.

(15) Land shown on a plan approved by resolutions of a local highway authority as land for a new highway or highway improvements.
(16) Land included on a plan published by the Secretary of State as land for a public highway improvement area and notified to the local planning authority.
(17) Land shown on plans resolved by the local highway authority for acquisition for mitigation purposes.
(18) Land shown in a written notice by the Secretary of State to be acquired for mitigation purposes.
(19) Land within the outer lines laid down in a new streets order, prescribing the minimum width of new streets, and upon which there is a dwelling erected before the making of the order (an owner may find his property is unsaleable, because such an order may be made long before the street widening takes place).

18.3.1.5. General improvement area

(20) Land in general improvement area (Part VIII of the Housing Act 1985) or indicated in a notice published for that purpose.

18.3.1.6. Compulsory purchase

(21) Land authorised to be acquired by a private Act or within limits of deviation permitted by such an Act.
(22) Land within a compulsory purchase order, including an order authorising the acquisition of rights (in respect of which the appropriate acquiring authority has not yet served a notice to treat), a compulsory purchase order submitted to the minister for confirmation or, where a minister is the acquiring authority, a draft compulsory purchase order (a resolution to make an order, but where no other steps are taken to make an order, is not within this paragraph).
(23) Land proposed in an application for, in a draft order under, within limits of deviation or authorised to be acquired by, an order under the Transport and Works Act 1992.
(24) Land falls within this paragraph if, for the purposes of the Planning Act 2008:

   (a) The compulsory acquisition of the land is authorised by an order granting development consent; or
   (b) The land falls within the limits of deviation within which powers of compulsory acquisition conferred by an order granting development consent are exercisable; or
   (c) An application for an order granting development consent seeks authority to compulsorily acquire the land.

(25) Land falls within this paragraph if the land is in a location identified in a national policy statement (for the purposes of the Planning Act

2008) as suitable (or potentially suitable) for a specified description of development.

Note: Land ceases to fall within this paragraph when the national policy statement:

(a) ceases to have effect; or
(b) ceases to identify the land as suitable or potentially suitable for that description of development.

18.3.2. *Persons entitled to serve a blight notice: section 149 of the 1990 Act*

Only certain persons may serve a blight notice, essentially those to whom having blighted land is an especial hardship:

(a) owner–occupiers of any hereditament, of which the net annual value for rating purposes does not exceed a prescribed limit (at present £34,800; see the Town and Country Planning (Blight Provisions) Order 2010) – this would include the owners of small business premises or shop;
(b) residential owner–occupiers of a private dwelling;
(c) owner–occupiers of agricultural units.

An owner–occupier is the owner of the freehold or a lease with at least three years unexpired tenure, who has been in occupation of the whole or a substantial part of the hereditament for the six months preceding the service of a blight notice, or a period of six months ending not more than 12 months before the blight notice; and if the latter, the hereditament or dwelling must have been unoccupied since the owner–occupier vacated (section 168). Where a property is let, it cannot be regarded as occupied: see *Aardvark SRE Ltd* v *Sedgefield BC* [2008].

Where the subsoil under an adjoining public highway is presumed to be owned with the adjoining land, and highway improvements are proposed which will affect the subsoil, it was held by the Lands Tribunal in *Norman* v *Department of Transport* [1996] that the subsoil is part of the hereditament for the purposes of serving a blight notice. Although this case was appealed, it is believed that the appeal was settled.

The 1990 Act also permits a mortgagee the right to serve a blight notice where he is entitled to exercise his powers of sale, can give immediate vacant possession and has been unable to sell the interest concerned except at a price substantially lower than he might reasonably have been expected to obtain had the land not been within the circumstances of blight (section 161). A personal representative of a deceased owner–occupier has a similar right to serve a blight notice and must also show that he has been unable to sell except at a substantially lower price (section 161 of the 1990 Act).

18.3.3. Procedure: sections 150–152 of the 1990 Act

A person who qualifies under part **18.3.2** above and has land within one of the circumstances of planning blight outlined in part **18.3.1** above may serve a blight notice on the appropriate authority. It is not good enough that the claimant's land is blighted and difficult to sell, because it will be close to a proposed highway: hardship alone will not found a blight notice; the land must be land that will be needed: *McDermott* v *Department of Transport* [1984]. In some circumstances, land may be blighted under more than one of the categories of blight and more than one authority may be the 'appropriate' authority. In *R* v *Secretary of State for the Environment, ex parte Bournemouth Borough Council* [1987], where land was blighted by a development plan indication of a proposed highway, it was decided that both the local planning authority and the highway authority were each the 'appropriate' authority, and the Secretary of State had no power to make a selection.

Except where the land falls within paras 21, 22 or 24(a) and (b) of Schedule 13 (see p. 221 above) and the powers of compulsory acquisition remain exercisable, the claimant must, before serving a blight notice, have made reasonable endeavours to sell his interest and in consequence of the blight has been unable to do so, or unable to do so except at a price substantially lower than he would have expected in the absence of blight. Individual circumstances will dictate what a reasonable endeavour to sell is. It would seem that in a case within para 23 of Schedule 13 – a transport and works order – there is no exception to this rule. It may be possible to argue that such an order is a 'special enactment' and therefore falls within paras 21 or 22 to which the exception does apply. Individual circumstances will dictate what a reasonable endeavour to sell is. It may include advertising or placing the property with an agent who makes attempts to seek a purchaser: see *Perkins* v *West Wiltshire District Council* [1975]. It is not sufficient to make no attempt to sell if an agent advises that no purchaser could be found in view of the planning blight. The attempt to sell need not be after a date when the land becomes formally blighted. The costs of any attempt to sell are not recoverable as compensation: see *Budgen* v *Secretary of State for Wales* [1985].

Where an owner owns the whole of a hereditament or an agricultural unit, a blight notice cannot be served in respect of part only; and where an owner is entitled to an interest only in part of a hereditament or an agricultural unit, a blight notice cannot be served in respect of less than the entirety of the respective part (section 150). However, if a blight notice is served in respect of less than the whole of the owner's land, and so in contravention of this provision, the invalidity of the notice must be raised in the authority's counter-notice or otherwise, the Upper Tribunal (Lands Chamber) will have no jurisdiction to determine the question: *Binns* v *Secretary of State for Transport* [1986].

This last case, together with the much earlier case of *Lake* v *Cheshire County Council* [1976], illustrates that it is strictly statutorily impossible to serve a valid blight notice in respect of part of a hereditament (as defined in the valuation list) or part of an agricultural unit. Even if a notice is served in respect of the whole property, and the authority is prepared to take part, it may be difficult to compel them to take the whole: *Lake* v *Cheshire County Council* [1976].

18.3.4. *Counter-notice by appropriate authority*

The appropriate authority may accept the blight notice and proceed to acquire the land: the purchase is then deemed to be a compulsory purchase. Alternatively, the authority may serve a counter-notice within two months objecting to the blight notice. The grounds of objection are set out in section 151 of the 1990 Act. They include the following:

(a) That no part of the hereditament or agricultural unit is comprised in blighted land.

(b) That, save in the case of land within para 25 of Schedule 13 (see the preceding text), the appropriate authority does not propose to acquire any of the land.

(c) That they only propose to acquire part of the land referred to in the blight notice for their purposes (they can only be made to acquire the whole of a person's property if there would be material detriment to any part not acquired – see Chapter 8).

(d) That in the case of land falling within paras 1, 3 or 13, but not 14, 15 or 16 of Schedule 13 (see the preceding text) the authority does not propose to acquire any of the land within the next 15 years, or such longer period as may be specified in the counter-notice.

(e) That on the date of the blight notice the person who served the blight notice is not entitled to an interest in any part of the land to which the blight notice relates.

(f) That the interest of the claimant does not qualify (the reasons must be stated in the counter-notice).

(g) That the person who served the blight notice has not made reasonable endeavours to sell, or he could only sell at a price that is depreciated by the planning blight.

An authority cannot rely on (d), that in appropriate cases they do not propose acquiring the land within the next 15 years, if ground (b) is appropriate (section 151(6)). Unfortunately, this does not apply if the authority is only proposing to take part of the land under ground (c); they may also serve a counter-notice relying on ground (d) that they do not propose acquiring within 15 years (*Parker* v *West Midlands County Council* [1979] LT).

The proper date for considering the validity of an objection that no part of the land is to be acquired is the date of the counter-notice, not the date of the blight notice: *Mancini* v *Coventry City* [1984] CA. However, if a local authority changes their policy, and abandons a scheme, after the date of the counter-notice, the Upper Tribunal (Lands Chamber) cannot consider the abandonment and is constrained to consider the facts as they were at the date of the counter-notice: see *Burn* v *North Yorkshire County Council* [1992]. This can mean that an authority is able to change their policy after receiving a blight notice and before they serve the counter-notice. As the Court of Appeal observed in the *Mancini* case, the Lands Tribunal has no discretion to deal with the hardship suffered by the owner. Further, the Upper Tribunal (now part of the Tribunals Service) can only deal with the grounds in any counter-notice, and it is irrelevant that at a later date the requirement for the land in question has ceased because of a new route for a road: see *Entwistle Pearson (Manchester) Ltd* v *Chorley Borough Council* [1993]. However, the Court of Appeal in the *Mancini* case left open whether events post the date of any counter-notice were relevant. This has been rejected by the Lands Tribunal in *Sinclair* v *Department of Transport* [1997] under ground (a).

Where an authority serves a counter-notice on grounds (b) or (d), that either they do not propose to acquire any of the claimant's land or they do not propose to acquire within the next 15 years, and they make a compulsory purchase order including that land, the order will have no effect in relation to the claimant's land in two circumstances. Either the counter-notice objection was referred to the Lands Tribunal and upheld, or no such reference was made within the time limit (section 155). This provision should deter authorities from improperly making use of objections based on grounds (b) or (d).

Where an authority has served a counter-notice objecting to a blight notice in respect of land falling within paras 1, 2, 3, 4 or 14 of Schedule 13 (see pp. 219–220), and the relevant plan, alterations, order or scheme comes into force, the authority may serve a further counter-notice specifying different grounds of objection (see section 152). The purpose of this provision is to allow the authority to serve a counter-notice containing objections based on the actual order, etc., which might have excluded the land in question. The second counter-notice must be served within two months of the appropriate plan, etc., coming into force (see section 152(2)). A second counter-notice cannot be served if an objection has been made and withdrawn, or the Upper Tribunal (Lands Chamber) has made a determination.

Where the land falls within paras 21, 22 or 24(a) and (b) of Schedule 13 (the categories of blighted land), and the powers of acquisition remain exercisable, a counter-notice cannot be served on ground (g) (that the owner has not made reasonable endeavours to sell).

18.3.5. *Counter-notice objections are referred to the tribunals service: section 153 of the 1990 Act*

A person who has received a counter-notice containing any of the grounds for objecting to a blight notice set out in part **18.2.4** above may require the objection to be considered by the Upper Tribunal (Lands Chamber) (Tribunals Service). A reference must be made within one month of the counter-notice: see the Tribunal Rules 2010. Where a blight notice has been served on two 'appropriate' authorities, and each has served a counter-notice, the Tribunal may review the validity of the objection in each notice and decide if the blight notice or notices can be upheld: *R* v *Secretary of State for the Environment, ex parte Bournemouth Borough Council* [1987]. The Tribunal Service will review any objection and will uphold it unless it considers the objection is not well founded. However, the position is different where the objection is that land is not required by the authority (see (b) and (c) on p. 224): the Tribunal is then required not to uphold the objection unless it is shown to the satisfaction of the Tribunal that the objection is well founded. It was said in *Perkins* v *West Wiltshire District Council* [1975] LT that the designation of land as public open space in the development plan was an indication that land was to be acquired. The House of Lords decided in *Essex County Council* v *Essex Congregational Church* [1963] that the authority could not raise an objection at the Lands Tribunal that was not in their counter-notice.

If the Tribunal Service does not uphold an objection, it will declare the blight notice to be valid; or, if it upholds an objection on the ground that the authority only proposes to acquire part of the land covered by the blight notice, it may declare the blight notice valid in respect of that part. Where a blight notice is declared valid in this manner, the authority concerned are deemed to have served a notice to treat and become obliged to acquire subject to compulsory purchase procedures and compensation (section 154).

18.3.5.1. Tribunal Service (Lands Tribunal) examples

Louisville Investments Ltd v Basingstoke District Council [1976]
Land belonging to L was included in a compulsory purchase order submitted to the minister for confirmation. After the service of the blight notice by L, the authority withdrew the compulsory purchase order but not before they had served a counter-notice that they had no proposals to acquire the land. In considering all the events up to the date of the decision, the Tribunal was not satisfied that the authority had abandoned their proposals to acquire L's land.

Charman v Dorset County Council [1986]
A service road was indicated on a local plan. The county council's objection to a blight notice was not upheld as they could not establish that the road was not required.

Kayworth v Highways Agency [1996]
The claimant's land was 'safeguarded' for a proposed road-widening scheme. The Agency refused to accept a blight notice on the ground that it did not intend to acquire any of the land; there had been a change of policy prior to the counter-notice, although it was not publicised. The blight notice was upheld as no action had been taken after the material date of the counter-notice.

Smith v *Kent County Council* [1995]

A blight notice was upheld in respect of a residential property. Regard was had to the broad issues of loss of privacy due to loss of trees; there would have been a serious effect on the amenity of the property under the scheme.

Head v *Eastbourne Borough Council* [2009]

Counter-notice not upheld where dwellinghouse was within a site of proposed regeneration.

18.3.6. Agricultural units

Sections 158–160 of the 1990 Act contain special provisions as to blight notices served by an owner–occupier of agricultural units. It is possible that not all of a farm is within one of the categories of planning blight specified in Schedule 13 to the 1990 Act (see part **18.3.1** above); an owner–occupier may, nonetheless, serve a blight notice in respect of the whole farm, or a part of the farm, if the 'unaffected area' (the area not subject to planning blight) is not reasonably capable of being farmed, either by itself or with other land in the same unit (if the blight notice does not cover the whole unit), or with another unit occupied by the person serving the blight notice (and of which he owns the freehold or a lease with at least three years unexpired: section 158).

The council receiving such a blight notice may, by counter-notice, object to the inclusion of the 'unaffected area' on the grounds that it can be farmed by itself or with other land occupied by the owner. The question can then be referred to the Lands Tribunal; the Tribunal can declare the original blight notice valid, in which case both affected and unaffected land has to be acquired; or uphold the council's objection and only the affected land is then acquired. The council may make the separate objection that they have no proposal to acquire all the land in the blight notice, but this can only be made if the objection that the unaffected land can be farmed viably is also made (section 159).

18.3.7. Withdrawal of blight notice

Any person who has served a blight notice may withdraw it at any time before compensation has been determined by the Lands Tribunal (or within six weeks of such determination). A blight notice cannot be withdrawn once the appropriate authority has entered and taken possession (see generally section 156 of the 1990 Act). A blight notice can be deemed to have been withdrawn where a claimant sells the property: see *Carrel* v *London Underground* [1995] and *Bennett* v *Wakefield Metropolitan District Council* [1997].

18.3.8. Deemed compulsory purchase

Where there is no counter-notice to a blight notice, or the Tribunals Service upholds its validity, the appropriate authority is deemed to have served a notice to treat on the expiration of two months from the date of service of the blight notice, or where a counter-notice is not upheld on a reference to the Tribunals Service, the date directed by the Tribunal: see section 154(3) of the 1990 Act. By section 67 of the Planning and Compensation Act 1991, such a deemed notice to treat will cease to have effect after three years in cases where the blighted land becomes blighted after 25 September 1991.

18.3.9. Right to sell the whole hereditament

Where part of an owner's property falls within one or more of the categories of blighted land, the owner can require that that his whole interest is acquired where there will otherwise be material detriment to the part retained. Section 8 of the Compulsory Purchase Act 1965 has application, and the Tribunals Service must have regard to material detriment or injury to amenity to the unblighted part in considering any counter-notice ground of objection. Thus, in *Smith* v *Kent CC* [1995], where a blight notice had been served in respect of the whole of the owner's property, the Tribunal rejected a counter-notice that the authority did not wish to acquire the whole property on the ground that there would be serious injury to amenity to the unblighted part.

19 Compensation for decisions

19.1. Introduction

Many decisions of public authorities affect the value of land. A decision to formulate plans and policies that increase the possibilities of development or the more profitable use of land may cause land value increases; other decisions, such as to revoke a planning permission or to prescribe a building line in connection with a highway, may cause land value decreases. Such a land value increase is known as betterment; it may be collected if capital gains tax becomes payable or if rateable values are increased. The described land value decrease is one form of worsenment. This chapter is concerned with, first, the legality of interfering with property rights by decisions, and second, with some statutory provisions enabling compensation to be claimed for land value decreases, or other losses, caused by decisions of public authorities.

Section 31 of the Planning and Compensation Act 1991 repealed several planning provisions for the payment of compensation. A brief description of the repealed provisions is included in this chapter for the sake of completeness.

19.2. Interference to property rights: is there a right to compensation?

In England, it is acceptable that Parliament, by virtue of its supreme authority, may authorise the Crown, or indeed any corporate or incorporate person, to take land by compulsory acquisition. Alongside this accepted right is a principle restated by Viscount Simonds in *Belfast Corpn* v *OD Cars Ltd* [1960] HL:

> It is no doubt the law that the intention to take away property without compensation is not to be imputed to the legislature unless it is expressed in unequivocal terms.

There is a full review of the cases on constitutional protection of property in the Privy Council decision in *Campbell-Rodriques* v *Attorney-General for Jamaica* [2007]. Article 17 of the Universal Declaration of Human Rights states that everyone has the right to own property and shall not be arbitrarily deprived of his property; and the European Convention on Human Rights, ratified by the United Kingdom, provides that every person is entitled to the peaceful enjoyment of his possessions and shall not be deprived of them except in the public interest and subject to the conditions provided for by law, but that the State may enforce such laws as it deems necessary to control the use of property in accordance with the general interest.

The taking of property by the acquisition of title, or by possession, is clearly within the principles just stated; and compensation is available for all such situations under enactments authorising compulsory acquisition. Nevertheless, is the right to develop land a property right which, if interfered with by decisions of planning or highway authorities, is to be compensated as of right in accordance with those principles? Clearly, every restriction imposed by authorities deprives an owner of rights that he previously enjoyed, and if those rights are restricted to prevent some use or development harmful or dangerous to public interest, the non-payment of compensation is probably justified. However, in an American case, *Pennsylvania Coal Co* v *Mahon* [1922] and cited in the *Belfast* case, Holmes J said:

> The general rule at least is, that while property may be regulated to a certain extent, if regulation goes too far it will be recognised as a taking.

These words were echoed by Viscount Simonds in the *Belfast* case itself when he had to consider whether certain restrictions on rights of development imposed without compensation amounted to a taking of property without compensation and therefore invalid under the constitution of Northern Ireland: he decided it was not a taking of property, and therefore valid.

In England, there is no written constitutional limit on Parliament's powers to authorise restrictions on property rights, and so any Act of Parliament imposing restrictions without compensation is perfectly valid in English law: but that is not to say it may not be contrary to the European Convention. See *Arrondelle* v *UK Government* [1982] for a case brought by a property owner harmed by the noise of aircraft at Gatwick Airport: the case was settled by an *ex gratia* payment made by the government before the Commission had to consider whether a breach of the Treaty had occurred.

In another case, *James* v *UK Government* [1986], the applicants argued that the Leasehold Reform Act 1967 was a breach of the Convention for two reasons: first because the compensation basis was unfair to landlords because landlords may receive less than full market value; and second because the legislation itself is indiscriminate in its application. Both grounds were dismissed: the Court deciding that there was a margin of appreciation for Parliament in balancing the burden imposed on an individual against the wider public interest in achieving the purpose of the legislation.

A person who is affected by a law that imposes such restrictions on his property as to amount to a breach of the articles or protocols of the Convention may lodge a complaint with the European Commission of Human Rights; or raise the matter in any proceedings under the Human Rights Act 1998. This matter is more fully considered in Chapter 20, where both Act and the Convention rights are discussed.

One further difficulty in English law is illustrated by *Westminster Bank Ltd* v *Minister of Housing and Local Government* [1970]. The bank had been refused planning permission for an extension because of a proposed road-widening scheme; the bank argued that it was *ultra vires* to use planning powers, because compensation was not payable for a refusal of planning permission. There was power under section 72 of the Highways Act 1959 to prescribe an improvement line to safeguard land for road widening: the frontager is entitled to compensation for injurious affection if he cannot build inside the improvement line. The bank's argument was founded on the principle that property rights should not be taken away without compensation and that where there was a choice of statutory powers, the power providing for compensation should be used. The House of Lords rejected the bank's argument: an authority may decide which of two powers it exercises, and choose the one that has no compensation burden to the ratepayers: a similar point arose in *Hoveringham Gravels Ltd* v *Secretary of State for the Environment* [1975].

19.2.1. Summary

The points discussed can now be summarised. Although Parliament may pass any laws so as to restrict property rights, and a restriction if severe enough may amount to a taking away of property rights, the courts assume that property rights are not taken away without compensation unless the legislation so provides in unequivocal terms. As the United Kingdom is a signatory to the European Convention on Human Rights, its legislation must satisfy the requirements of that convention. This matter is more fully discussed in Chapter 20.

19.3. Compensation for refusal of planning permission

Most of the provisions for this compensation were repealed by the Planning and Compensation Act 1991. (For details, see the third edition of this book.) What is left is a very limited claim.

19.3.1. Compensation for refusal of planning permission for development specified in a development order or local development order

This right to claim compensation for the refusal of planning permission, or its grant subject to conditions, seems at odds with the general philosophy of the 1990 Act that compensation is not payable for a planning refusal. However, section 108 provides that if planning permission has been granted by a development order, or a local development order, and that permission is revoked or modified, or withdrawn by the issue of a direction (such as an article 4 direction under the

General Permitted Development Order 1995), then, if a person is subsequently refused planning permission for that development or it is granted subject to conditions, he is entitled to compensation. However, the 1991 Act amended section 108 to provide power to make regulations to exclude the right to compensation; certain classes of development are so specified in the Town and Country Planning (Compensation) (No 3) (England) Regulations 2010. For recent claims, see *Slot* v *Guildford Borough Council* [1993] and *Bolton* v *North Dorset District Council* [1997].

As the claim for compensation under section 108 is in respect of the refusal of planning permission, not the withdrawal of development rights by an article 4 direction, a claim for compensation can be made after the last of a number of planning applications, even years after the withdrawal of development rights: see *Green* v *Durham City Council* [2007].

19.4. Compensation for revocation, modification and discontinuance under the Planning Acts

Under the Town and Country Planning Act 1990, section 97, a planning permission already granted may be revoked or modified by an order confirmed by the Secretary of State before the building or other operations to which it relates have been completed, or, if it permits a change of use, before that change has taken place.

Section 102 of the same Act enables a local planning authority to serve a discontinuance order, which must be confirmed by the Secretary of State; such an order can require the discontinuance of any use of land, or impose conditions as to its further use, or require the alteration or removal of any buildings or works.

These powers clearly interfere with rights already enjoyed by persons with interests in the land: the 1990 Act provides for compensation.

19.4.1. Compensation for the revocation or modification of a planning permission

Section 107 provides that a person who has an interest in land affected by an order for the revocation or modification of a planning permission may claim from the local planning authority compensation for the following matters: expenditure incurred on work rendered abortive by the order and other loss or damage directly attributable to that order.

The preparation of plans and other similar preparatory matters can be included in the claim, if such work is also abortive; but any other work done before the grant of planning permission is disregarded. The claim must be made within 12 months of the date of the decision, although the time can be extended: see Town and Country Planning General Regulations 1992.

In *Pennine Raceway Ltd* v *Kirklees Metropolitan Council* [1982], the Court of Appeal decided that a person who had a licence to use land for motorcar and motorcycle racing was 'a person interested in land' and therefore entitled to claim compensation when an article 4 direction was issued which stopped the use of an old airfield for these purposes. In *Bond* v *Dorset County Council* [2010], it was held that a tenant holding a reversionary lease at the relevant valuation date (date of confirmation of the modification order) was entitled to claim compensation.

Apart from compensation for abortive work already carried out, a claimant, with an interest in the land, is entitled to compensation for the depreciation in the value of his interest by reason of the loss of development rights; but in this connection, it is assumed that planning permission would be granted for development of any class specified in the Third Schedule: see the preceding text, and further later on. This presents a problem if the revocation order in fact concerns any such development. In *Canterbury City Council* v *Colley* [1993], the House of Lords were concerned with a claim for compensation for the revocation of planning permission to rebuild a house which had been earlier demolished. It was held that in valuing the land what is now section 107(4) of the 1990 Act required the assumption that there was planning permission to rebuild. Accordingly, it could be the case that no compensation is payable if the assumed permission is the same as that revoked. Where the claim is for diminution in value of the interest of the claimant, then section 117 of the 1990 Act applies the rules for the assessment of compensation in section 5 of the Land Compensation Act 1961.

The *Pointe Gourde* principle cannot apply to the assessment of compensation in the case of a modification order, as an order is an act not linked to or in furtherance of any scheme: see *Land & Property Ltd* v *Restormel Borough Council* [2004]. However, see **19.4.3** below.

Compensation is not restricted to a depreciation in land value; it includes any loss or damage:

> *Hobbs (Quarries) Ltd* v *Somerset County Council* [1975] LT
> A planning permission, granted in 1947, to work limestone in a quarry, was revoked. At the time, the reserves of limestone amounted to 2 million tons and it was accepted as almost certain that the claimants, but for the revocation, would have obtained a subcontract for the supply of material for the construction of the M5 motorway. Had that contract been obtained, they would have earned nearly £200,000 in profits; but without that contract, they would have earned £84,000 supplying the general market.

The depreciation in the market value of the quarry due to the revocation order was agreed to have been £72,000. The Tribunal decided that the loss of the motorway contract was not too remote; that, although £72,000 was the depreciation in the market value of the quarry, the claimant could not have purchased another quarry to earn the same profits; and, as the company did not claim for the depreciation in value of the land, they could have loss of profits instead, suitably deferred.

See **19.4.3** below. for the present position in relation to minerals and mining operations.

In *Cawoods Aggregates (South Eastern) Ltd* v *Southwark London Borough Council* [1982], the Tribunal concluded that a loss is recoverable if it is not too remote. It is arguable that the costs of obtaining the planning permission may be recoverable. In *Bond* v *Dorset County Council* [2010], where the county council had modified a planning permission for mineral extraction in consequence of certain European habitat designations, one of the claimants (H) contended that, in the absence of the designations and the resulting modification order, the lease it had held of the quarry would have been renewed, and it should be compensated accordingly. The Tribunal held that section 9 of the Land Compensation Act 1961 (disregard that land likely to be compulsorily acquired) did not apply to section 107 of the 1990 Act. It also held that the failure of H to be granted a lease renewal was not 'directly attributable' to the modification order in terms of section 107.

There is now provision for the payment of interest from the date of the revocation order, although, if a reference is made to the Tribunals Service, that body has discretion to award interest from the date of its award: see also *Knibb* v *National Coal Board* [1986]. In *Loromah Ltd* v *Haringey London Borough Council* [1978] LT, it was held that the payment of Development Land Tax on the compensation sum was not attributable directly to the revocation; accordingly, any such liability could not be added.

In assessing any part of the compensation for depreciation in land values, the rules in section 5 of the Land Compensation Act 1961 apply: see section 117 of the 1990 Act.

Where compensation in excess of £20 is paid, it is registrable as a local land charge and repayable if development is subsequently allowed.

19.4.2. Compensation in respect of a discontinuance order: section 115

A claim for compensation can be made if any person has suffered damage, in consequence of a discontinuance order, by a depreciation of the value of his interest in the land; he is also entitled to be compensated for damage attributable to 'being disturbed in his enjoyment of the land'.

Additionally, the costs of carrying out any work to comply with the order which have been reasonably incurred for that purpose can also be recovered. In *K & B Metals Ltd* v *Birmingham City Council* [1977], it was held that the acquisition of a scrap bailer between the date of the making of an order and its confirmation could be the subject of a claim.

The rules in section 5 of the 1961 Act will apply to the assessment of compensation for a depreciation in the value of the land.

19.4.3. Compensation and mining operations

The Town and Country Planning (Minerals) Act 1981 introduced amendments to the control of mining operations. There are now amended provisions dealing with the revocation or modification of a planning permission for mining and in the use of discontinuance orders. There are also prohibition orders to prohibit the resumption of winning and working minerals, and suspension orders for temporary suspension: see now sections 102 and 116 and Schedules 5, 9 and 11 to the Town and Country Planning Act 1990. Under section 96 of, and Schedule 14 to, the Environment Act 1995, there can be a periodic review of mining planning permissions. If a restriction on working rights is imposed, this is treated as if a modification order had been made under section 97 of the 1990 Act, as in the preceding text.

In circumstances where the 'minerals compensation modifications' apply, there is a reduced basis of compensation: see Town and Country Planning (Compensation for Restrictions on Mineral Working and Mineral Waste Depositing) Regulations 1997.

A number of minerals' cases are explained in the preceding text under compensation for modification orders. In *MWH Associates Ltd* v *Wrexham County Borough Council* [2012], a case concerning a claim for compensation for loss of the ability to work minerals in consequence of an order that was treated as a modification order, the claim failed. The claimant failed to show any depreciation in value of the affected land, as it had not intended to work the area, and would not have got a licence to relocate some great crested newts.

19.4.4. Measure of compensation

Section 117 of the 1990 Act applies the rules set out in section 5 of the Land Compensation Act 1961 to the assessment of compensation in relation to the depreciation in the value of an interest in land. In *Loromah Estates Ltd* v *Haringey London Borough Council* [1978], the Tribunal decided that a principle analogous to the *Pointe Gourde* principle must be implied so that the market value of land is determined ignoring the effect of the revocation or other order.

19.5. Compensation in connection with listed buildings and ancient monuments

A building of special architectural or historic interest may be listed under section 1 of the Planning (Listed Buildings and Conservation Areas) Act 1990. It then becomes an offence if any works are carried out for the demolition of a listed building or for its alteration or extension in any manner which would affect its character as such a building, without obtaining a listed building consent. Compensation is not payable upon the listing of a building. In respect of applications made for listed building consent made before 16 November 1990, compensation may be payable if listed building consent is refused. There is no compensation for applications made after this date. Compensation is payable for the revocation of listed buildings consent and in connection with building preservation notices. There are similar provisions in the Ancient Monuments and

Archaeological Areas Act 1979 in respect of scheduled monuments: these are considered in the following text.

19.5.1. Compensation for the refusal of listed building consent

In respect of an application made before 16 November 1990, compensation was payable if listed building consent is refused by the Secretary of State, or granted subject to conditions, for the alteration or extension of a listed building: see section 27 of the Planning (Listed Buildings and Conservation Areas) Act 1990. The work proposed must not constitute development within the meaning of the 1990 Act, or, if it does, the development is permitted by a development order; see *Burroughs Day* v *Bristol City Council* [1996]. There is no compensation for a refusal of consent to demolish a listed building: see *Shimizu* v *Westminster City Council* [1996].

Compensation shall equal the difference between the value of the claimant's interest had the consent been granted, and its value subject to the decision of the Secretary of State (including any alternative consent he may have granted or undertaken to grant).

Compensation is not payable for a refusal of listed building consent if an application is made after 16 November 1990: see section 31 of the Planning and Compensation Act 1991.

19.5.2. Compensation for the revocation or modification of a listed building consent

Compensation for the revocation or modification of a listed building consent is payable in the same circumstances as the revocation of a planning permission: see part **19.4.1** above and also section 28 of the Planning (Listed Buildings and Conservation Areas) Act 1990.

19.5.3. Compensation in respect of a building preservation notice

Under section 3 of the Planning (Listed Buildings and Conservation Areas) Act 1990, a local planning authority may serve a 'building preservation notice' on the owner and occupier of a building. It will do so where it is intended to seek the listing of the building and the building requires temporary protection, because it is in danger of demolition or alteration.

If the Secretary of State decides not to list the building, the building preservation notice, and the temporary protection it afforded, lapses. Any person with an interest in the building at the time of the notice may then claim compensation in respect of any loss or damage directly attributable to the effect of the notice: see section 29. The section permits the costs involved in terminating a contract to be claimed where the proposed work, say demolition, is prevented by the notice. However, such expenses cannot be claimed if the building is eventually listed. This can produce severe hardship: in *Amalgamated Investment & Property Co*

Ltd v *John Walker & Sons Ltd* [1976] CA, a building sold for development was listed one day after the plaintiff signed a contract to purchase for nearly £1.75 million. The effect of listing was to reduce the value of the building to £200,000. Amalgamated were bound by their contract and could not seek a reduction in price: the planning legislation provided no right to claim, as compensation, their loss of £1.33 million.

19.5.4. Compensation in respect of ancient monuments

Under the Ancient Monuments and Archaeological Areas Act 1979, the Secretary of State may compile a schedule of monuments; it then becomes an offence to carry out certain works to a 'scheduled monument' without consent. A scheduled monument may include a building, structure or work, or a site comprising the remains of such things. The proscribed work includes work of demolition, destruction, removal, repair, alteration or addition, or flooding or tipping operations: section 1.

If scheduled monument consent is refused, or granted subject to conditions, compensation is payable to any person who has an interest in the monument and who 'incurs expenditure or otherwise sustains any loss or damage in consequence': section 2.

Compensation is only payable for a refusal of consent for the following works:

(a) Works reasonably necessary for development for which planning permission was granted before the monument was scheduled.
(b) Works which do not constitute development (or, for which permission is granted by a development order), other than works for the demolition or destruction of the monument.
(c) Works which are reasonably necessary for the continuation of any use of the monument for any purpose for which it was in use immediately before the application for scheduled monument consent: section 7.

However, in relation to (a) above, the compensation is limited to expenditure incurred or other loss or damage sustained which is due to the fact that development for which planning permission has been granted cannot be carried out because of the scheduling of the monument: section 7(3). Compensation is not available where planning permission is granted after a monument has been scheduled and scheduled monument consent is refused.

In relation to (b) above, there is a potential problem where agricultural development enjoys permitted development rights, but which are subject to the prior notification procedure in the General Permitted Development Order 1995. It could be argued that the development would never be permitted and therefore no compensation is payable. That seems unjust if the only reason for refusing permission for the development is the ancient monument, as it defeats the purpose of the compensation provisions.

The rules in section 5 of the 1961 Act apply for the purpose of assessing compensation in respect of any loss or damage consisting of depreciation of the value of an interest in land: section 27.

Compensation is to be assessed as at the date of refusal: *see Currie's Exors* v *Secretary of State for Scotland* [1993].

19.6. Compensation in respect of tree preservation orders, control of advertisements and stop notices

19.6.1. *Tree preservation orders: section 203–204 of the 1990 Act*

Section 198 of the 1990 Act empowers a local planning authority to make a tree preservation order for the preservation of specified trees, groups of trees or woodlands. The order may prohibit the cutting down, topping, lopping of trees without the authority's consent; it may also provide for replanting when any part of a woodland is felled in the course of permitted forestry operations.

If consent is refused for any matter prohibited by a tree preservation order, then compensation in respect of loss or damage as a result of that refusal is payable if the order itself permits of this: see section 203 of the 1990 Act. The standard form of order in the Town and Country Planning (Trees) Regulations 1999 restricts compensation to claims made within 12 months and which exceed £500. The order contains a number of other restrictions on the right to compensation. The model order in force pre-1999 contains fewer restrictions. For that reason, the terms of the order should always be checked. In the case of TPOs made under the 2012 Tree Preservation Regulations, these make provision for compensation

The cases now discussed mainly relate to the orders used pre-1999, and may not have relevance to the 1999-model order. When payable, compensation may equal the depreciation in the value of the trees: *Cardigan Timber Co* v *Cardiganshire County Council* [1957] LT. It would seem that the capital value of the profits that could have been made from the land by growing Christmas trees had felling taken place cannot be recovered: *Bollans* v *Surrey County Council* [1968] LT: see also *Fletcher* v *Chelmsford Borough Council* [1991] (costs of expert awarded). Nevertheless, the Court of Appeal approved a claim based on the difference in the value of land subject to an order, and its value as agricultural land in *Bell* v *Canterbury City Council* [1988]. However, the Town and Country Planning (Tree) Regulations 1999 standard tree preservation order excludes compensation for loss of development value.

Where consent was initially refused, and later granted, to prune a tree, compensation awarded to a developer included increased borrowing costs attributable to the delay: see *Factorset Ltd* v *Selby District Council* [1995]. In addition, where a property was damaged by the roots of a tree which could not be felled, the owner was entitled to compensation for remedying that damage: see *Buckle* v *Holderness Borough Council* [1996]. In *Mooney* v *West Lindsey District Council* [2000], compensation of £5,739 was awarded for stress, inconvenience and loss of earnings in seeking consent to fell two trees which was only granted on appeal. In *Wright* v *Horsham District Council* [2011], compensation was awarded for the cost of underpinning works and for distress and inconvenience.

The cost of repairing damage caused by a tree prior to the date of a refusal of consent to fell is not recoverable as compensation, but if the effect of refusal was to increase the cost of repairs to past damage, the extra cost is compensatable: see *Duncan* v *Epping Forest District Council* [2004].

A licence is required for the felling of timber, with some exceptions, under the Forestry Act 1967: there is provision in that Act for compensation if such a licence is refused.

In respect of a direction made under the provisions of a tree preservation order that replanting should take place, compensation for the loss or damage incurred complying with the direction is payable. However, such compensation is only payable where the Forestry Commission consider the replanting would not be in the interests of commercial forestry, and the local planning authority requires such replanting in the interest of amenity.

19.6.2. Restriction on advertisements

The Town and Country Planning (Control of Advertisement) Regulations 2007 are made under section 220 of the 1990 Act. These regulations provide for restricting and regulating the display of advertisements in the interests of amenity or public safety.

No compensation is payable for a refusal of consent to display an advertisement. However, section 223 of the 1990 Act does provide for compensation to any person who carries out work to comply with the regulations in the following circumstances: he is entitled to any expenses reasonably incurred in complying with an order for removing an advertisement displayed on 1 August 1948; or for discontinuing the use of a site used for an advertisement displayed on that date. A claimant must submit his claim to the local planning authority within six months of completing the necessary work.

19.6.3. Stop notices

Section 184 of the 1990 Act empowers a local planning authority, who have issued an enforcement notice requiring a breach of planning control to be remedied, to serve a further notice called a 'stop notice'. The stop notice may prevent the carrying out of the activity alleged to constitute the planning breach before the expiration of the period of compliance allowed in the enforcement notice.

If the enforcement notice takes effect, there is no compensation for loss or expense incurred in complying with such a notice, or a stop notice that may have been served with it. However, section 186 of the 1990 Act does provide compensation for loss or damage directly attributable to a stop notice in the following cases: the enforcement notice is quashed on appeal; the enforcement notice is varied so that the matter covered by the stop notice is no longer part of the enforcement notice; and, either the enforcement notice or the stop notice is withdrawn. In such cases, any sum payable in respect of a breach of contract made necessary by compliance with a stop notice is considered to be loss or damage attributable to the prohibition in a stop notice.

A claim can only be made by a person with an interest in the affected land; a mere licence to use the land is not sufficient: see *International Ferry Traders Ltd* v *Adur District Council* [2004]. A claim must be made within 12 months of the decision of the Secretary of State and an informal letter may constitute such a claim: *Texas Home Care Ltd* v *Lewes District Council* [1986]. In *J Sample (Warkworth) Ltd* v *Alnwick District Council* [1984] LT, a stop notice and enforcement notice were served on the same day. The enforcement notice was quashed on appeal 22 months later without any order as to costs. The parties disputed the meaning of directly attributable in connection with the compensation claim, but the Tribunal did not accept that the words were qualified by the concept of reasonable foreseeability. Building delays were directly attributable to the stop notice and the Tribunal awarded as compensation loss of interest on purchase money and extra labour costs. The costs of the enforcement notice appeal were disallowed.

Before serving a stop notice, and therefore incurring the unknown liability for compensation, the local planning authority ought to be reasonably certain that a breach of planning control has taken place. It was held by the Court of Appeal in *Malvern Hills District Council* v *Secretary of State* [1982] that the marking out of an estate road with pegs was a 'specified operation' which kept alive a planning permission under what is now section 56 of the 1990 Act. The cost of delay caused by the stop notice was considered by the Lands Tribunal in *Robert Barnes & Co Ltd* v *Malvern Hills District Council* [1985]; interest on unpaid compensation between the date the stop notice ceases to have effect and the date of the tribunal's award could not then be awarded. It now can be.

Where a loss arises substantially due to the impecuniosity of the claimant, it may be rejected under the *Liesbosch* principle as being too remote: see *Graysmark* v *South Hams District Council* [1989].

Compensation is excluded in two circumstances. First, where loss arises due to a stop notice which prohibited anything which was a breach of planning control during the period the notice was in force: see section 186(5)(a) of the 1990 Act. Second, for losses which could have been excluded had the claimant provided information properly requested by the local planning authority: see section 186(5)(b) of the 1990 Act.

20 Judicial supervision

20.1. Introduction

Most readers will have observed by now that the development control process has received a good deal of judicial interest. The purpose of this chapter is to describe, in outline only, how cases are brought before the courts, and the principal legal grounds upon which they are founded. This chapter is written primarily for the non-lawyer reader. The Town and Country Planning Act 1990 contains procedures for the legal review of certain decisions and actions of local planning authorities and the Secretary of State. Apart from these statutory procedures, there is also a right to judicial review under the Civil Procedure Rules 1998. In many circumstances, this latter right, which was in origin a common law right, is precluded by the statute, but where it is not precluded, it has provided a useful addition to the rights of aggrieved parties, as judicial review may be available in circumstances outside the scope of the statutory right to review.

Although the scope of the statutory and judicial review rights may differ, the principles upon which the courts have exercised these rights are similar. The Administrative Court of the Queen's Bench Division of the High Court is concerned with the judicial supervision of authorities or ministers to ensure that they act within the law; that their administrative and discretionary decisions are decisions that can properly be made under the law; and that they follow the statutory procedures as well as the rules of natural justice to ensure openness, fairness and impartiality where these are considered appropriate. The court has developed a number of principles to test the legality of administrative decisions; they possess a mystery and meaning understandable only to the initiated! The courts often refer to the Secretary of State, or inspector or a local planning authority as the decision-maker.

The expectations of the court are that administrative decisions are made in accordance with these principles, and that the decision-makers think appropriately. It is doubtful whether this really happens. Administrative decisions are made against a wide background of facts, policies and political views; the principles of this branch of the law are not always understood by decision-makers.

In any event, these principles are often satisfied without in any way altering a decision. For example, the decision-maker must have regard to all the relevant considerations when making his decision (see below). If the court later identifies a matter that was omitted from the decision-maker's consideration, then provided that matter is then considered, the new decision could well be the same as the original decision. It seems that often the right incantation in a written decision is all that is required.

These remarks may suggest a certain impatience with this aspect of the subject. That is not entirely the case as blatant unfairness, procedural irregularity or illegality should be remedied, although few cases seem to involve this. Anyway, the large number of cases are extremely useful in determining the scope and meaning of the statutes and the legal boundaries of decisions.

20.2. The statutory rights of challenge to the courts

The right to challenge the validity of certain schemes, orders, decisions and directions is precluded except in the circumstances outlined in the following text (section 284 of the 1990 Act). Section 284 continues to have effect for a transitional period in relation to any structure or local plans, although this will now be very limited with the abolition of structure plans and the replacement of local plans with local development documents under the Planning and Compulsory Purchase Act 2004. Under section 284, the orders include revocation, discontinuance and tree preservation orders; and the decisions are those of the Secretary of State, either where some planning matter is referred to him, or following an appeal. These orders, plans, and various decisions may be challenged under section 288 of the 1990 on two grounds: that the order, decision, etc., is not within the powers of the Act, or that some procedural requirement has not been complied with. In the latter case, there must be substantial prejudice.

In relation to strategies, plans and documents that constitute the development plan, there is a separate right of challenge under section 113 of the Planning and Compulsory Purchase Act 2004 where a document is not within the appropriate power or a procedural requirement has not been complied with.

These statutory rights to challenge are subject to a six-week time limit. The purpose of restricting the challenge of these matters is to ensure that action can be taken only within the limited period of six weeks so as to ensure that after that period there is no continuing legal uncertainty. Even if the grounds for challenging the legal validity of some scheme, order, decision or direction are not apparent until after that period, the expiration of the six weeks will prevent any appeal to the High Court (*R* v *Secretary of State for the Environment, ex parte Ostler* [1976]), even where the claimant did not know about an application for permission near his home, or of the planning appeal decision on it: see *R* v *Secretary for the Environment, ex p Kent* [1990]. In the case of a challenge to a development plan document, the six-week period starts on the date of the adoption of the plan document, not the next day: see *Hinde* v *Rugby Borough Council* [2011].

The right to question the validity of an enforcement notice or a listed building enforcement notice is separately considered at **20.4** below.

20.2.1. Challenging development plans

Where section 285(1) of the 1990 Act still has transitional application, a structure or local plan, or any alteration, repeal or replacement of a plan, or any order made under the 1990 Act in respect of highways, may be challenged in the High Court within six weeks (section 287). The party making the application must be a 'person aggrieved' and must either show that the development plan or order is outside the powers of the Act or that he has been 'substantially prejudiced' by a failure to comply with the statutory procedures.

Section 113 of the Planning and Compulsory Purchase Act 2004 provides that the validity of the following strategies, plans and documents may be challenged by 'a person aggrieved' within six weeks on the grounds that the document is not within the appropriate power, or some procedural requirement has not been complied with: revision of the regional spatial strategy, Wales spatial strategy (or revision), a development plan document (or revision), a local plan (or revision), the Mayor of London's spatial development strategy, and an alteration or replacement of the spatial development strategy. (Few spatial strategies remain.) The foregoing 'relevant documents' may not otherwise be challenged. The High Court may quash or remit the document with directions. Part of a core document restricting house building affected by aircraft noise was quashed in *Taylor Wimpey UK Ltd* v *Crawley BC* [2008] as it was inconsistent with a structure plan policy and government guidance.

In *Edwin H Bradley & Son* v *Secretary of State for the Environment* [1982], the court refused to quash a plan as the Secretary of State had sufficient information, and had given adequate reasons for excluding the applicant's land for development. A plan that makes express reference to guidelines that are not part of the plan may be quashed as not complying with the law to formulate 'proposals for the development and other use of land in the area': *Great Portland Estates* v *Westminster City Council* [1984]. The local plan in *Fourth Investments Ltd* v *Bury Borough Council* [1985] was quashed as the inspector could not have reached a conclusion on the adequacy of land for housing on the evidence before him.

20.2.2. Challenging other orders and decisions

In respect of certain orders, and in respect of certain actions of the Secretary of State, a 'person aggrieved' by the order or the action may question the validity of that matter in the High Court within six weeks on the ground that the order or action is outside the powers of the Act or there has been a non-compliance with some statutory procedure (section 288). The orders include those of revocation or modification, discontinuance or tree preservation, the actions include decisions of the Secretary of State on planning applications referred to him, decisions on a planning appeal, a decision to grant planning permission following an enforcement notice appeal, and other decisions in relation to such matters as tree preservation orders, advertisement control, listed building consents and purchase notices (section 284(2)). The party making the application to the High Court must show that the order or action is outside the powers of the Act or that he has been 'substantially prejudiced' by a failure to comply with statutory procedures.

It was held in *Co-operative Retail Services* v *Secretary of State for the Environment* [1980] that a refusal to adjourn a local inquiry was not a decision within the scope of the statutory right of appeal.

20.2.3. Person aggrieved

The statutory right of appeal to the High Court can only be brought by a 'person aggrieved'. Although a restricted meaning was once attached to this, so that adjoining landowners to a development, who had no legal right to have their representations considered by the Secretary of State, were held not to be entitled to appeal (*Buxton* v *Ministry of Housing and Local Government* [1961]). A wider view of this is now taken by the courts. In *Turner* v *Secretary of State for the Environment* [1973], Ackner J considered that persons who had been permitted to appear at the local inquiry could be 'persons aggrieved'. A wider view was taken by the Court of Appeal in *Times Investment Ltd* v *Secretary of State for the Environment* [1991].

20.2.4. The statutory grounds of appeal

The two principal grounds, of outside the statutory powers and procedural irregularity, overlap to some extent. They also embrace a number of other grounds; the broad principles upon which the courts are prepared to quash decisions and orders are really the same as where law judicial review procedure applies, and these are considered in the following text. Although there is a specific statutory ground of a failure to comply with the relevant legislative requirements, and where substantial prejudice must be shown. This is also considered at **20.4.2** below.

20.2.5. The court's discretion and basis of decision

The court has a discretion under section 288 to quash a decision. In *Bolton Metropolitan Borough Council* v *Secretary of State for the Environment* [1990], the Court of Appeal gave guidance as to how that discretion should be exercised where the decision-maker has failed to take into account a material consideration. The decision-maker ought to take into account a matter which might have caused him to reach a different conclusion to that which he would reach if he did not take it into account. If the matter is trivial or of small importance in relation to the particular decision, it would follow that if it were taken into account, there would be a real possibility that it would make no difference to the decision; accordingly, it is not a matter which the decision-maker ought to take into account. If the court concludes that the material consideration was fundamental to the decision, that it is clear that there is a real possibility that the consideration of the matter would have made a difference to the decision, the decision will not have been validly made. However, if the court is uncertain whether the matter would have had this effect or was of such importance in the decision-making process, then the court does not have the material necessary to conclude that the decision was invalid.

In *Seddon Properties Ltd* v *Secretary of State for the Environment* [1978], Forbes J set out a number of principles in relation to the validity of the decision of a decision-maker. He said that the decision-maker must not act perversely; he must not take into account irrelevant material or fail to take into account that which is relevant; the decision-maker must follow statutory procedures, and in the case of the Secretary of State, these would mean the Inquiry Procedure Rules, and give proper and adequate reasons for any decision; and the decision-maker must not depart from the principles of natural justice.

20.2.6. *Decision of the court*

The power of the court is usually limited to quashing the decision and remitting it to the Secretary of State. This would mean that the Secretary of State would have to remake his decision, avoiding the defects found by the court. The court is not concerned with the merits of planning or policy; these are for the decision-maker (minister or local planning authority): see *Tesco Stores* v *Secretary of State for the Environment* [1995]. However, where a development plan document is remitted in whole or part under section 113 of the 2004 Act, the Court may give directions in relation to the remission.

20.3. Judicial review

The jurisdiction of the Administrative Court of the Queen's Bench Division of the High Court goes back to the close association the original King's Bench had with the Council of the King, and the power the court possessed over officials and other courts. The principles of judicial supervision, often referred to today as 'administrative law', have been substantially developed in the course of the last few decades. Judicial review is only available against public bodies and in the exercise of powers involving a public element. It cannot be exercised to enforce private law rights.

The present procedure is known as a claim for judicial review (rule 54 of the Civil Procedure Rules 1998). The claim means a claim to review the lawfulness of an enactment or a decisions, action, or failure to act in relation to the exercise of a public function (rule 54.1). Judicial review may be given on the following grounds: error of law, procedural impropriety, irrationality, and abuse of power. The Court will apply the Human Rights Act 1998, and the rights contained in the European Convention on Human Rights. Judicial review must be used where the claimant is seeking one of the prerogative orders, of which the most usual are now known as: a mandatory order (to compel performance), a prohibiting order and a quashing order (rule 54.2). Judicial review may be used where claimant is seeking a declaration or an injunction, and a claim may include a claim for damages, restitution or the recovery of a sum (rule 54.3).

A claimant must first obtain the permission of the court to make a claim for judicial review (rule 54.4). If the court accepts the matter as being arguable, and gives permission, there will be a full hearing of the substantive claim. Application for permission should be made as soon as possible and in any event, within three months of the decision to be reviewed. The decision in *R* v *Ceredigion County Council, ex parte McKeown* [1998], that an application must normally be made within six weeks – the same as the time limit for statutory appeals, was overruled in *R* v *Hammersmith & Fulham LBC, ex p Burkett* [2002], where the House of Lords also decided that where a planning permission was the subject of judicial review, the three-month period runs from the date of the grant of permission, and not the resolution to grant.

The statutory rights of challenge under the provisions of the 1990 or 2004 Acts, considered at **20.2** above, are not available for, and do not preclude, a claim for judicial review in the following examples: where planning permission has been granted by a local planning authority or the Secretary of State, a claim made by a third party seeking to challenge the lawfulness of the grant; or where development plan documents are in preparation, but no final document has come into existence, a claim that the material considerations are being ignored in the plan making procedure.

As commented in the preceding text, a claim for judicial review may include a claim for a declaration or an injunction. A declaration was used in *Pyx Granite Co* v *Ministry of Housing and Local Government* [1958] to determine the validity of certain planning conditions; and in *Heron Corporation* v *Manchester City Council* [1978] to declare invalid the refusal of the authority to consider an application for the approval of reserved matters. In *Irlam Brick Co* v *Warrington Borough Council* [1982], Woolf J considered that judicial review under what is now rule 54 was preferable to a declaration where the issues concerned the conduct of a public body, and persons beyond the immediate parties were affected. The interaction between public law and private law remedies is considered separately at **20.5** below.

A person must have standing, that is, a sufficient interest in the matter, to claim judicial review. In *Covent Garden Community Association Ltd* v *Greater London Council* [1981], a local residents' association were held to have standing, as representations of such bodies would be taken into account in planning appeals and inquiries; and in *Steeples* v *Derbyshire County Council* [1981], a person was said to have standing if he were a ratepayer or an adjoining owner as in *R* v *Castle Point Urban District Council, ex parte Brooks* [1985]. However, a more restrictive view was taken in *R* v *Secretary of State, ex parte Rose Theatre Trust* [1990] where it was held that a group concerned with the preservation of the theatre did not have standing, and in *R* v *North Somerset DC, ex p Garnett* [1997], it was held that the applicants, who lived three miles from the site in issue, and who had not objected to the planning application, did not have standing to claim judicial review of the decision to grant planning permission.

A final and important feature of the claim for judicial review is that, even where the requisite grounds are made out, the Court retains a discretion as to whether to grant any relief.

20.4. Grounds of challenge

The grounds upon which the High Court will accept a challenge to an administrative or policy decision, whether under the statutory appeals, or under the common law judicial review procedure, are broadly the same: the decision is *ultra vires*, or there has been some procedural irregularity. The role of the court is to ensure that the process of decision-making is fair; it does not re-consider the planning merits and does not replace a decision on merits with one of its own.

Three cases form the bedrock upon which all later developments in this branch of the law have evolved. In *Cooper* v *Wandsworth Board of Works* [1863], the court decided that the Board of Works could not pull down a house, of which no notice of construction had been received, without first giving the owner an opportunity of explaining his case. In *Associated Provincial Picture Houses* v *Wednesbury Corporation* [1948], the judgment of Lord Green MR in the Court of Appeal, concerning an authority exercising some discretion, can be summarised in the following two principles:

 (i) The authority must direct itself properly in law, and should not refuse to take into account matters that are relevant, nor take into account matters which ought not to be taken into account.
(ii) Even if the first principle has been followed, the decision must not be so unreasonable that no reasonable authority would have made it.

Those principles, the *Wednesbury* principles, were enlarged on by Lord Denning MR in *Ashbridge Investments* v *Minister of Housing and Local Government* [1965] CA. He added that the court could interfere with a decision if the decision-maker had acted on no evidence, or come to a conclusion to which on the evidence one could not reasonably come.

The grounds for judicial review of administrative decisions, and it is not always clear whether a planning decision is merely a policy matter or an administrative one, have been much developed in recent years. In *Council of Civil Service Unions* v *Minister for the Civil Service* [1984], HL Lord Diplock said the grounds were:

(a) *Illegality*: '... I mean that the decision-maker must understand correctly the law that regulates his decision-making power and must give effect to it ...'
(b) *Irrationality*: '... I mean what can now be referred to as "*Wednesbury* unreasonableness". It applies to a decision which is so outrageous in its defiance of logic or of accepted moral standards that no sensible person who had applied his mind to the question to be decided could have arrived at it.'

(c) *Procedural impropriety*: This covers a 'failure to observe basic rules of natural justice or failure to act with procedural fairness towards the person who will be affected' as well as 'failure … to observe procedural rules laid down in the legislative instrument by which … jurisdiction is conferred, even where such failure does not involve any denial of natural justice …'. This would also cover abuse of power.

Today, the grounds can be considered under the following headings, and examples are given. However, in the broader judicial review context, the idea of failing to give effect to a legitimate expectation is part of the notion of an error of law. Many of the cases in these examples arose under the statutory grounds of challenge considered at **20.5** above, rather than by way of judicial review.

20.4.1. Error of law

This arises where decision is contrary to law. That could include a misinterpretation of planning policy, as in *South Somerset DC* v *Secretary of State for the Environment* [1993]. It would also be an error of law to fail to give effect to a legitimate expectation, although the Courts recognise that a legitimate expectation cannot always be given effect to where there are greater public interests that would preclude this; compensation could remedy any failure to give effect: see *Rowland* v *Environment Agency* [2005].

20.4.2. Procedural impropriety

Where a decision is being appealed under the statutory right of appeal in the planning Acts, rather than by common law judicial review, the statute requires proof of 'substantial prejudice' if a procedural irregularity is being alleged. Although a serious procedural irregularity probably amounts to *ultra vires*. This separate ground is presumably to provide for the more trivial procedural irregularity. In *Performance Cars Ltd* v *Secretary of State for the Environment* [1978], the applicant failed to show that he had been substantially prejudiced by a refusal to give an adjournment of a length necessary to consider new evidence, the evidence made no difference to his case. Although, confusingly, that refusal was in breach of the rules of natural justice, and the decision was as a consequence *ultra vires* under the first ground.

In *Miller* v *Weymouth and Melcombe Regis Corporation* [1974], Kerr J considered that if there was no substantial prejudice to the applicant as the result of a clerical error, the court had a discretion not to quash the decision. The fact that the applicant did not make some early objection to a procedural irregularity may be accepted as showing he was not substantially prejudiced (*Davies* v *Secretary of State for Wales* [1977]).

In *Reading Borough Council* v *Secretary of State for the Environment* [1986], it was said that even if the inquiry procedure rules have been satisfied, these did not exhaust the rules of natural justice. For, Parliament could not have intended that statutory powers should be used in breach of the rules of natural justice. Second, it was said there is no such thing as a technical breach of the rules of natural justice, and therefore substantial prejudice must be shown as in *George* v *Secretary of State for the Environment* [1979].

The issue of bias, or the appearance of bias, on the part of an inspector, could also be treated as an error of law. Thus, in *R (Ortona)* v *Secretary of State for the Communities and Local Government* [2009], a decision was quashed, because the inspector had previously been employed in the local planning authority whose transport plans were in issue, and he had worked on those policies. Where the landowner fails to appear at a pre-arranged site visit, it was an error of law by the inspector to speak to the planning officer, as justice must be seen to be done: see *R(Tait)* v *Secretary of State for Communities and Local Government* [2012].

For further examples, see:

Fairmount Investments Ltd v *Secretary of State for the Environment* [1976]
   Basing a decision on evidence obtained by an inspector on his site visit may involve a breach of the rules of natural justice, as a party is entitled to hear all the evidence against him. A breach of the rules may amount to *ultra vires*.

The *Covent Garden* case (see p. 246 above)
   A residents' association sought certiorari (now known as a quashing order) to quash a decision of the local planning authority to grant planning permission for office use contrary to a previous policy that favoured residential use of the property in Covent Garden. However, the order was not granted, as there had not been a breach of statutory requirements: a modification to the local plan did not have to be made.

The *Steeples* case (see p. 246 above)
   A member of the Friends of Shipley Park sought, and was granted, a declaration that planning permission granted by the county council for the development of a 'Wonder Park' in partnership with private developers was invalid, as certain statutory procedures had been properly complied with.

R v *North Yorkshire County Council, ex parte Brown HL* [1999]
   A planning permission granted for mining operations by the respondent local planning authority was quashed because of a failure to undertake an environmental assessment in compliance with European Union directives.

Berkeley v *Secretary of State for the Environment, Transport and the Regions* [2001]
   Where an application for a project required compliance with certain regulations, and those regulations were not complied with, the resulting planning permission would have to be quashed.

20.4.3. Irrationality

R v Hillingdon London Borough Council, ex parte Royco Homes [1974]
    Certiorari (now known as a quashing order) was granted to quash a planning decision containing conditions which were so unreasonable no reasonable authority could have imposed them.

20.4.4. Abuse of power

Co-operative Retail Services Ltd v *Taff-Ely Borough Council* [1979]
    A declaration and injunction was granted with the effect that a planning permission purported to have been granted for a supermarket site was invalid: the Borough Council had no power to grant, or to ratify the decision of their clerk purporting to grant the planning permission that was issued.

20.4.5. Human rights

The Courts are required by the Human Rights Act 1998 to apply the application of the European Convention on Human Rights. The rights most likely to be engaged are Article 1 of the First Protocol (protection of private possessions), Article 6(1) (right to a fair trial) and Article 8 (right of respect for the home). In relation to planning, there are two stages that must be considered in relation to Human Rights: (1) Is any Article of the Convention engaged by some aspect of the domestic law or some decision, order or plan done under it? (2) If engaged, is the decision, order or plan compliant with the Article engaged, having regard to the proportionality of balancing the policies and decisions of the state against that of the individual? Where a state has a system of planning controls and zoning, this will affect the private property (private possessions) of an individual. Article 1 of the First Protocol will be engaged, but in broad terms, such a system is proportionate and is not inconsistent with Human Rights, although there could be a breach if compensatory provisions are discriminatory against certain persons: see *Pine Valley Developments Ltd* v *Ireland* [1993].

    In England and Wales, many planning decisions are appealable to the Secretary of State. He appoints the inspector, and the decision of the inspector is the decision of the minister. As the Secretary of State is concerned with policy and law making, as well as the inquiry/dispute decision-making process, it has been argued that the inquiry process is in breach of Article 6, as the individual would not get a fair hearing. In three conjoined appeals, *R* v *Secretary of State for the Environment, Transport and the Regions, ex p Alconbury Developments Ltd*, *R* v *Secretary of State for the Environment, Transport and the Regions, ex p Holdings & Barnes plc, R* v *Secretary of State for the Environment, Transport and the Regions, ex p Legal & General Society Ltd* [2001], the House of Lords decided that planning processes engaged Article 6(1), and individuals are entitled to the protection of that Article, but that there was sufficient judicial control by the courts to ensure that determinations were carried out independently and impartially in relation to the planning appeal and inquiry process.

However, the *Alconbury* case will not necessarily apply to decisions made by local planning authorities that affect individuals. Here, an authority may have to consider hearings, to ensure that an individual has an opportunity to put his case for the purposes of Article 6(1), in appropriate cases that particularly affect the individual: see *R (Adlard)* v *Secretary of State for Transport, Local Government and the Regions* [2002].

In relation to Article 8, and respect for a person's home, the decision of the European Court of Human Rights in *Buckley* v *United Kingdom* [1996], shows that an individual's preference as to a place of residence is subject to the overriding general interest in planning terms.

20.4.6. Duty to give reasons

The duty to give reasons for a planning decision is imposed on the Secretary of State, or the inspector, by the inquiries procedure rules, and in practice, reasons are always given. Woolf J in *Grenfell-Baines* v *Secretary of State for the Environment* [1985], inferring from the general duty under the *Wednesbury* principles, said it was implicit on a decision-maker to give reasons. Where the reasons are obscure, there will be a breach of this duty and the decision may be quashed: *Givaudan* v *Minister of Housing and Local Government* [1967]. Guidance on the giving of reasons is found in *Bolton Metropolitan Borough Council* v *Secretary of State for the Environment* [1995] and *MJT Securities* v *Secretary of State for the Environment* [1997].

The increasing tendency in recent years to scan decisions of inspectors with the most precise scrutiny, seeking out minor issues which an inspector might not have dealt with fully, has been criticised by the Court of Appeal. In *Weitz* v *Secretary of State for the Environment* [1987], Russell LJ said that one must read a decision letter as a whole, and then ask whether the inspector did not or may not really have come to grips with the conflicts presented to him.

An inspector's decision letter will be regarded as disclosing inadequate reasons if it does explain the basis of the inspector's decision to the local planning authority: *Stephenson* v *Secretary of State for the Environment* [1985]. Finally, in *South Buckinghamshire CC* v *Porter (No 2)* [2004], the House of Lords addressed the issue of the duty to give reasons. The reasons need only deal with the main points in issue, and not every material consideration. Lord Brown said that a reasons challenge will only succeed where the claimant can show that he genuinely been substantially prejudiced.

20.5. Private law remedies

Examples of the use of the private law remedy of an injunction were given on an earlier page. In some circumstances, there will be a cause of action in private law, such as an allegation of breach of contract, or a claim for damages for an alleged negligent statement. In *Avon County Council* v *Millard* [1986], it was said that where a landowner was in breach of a planning agreement, there was no obligation on the planning authority to exhaust the public law procedures before seeking an injunction to restrain the breach of what was a contract.

Confusion can arise where a party is seeking both to challenge an alleged unlawful decision, and to recover damages for any harm suffered. In the House of Lords case of *O'Reilly* v *Mackman* [1983], it was said it would be an abuse of court procedures '... to permit a person seeking to establish that a decision of a public authority infringed rights to which he was entitled to protection under public law by way of an ordinary action, and by this means evade the provision of Order 53 for the protection of such authorities'. The provisions of the order (now rule 54) filter out claims with little merit, or which do not substantially concern matters of wider public interest.

This check on private law actions did not prevent the House of Lords from allowing a claim based on negligence to be brought in *Davy* v *Spelthorne Borough Council* [1983]. The claimant alleged that he had foregone appealing an enforcement notice because of negligent advice he said he received from the planning department. His action for damages could be brought as he was not seeking to challenge the validity of the enforcement notice, and his private law claim did not really concern issues of wider public interest. It was also said that the privative clause in the planning Act (section 243(1)), that excluded a challenge of an enforcement notice otherwise than on the grounds provided by that section, did not prevent an action in private law.

In recent years, actions by individuals against local planning authorities, based on private law remedies, have tended not to be successful. In *Tidman* v *Reading Borough Council* [1994], a landowner sought advice from a planning authority and later contended that the authority ought to have advised him to make an application as to whether or not planning permission was required for a particular proposal. His proceedings alleging negligence were dismissed. In *Lambert* v *West Devon Borough Council* [1997], the plaintiff was given the 'go ahead' to carry out certain works by a planning officer. The local planning authority later served an enforcement notice and the plaintiff brought proceedings alleging negligent advice by the planning officer. It was held that the local planning authority was liable in negligence; they owed a duty of care in respect of the advice of their officer. However, in *Lam* v *Brennan* [1997], the plaintiffs owned a restaurant and the local planning authority granted planning permission for a change of use of adjoining premises to industrial processes. The industrial processes affected the use of the restaurant. The Court of Appeal held that the local planning authority were not in breach of any statutory or common law duties, nor were they in breach of a duty to prevent the creation of a nuisance.

20.6. Challenging the validity of enforcement notices

The right to appeal to the Secretary of State against an enforcement notice on the grounds in section 174 of the 1990 Act has already been dealt with in Chapter 15.

Apart from this right of appeal, the validity of an enforcement notice cannot be challenged on any of the grounds in section 174 (see the preceding text) in any other legal proceedings: see section 285 of the 1990 Act. This would include any prosecution proceedings for failing to comply with an enforcement notice in the magistrate's court. Where a person is prosecuted for breach of a stop notice, the defendant may challenge the validity of the stop notice in the magistrate's court: *R* v *Jenner* [1983]. Section 285 does preclude an action for tort claiming damages in nuisance: *Davy* v *Spelthorne Borough Council* [1983].

In *Square Meals Frozen Foods* v *Dunstable Corporation* [1974], a declaration was first sought to determine whether an intended user was lawful; an enforcement notice was then issued to stop the continuance of that user. It was held that as the application for a declaration was a proceeding in anticipation of enforcement, section 285 preventing that legal action from continuing.

However, it was said in *Miller-Mead* v *Minister of Housing and Local Government* [1963] that if an enforcement notice is *ultra vires* (see the preceding text) or is incurably bad in its content, it is a nullity and void. It may then be challenged by the ordinary common law right of judicial review (*Stringer* v *Minister of Housing and Local Government* [1970]), either for a quashing order to have the notice quashed, or for a declaration that the notice is a nullity, as confirmed in *Rhymney Valley District Council* v *Secretary of State for Wales* [1985].

The alternative to judicial review of an enforcement notice that is a nullity is to await a prosecution for non-compliance, and then raise the validity of the notice as a defence to the prosecution.

If an enforcement notice is first appealed to the Secretary of State on one or more of the grounds in section 174, and the Secretary of State gives a decision in the proceedings, the appellant, the local planning authority or any other person served with a copy of an enforcement notice may appeal to the High Court on a point of law, or have a case stated for the opinion of the court (section 289). In accordance with the rules of the court, permission is first required. The application must be made within 21 days.

For a fuller description of the right to appeal and enforcement, see Chapter 15.

20.7. The local ombudsman

The Local Government Act 1974 introduced the local ombudsman. Strictly, the appropriate body is called the Commission for Local Administration. There are local commissioners who have powers to investigate written complaints made by or on behalf of members of the public. The complaint can only concern injustice in consequence of maladministration.

Maladministration includes taking too long to take action without good reason; not following the authority's own rules or the law; breaking their promises; giving wrong information; or not making a decision in the correct way. The local ombudsman cannot investigate the substance of a planning decision.

Accordingly, if the decision of a local planning authority is simply disagreed with, that is not a matter the local ombudsman can be concerned with.

In practice, the local ombudsman will expect the local authority concerned to be first given an opportunity to explain the background to the complaint of maladministration before the ombudsman investigates and makes a determination.

The local ombudsman will prepare a report setting out his recommendations. The report might include a recommendation that compensation should be paid to the complainant in appropriate circumstances. In many cases, the compensation is recommended as the depreciation in value of the complainant's property caused by the maladministration in making the planning decision in question.

Part VI
Services and highways

21 Water, sewers, gas and electricity

21.1. Introduction

The development of land will require connection to and the supply of services. Sometimes the inadequacies of a particular service, such as drainage, might be a reason for refusing planning permission. In any event, a developer may be constructing sewers and other services within the scheme and will want an assurance that the facilities will be adopted by the relevant undertaker and appropriate connections made with the mains.

This chapter provides a brief outline of the principal powers and rights in relation to the supply of services. All these services are now provided by the privatised industries.

21.2. Supply of water

The Water Industry Act 1991 sets out the powers and duties of the water undertakers. Under section 37 of the 1991 Act, it is the duty of every water undertaker to develop and maintain an efficient and economical water supply within its area and to ensure that all such arrangements have been made for providing supplies of water to premises in that area and for making such supplies available to persons who demand them, and for maintaining, improving and extending the water undertaker's water mains and other pipes. The Act makes a distinction between a 'water main' and a 'service pipe'. A water main means any pipe used by the water undertaker for the purpose of making a general supply of water available to customers, as distinct from the supply of a particular customer (see section 219). A service pipe is that part of a pipe that provides the connection between the water main and any particular premises and is subject to the water pressure from that main (see section 219). The obligations of a water undertaker are generally in relation to the provision of a water main and the appropriate water pressure in that main. The provision of a service pipe is a matter entirely for the developer or, in due course, the customer or landowner.

21.2.1. Duty to make connections

Under section 45 of the 1991 Act, the water undertaker has a duty to connect a service pipe to one of the undertaker's water mains in certain circumstances. The duty arises in respect of domestic properties that either exist or are proposed to be erected, and the owner or occupier serves a notice requiring a connection. Where a service pipe has to be provided in a street between the main and the premises to be connected, the duty of the water undertaker includes an obligation to provide the service pipe (see section 46).

21.2.2. Connection for non-domestic purposes

The owner or occupier of premises may serve a connection notice requiring connection to a water main to supply water for non-domestic purposes. However, there is no obligation on the water undertaker to provide a supply if that would require the undertaker to incur unreasonable expenditure (see section 55).

21.2.3. Requisitioning a water main

Where an owner of land proposes to erect buildings to be used for domestic purposes, a water main requisition notice may be served on the water undertaker. The undertaker is then required to provide a water main (see section 41). Sections 42–43 of the 1991 Act make provision for the financial requirements that must be satisfied by the person serving the water main requisition notice. In principle, the requisitioner must reimburse the water undertaker for a deficit that has regard to the costs of complying with the requisition and the increase in income that will result from the supply of water.

21.2.4. Completion date and route for requisitioned water main

The water undertaker must provide the requisitioned water main within a period of three months of the water main requisition notice. The route for the requisition main will be a matter for agreement or as may be determined by arbitration (see section 44).

21.2.5. The right to lay a water main

The water undertaker has powers in sections 158–159 of the 1991 Act to lay pipes in streets or through private land. In the latter case, the water undertaker is obliged to give notice to the owner and occupier of any land through which a pipe is laid. Those persons are then entitled to compensation under Schedule 12 to the Act. The compensation principles are explained in the author's book, *Compulsory Purchase and Compensation* (Tenth Edition, *Estates Gazette*, 2013).

21.2.6. Disputes

The powers and duties set out in the preceding text make a number of provisions for the reference of disputes to the director-general of water services, the relevant watchdog, under section 30A of the 1991 Act.

21.3. Sewers

The Water Industry Act 1991 deals with the rights and obligations of the sewerage undertakers. Under section 94, there is a duty on every sewerage undertaker to provide, improve and extend a system of public sewers and to make provision for the emptying of those sewers.

Section 219 of the 1991 Act defines a sewer as all sewers and drains used for the drainage of buildings and yards appurtenant to buildings. A pipe may be a sewer whether it drains foul water from buildings or drains storm or surface waters. A public sewer is a sewer that is vested in a sewerage undertaker in its capacity as such. Public and private sewer and lateral drains, are the responsibility of the sewerage undertaker who are regarded as the owners of, at least, the public sewer. Drains serving one building or curtilage are not the responsibility of the sewerage undertaker.

21.3.1. Right to requisition a public sewer

An owner or occupier of premises, or the local authority, may serve a requisition notice requiring the provision of a public sewer to drain premises used, or to be used, for domestic purposes (see section 98). The sewerage undertaker then has six months within which to comply with the requisition notice (section 101). As with the requisition of a water main, there are financial requirements that the person serving the requisition notice must satisfy. Any disputes relating to the specifications or financial contributions are referable to a single arbitrator under section 99(6). There is also a right to requisition lateral drains, being drains serving one curtilage, but outside its boundaries: see sections 95–98 of the Water Act 2003.

The right of a sewerage undertaker to lay pipes in streets or through private land is the same as the right of a water undertaker: see at **21.2.5** above – The right to lay a water main – in the preceding text.

21.3.2. Sewer agreements

Under section 104 of the 1991 Act, a sewerage undertaker may agree with any person constructing or proposing to construct any sewer that if the sewer or works are constructed in accordance with the terms of the agreement, the undertaker will, upon the completion of the work, declare that the sewer or works be vested in the undertaker. The purpose of such an agreement is to enable a developer to have the assurance that if he lays out a development scheme, and constructs sewers within that scheme, the sewers will be adopted and will be vested in the sewerage undertaker provided he constructs them to an appropriate and agreed specification, thereafter. The sewers will then be public sewers for which the sewerage undertaker will have responsibility.

21.3.3. Adoption of sewers

Although most developers will normally proceed by way of a sewer agreement under section 104 of the 1991 Act, section 102 makes provision for the adoption of sewers by a sewerage undertaker. Thus, an owner of land containing a sewer may request that the sewerage undertaker make a declaration under section 102 that the sewer shall become vested in the undertaker. If that declaration is made, then the sewer becomes a public sewer and the responsibility of the undertaker.

21.4. Supply of gas

The Gas Act 1986 made provision for the privatisation of British Gas. The principal supplier was originally British Gas, although there are now many other competitors. As amended, the Act provides for public gas transporters, who provide the pipes and principal installations, and public gas suppliers, who purchase gas on the wholesale market and sell it to the consumers; both types require licences: see sections 7 and 7A of the 1986 Act.

The general obligations of public gas suppliers and transporters are set out in sections 1–11 of the Gas Act 1986. Under section 10, the owner or occupier of any premises requiring a supply of gas may serve a notice on the public gas transporter. The public gas transporter must comply with the request, if the premises are within 23 metres of the transporter's gas main. The public gas transporter is entitled to recover the costs of laying a pipe within the property to be supplied, and the cost of any distribution pipe. A deposit as security for these costs may be demanded. A public gas transporter cannot be required to provide a connection to its gas main where a consumption in excess of 75,000 therms per year is expected. There are a number of other exceptions set out in section 10.

If a request has been made for a supply of gas likely to exceed 2,500 therms annually, and that supply cannot be given without a new main, or the enlarging of an existing main, or the constructing or enlarging of other works, the public gas transporter is not required to comply with the request unless certain matters can be satisfied. The first of these is that the person requesting the supply enters into a written contract to take and pay for minimum quantities of gas for a minimum period having regard to the expenditure necessary on the part of the public gas supplier and to make any payment towards that expenditure as may be reasonably required. A deposit as security may be demanded.

The developer will have to secure any necessary easements to lay pipes through any land not in his ownership. The public gas transporter has power to acquire land or rights over land for the purpose of discharging their functions (Schedule 3 to the Gas Act 1986). It may be possible to persuade the public gas transporter to exercise this power if legal rights cannot be obtained over land not owned by a person requiring a gas supply, the public gas transporter is under a duty to provide the supply as described in the preceding text.

21.5. Supply of electricity

The Electricity Act 1989 makes provision for the generation, transmission, distribution and supply of electricity by operators with the appropriate licences. A public electricity distributor distributes electricity, through lines and cables, within a specified area, and a public electricity supplier distributes electricity to any consumers, and is not necessarily restricted to a specified area. A public electricity distributor requires a licence and has a duty to develop and maintain an efficient, co-ordinated and economical system of electricity supply (see section 9).

Under sections 16 and 16A of the 1989 Act, the owner or occupier of any premises may request a supply of electricity and the provision of electric lines or electrical plant from a public electricity distributor. A notice requesting a supply of electricity must give certain details including the description of the premises in respect of which the supply is required, the day on which the supply is required to commence, the maximum power which may be required at any time and the minimum period for which the supply is required to be given: section 16A. Under section 16A, the public electricity distributor may serve a counter-notice setting out any counter-proposals and stating the prices and any payments that are required to be made.

Section 17 of the 1989 Act sets out the exceptions from the duty to make a supply of electricity:

(a) The public electricity distributor is prevented from providing a supply by circumstances not within its control.
(b) Providing a supply would be a breach of regulations that relate to securing a safe supply of electricity on a regular and efficient basis to the public.
(c) It is not reasonable in all the circumstances for the public electricity supplier to be required to provide a supply.

These exceptions, to the duty to make an electricity supply, are rather vague. A person requisitioning an electricity supply might therefore be in some difficulties in disputing any contention of a supplier based on one or other of these grounds.

Under section 19 of the 1989 Act, the public electricity supplier has powers to recover any expenditure incurred in providing a supply in accordance with the terms upon which that supply was to be made.

The Act does not make any provision for the route of an electricity supply or the time limit within which the supply must be made. It would seem that the route will be a matter of agreement following the service of the requisition notice and any counter-notice under section 16A. Any dispute arising out of the requisition of an electricity supply may be referred to the Gas and Electricity Markets Authority under section 23. On such a reference, the dispute may be determined by order made either by the Authority or, if it thinks fit, by an arbitrator.

Public electricity distributors have powers under Schedules 3 and 4 to the Act to take land or rights over land compulsorily or to acquire a statutory way leave.

22 Highways

22.1. Introduction

A highway is a strip of land over which the public at large have the right to pass and repass. There are various categories of public rights over highways; on carriageways, there is the right to pass and repass on foot, horse or by vehicle; on bridleways, foot, horse or by bicycle; on footpaths, on foot only. There are, in addition, special types of highways such as motorways, walkways and cycle paths.

The land over which the highway passes may be privately owned by the adjoining frontagers. Subject to any evidence to the contrary, ownership extends to the centre of the highway. This may mean that if a person on a highway conducts an activity that is not simply the exercise of his legal right to pass or repass, and matters incidental and ancillary thereto, he may be trespassing. Causing a disturbance to the landowner's grouse shooting (*Harrison* v *Duke of Rutland* [1893]), or picketing outside premises which exceeds the immunity normally applicable in connection with a trade dispute (*Hubbard* v *Pitt* [1975]), would amount to a trespass. However, conducting a peaceful non-obstructive assembly on the public highway is part of the public's right over a highway, provided it is reasonable: see *DPP* v *Jones* [1999].

When land is privately owned, the highway over it is only maintainable at public expense if it was created before 1836, or, if created after 1835, has been adopted by the local highway authority. The interest of the highway authority in a highway maintainable at public expense is in 'the top two spits of the road': *Tithe Redemption Commission* v *Runcorn Urban District Council* [1954]. It is a fee simply determinable when the highway ceases to exist, and extends between the adjoining fences, hedges or ditches.

Where a highway passes over private land, and is not maintainable at public expense, it may be privately repairable, or there may be no private liability for repairs. A private liability to repair may arise by prescription, either *ratione tenurae*, or *ratione clausurae*. Prescription under *ratione tenurae* occurs where it has been a condition of landownership to keep a highway in repair: *Esher and Dittons Urban District Council* v *Marks* [1902]. An owner is liable to repair *ratione clausurae* where the owner fences open land over which the public had rights of passage to avoid a highway when it becomes impassable, and confines the public to the highway. If a highway passes over private land, and is not maintainable at public expense, the responsibility of the owner is limited to any liability he may owe under the law of public nuisance – he must not interfere with or obstruct the rights of the public.

The highway authorities are the Department of Transport for motorways and trunk roads (through the Highways Agency), and the county and unitary councils for all other highways. In practice, many of the functions of a highway authority are delegated to district and borough councils. The highway authorities will own the land over which a highway passes only if the land was acquired and used for the purpose of constructing and creating a new or improved highway.

Highways are created at common law or by statute. At common law, a highway may be created if the landowner 'dedicates' the right for the benefit of the public at large, and the public 'accept' this right by use of the highway. Dedication and acceptance are the two essentials. Dedication must be intended, expressly or impliedly. If the public uses a strip of land without consent or hindrance for a sufficient period of time (20 years), dedication will be implied on the part of the landowner. However, public rights of way cannot be acquired for vehicular traffic: see section 66 of the Natural Environment and Rural Communities Act 2006.

Highways may also be created by statute. The Highways Act 1980 contains the principal powers enabling highway authorities to construct new highways or to make private roads into highways.

In connection with the creation of new highways, the Secretary of State is required to give consideration to an Environmental Assessment Statement before giving consent to its construction: see the Highways (Assessment of Environmental Effects) Regulations 1999, which amend and insert into the Highways Act 1980 a new Part VA directed to such matters.

Although there is a common law right of access onto a highway by adjoining landowners, the making of such an access may involve development and require planning permission, or may be restricted by the provisions of the Highways Act 1980 on the grounds of danger to traffic.

Certain features of highway law are dealt with in this chapter. The development of land may involve obstruction or interference to a highway. The developer may be constructing new streets and roads and may expect that these become highways maintainable at public expense. Although the common law rule 'once a highway, always a highway' means that a highway can never cease to exist, there are several statutory provisions concerning the stopping up and diversion of highways to facilitate development or other purposes.

22.2. Obstructions and interferences to highways

22.2.1. Public nuisance

If in the course of the development of land, the developer obstructs the highway or carries out his activities so as to endanger the use of the highway, that will be a public nuisance. If a vehicle is parked on a highway for long enough, that removes that part of the highway from public use and will be an obstruction: *Dymond* v *Pearce* [1972]. The erection of a fence, hoarding or scaffolding would similarly constitute an obstruction: *Harper* v *Haden & Sons* [1933]. Unguarded excavations on land immediately adjoining the highway may endanger users of the highway (*Jacobs* v *London County Council* [1950]); the same point would also apply to excavations in the highway (*Haley* v *London Electricity Board* [1965]).

If an obstruction or danger constitutes a public nuisance, either the Attorney-General or the local authority may seek an injunction to remove the cause of the nuisance. It is also a criminal offence (section 137 Highways Act 1980). This also applies to crops other than grass (see section 137A of the Highways Act 1980).

A private individual may bring a civil action in respect of a public nuisance if he can show special and particular damage over and above that suffered by the public at large. This will include loss of business due to the obstruction: see the *Harper* case. The protest camp on the public highway outside St Paul's Cathedral in 2011 gave rise to proceedings in public nuisance and orders and injunctions were granted under section 130 and 143 of the Highways Act 1980. It was held in *City of London* v *Samede* [2012] that the granting of such remedies did not interfere with the protestors' rights under the European Convention of Human Rights.

However, any owner of land is entitled to erect hoarding or scaffolding in the highway to effect repairs or improvements to his premises: a temporary obstruction would not constitute a public nuisance, provided it was reasonable in extent and duration (*Herring* v *Metropolitan Board of Works* [1865] – six months obstruction of a business by hoardings). If the highway authority has given a licence to erect scaffolding or hoarding in the highway (see the following text), this will not make legal an obstruction that is otherwise illegal, because it is a public nuisance by reason of its extent or duration.

22.2.2. Lawful and unlawful interference: Highways Act 1980

This Act makes provision for the protection of public rights on the highway by penalties. It also provides for precautions to be taken in connection with building works including the issue of licences for some activities.

22.2.2.1. Builders skips: sections 139–140A

These may only be deposited on the highway with the permission of the highway authority. Permission may be granted subject to conditions as to siting, dimensions, visible colouring, care of contents, lighting and removal; a charge may be imposed. A skip must be properly lit during darkness and must be marked with the owner's name and telephone number or an address. The Builders' Skip (Markings) Regulations 1984 (SI 1984/1933) requires the marking of skips with the owner's details. Where a skip is placed in the road with the permission of the highway authority, an offence is committed if the skip is not lit, and the fact that lights are vandalised will only be a defence if all reasonable steps are taken to prevent the loss of the lights: *PGM Building Ltd* v *Kensington Chelsea Royal Borough Council* [1982].

22.2.2.2. Building operations affecting public safety: section 168

It is an offence to carry out building operations which give rise to the risk of serious injury to persons on a highway or would have given rise to such a risk if the local authority had not used their powers to remove the risk.

22.2.2.3. Control of scaffolding on the highway: section 169

It is an offence to erect scaffolding or other structure in a highway without permission of the highway authority. A licence to erect the scaffolding must be granted by the authority unless the scaffolding will cause an unreasonable obstruction or there is some other way of arranging the scaffolding than that proposed by the applicant. The person to whom the licence is granted must ensure that the scaffolding or other structure is properly lit during darkness; he must also comply with any directions given concerning the erection of traffic signs.

22.2.2.4. Materials on the highway: sections 170–171

It is an offence to damage a highway by mixing mortar which will stick to the surface or enter drains or sewers. Consent may be given for the temporary deposit of building materials on a highway. Consent may be subject to conditions, and the materials must be properly lit and guarded.

22.2.2.5. Hoardings: sections 172–173

Unless this requirement is dispensed with by the highway authority, hoardings must be erected between the highway and land where the erection or demolition of a building is to take place. If necessary, a covered platform and handrail must be provided for pedestrians. The hoarding must be properly secured and lit.

22.2.2.6. Payment of charges: section 171A

Under section 171A, provision is made for regulations to charge for hoardings, structures or materials permitted on the highways.

22.2.2.7. Bridges and cellars: sections 176 and 179

If a landowner owns the subsoil under a highway, or owns land on both sides of a highway as well as the subsoil, he may build a cellar under the highway, or a bridge over it if he obtains the consent of the highway authority. That consent or licence may contain conditions. There is an appeal to the Crown Court against a refusal of a licence for a bridge, and to the magistrate's court for a refusal of consent for a cellar.

22.2.2.8. *Buildings over the highway: section 177*

A licence is required to construct a building over a highway. The licence may contain conditions relating to headway, maintenance, lighting and use of the building. There is an appeal to the Crown Court against a refusal of consent. No appeal can be brought if the land on which the highway lies is owned by the highway authority.

22.2.3. *Trees and highways*

Two matters arise in connection with trees. What is the liability for trees growing within the boundaries of a public highway and what is the liability for trees growing on private land adjoining the public highway.

In *Hurst* v *Hampshire County Council* [1997], it was held that where a public highway is adopted as maintainable at public expense the highway authority became liable for all trees within the boundaries of the highway, whether planted pre- or post-adoption. Accordingly, a highway authority is liable to users of the highway for dangerous trees and to adjoining landowners for damage which is reasonably foreseeable.

In the case of trees on private land adjoining a public highway, the owner of that land will be liable in nuisance for any damage caused by a tree provided the owner has, or ought to have, knowledge of the problem and the danger: see *Leakey* v *The National Trust* [1980].

22.3. Adoption of highways

Although any highway created before 1836 is maintainable at public expense, highways created since that date are only so maintainable if they have been constructed by the highway authority, or adopted by the highway authority under the statutory provisions now found in the Highways Act 1980 (section 36). A developer laying out new roads and streets will usually comply with the appropriate procedures to ensure that these are adopted.

22.3.1. *Adoption following notice: section 37*

Where a person proposes to dedicate a way over land as a highway, he may give three months notice of his intention to the local highway authority. The authority may complain to the magistrate's court that the way will not be of sufficient utility to justify its being maintained at public expense. If the way has been properly made up in a satisfactory manner (to a standard specified by the authority), the highway authority will issue a certificate to that effect. Then, provided the way is kept in repair for 12 months from the date of that certificate, and the magistrate's court do not make an order that the way will have insufficient utility, the highway will thereafter become maintainable at public expense.

22.3.2. Adoption by agreement: section 38

This procedure is usually more satisfactory to a developer proposing to construct new roads. An agreement is made between the local highway authority and a person who either proposes to dedicate an existing private carriage or occupation road, or who proposes to construct new roads and then dedicate them as highways. The agreement will contain the specifications of the new highway, the date on which it will become a highway maintainable at public expense; it may contain a bond and surety to pay the costs of making up the road if the developer defaults; it may also contain a financial contribution from the developer to the authority.

A walkway is a special type of highway over, through or under buildings or structures. Walkways are created by an agreement, under section 35 of the 1980 Act, which will make particular provision for maintenance, lighting, support, making of payments, and the termination of the public rights of way in specified circumstances. By-laws may be made for the proper regulation and use of walkways.

22.3.3. Agreements and highways

There are provisions in section 278 of the Highways Act 1980, as amended by the New Roads and Street Works Act 1991, that provide for agreements between developers and the highways authority for contributions to highway works necessary for the development of land. In *R v Warwickshire County Council, ex parte Powergen plc* [1997], it was held that a highway authority may have no option but to co-operate in implementing a planning permission by entering into a section 278 agreement.

22.3.4. Adoption under the advance payments code

This is dealt with below at *22.4.2*.

22.3.5. The duty to maintain: section 41

The 1980 Act imposes a duty on highway authorities to maintain a highway maintainable at public expense. The authority may be liable in damages to any user of the highway if they fail to discharge this duty. There are some special defences (section 58).

22.4. Making up private streets

The purpose of the two codes briefly described here is to ensure that existing private streets can be made up to a satisfactory standard, and that certain proposed private streets will be constructed to a proper standard, and then adopted (Part XI of the Highways Act 1980).

22.4.1. Private street works code

If the highway authority is satisfied that a private street is not properly sewered, levelled, paved, metalled, flagged, channelled, made good and lighted, they may resolve that the necessary street works are carried out by the authority at the expense of the frontagers (section 205(1)).

A specification, estimates and provisional apportionment of the costs are prepared, and if approved by the authority, notice to frontagers and publicity will be given. Any frontager may object to the magistrate's court (sections 208–209) on the grounds that the road is not a private street, there has been some procedural irregularity, the proposed works are insufficient or unreasonable, the estimated cost is excessive, that certain premises ought to be excluded from the apportionment, or that the provisional apportionment is incorrect (section 208).

When the street works have been executed, the authority will make a final apportionment of the costs. Any frontager may then appeal within one month to the magistrate's court if there has been an unreasonable departure from the specification, the actual expenses have without sufficient reason exceeded the estimated expenses by more than 15 per cent or the apportionment has not been properly made (section 211).

The sums due from the frontagers under the final apportionment are a charge on their frontage land until paid off. There is power to allow the sums to be paid, by way of annual instalments within a period not exceeding 30 years (section 212).

22.4.2. Advance payments code

This code ensures that a sum of money is deposited with the local authority as security for the proper construction of private streets; it also ensures that the frontagers can require that the private street be adopted as a highway maintainable at public expense.

Where it is proposed to erect a building or buildings for which plans must be deposited with the local authority in accordance with building regulations and the buildings will front a private street (existing or proposed), no building work may commence until a sum of money is deposited with the local authority (section 219(1)). The sum of money represents security for the costs of the necessary street works.

There are exemptions to this requirement of an advance payment, of which the following are the more important (section 219(4)):

(a) The proposed building will be in the curtilage of, or appurtenant to, an existing building.
(b) The advance payments code does not apply to the particular parish or community council area (it is extended to any such area by the county council, if not already an area to which the code applies).
(c) The developer has made an adoption agreement with the local highway under section 38 (see part **22.3.2** above).
(d) The local highway authority is satisfied that more than three-quarters of the aggregate length of the frontages, are or will consist of the frontages of industrial premises, or the proposed building will be industrial and has a frontage of at least 100 yards.

Unless any of the exemptions apply, the highway authority will serve a notice, within six weeks of the passing of the plans by the district council under the building regulations, specifying the sum to be deposited (section 220).

If the developer makes up the private street, the sum deposited will be refunded to that person, subject to any representation of the frontagers (section 221). If the highway authority makes up the street, they may recover any excess cost above the advance payment deposited and held by them, and must refund any excess if the advance payment exceeds the cost of the works (section 222). There is provision for a return of the advance payment if the proposed building work is abandoned (section 223). Interest is paid on the sum deposited at rates stipulated from time to time (section 225).

When the street works have been carried out, the highway authority may, by notice, declare that the highway shall be maintainable at public expense. Any objections are considered by the magistrate's court (section 228).

A majority of the frontagers, or those owning more than half the aggregate length of the street, may by notice require the highway authority to carry out the necessary street works and declare the highway to be maintainable at public expense (section 229). This right may only be exercised if more than half of the length of the street on both sides has been built up and the length of the street concerned is not less than 100 yards long.

22.5. Stopping up and diversion of highways

A highway may only be diverted or stopped up under statutory powers. Powers are available under the Highways Act 1980, the Town and Country Planning Act 1990, the Acquisition of Land Act 1981, and the Housing Act 1985.

22.5.1. *Stopping up and diversion of highways: Highways Act 1980*

The highway authority may apply to the magistrates court for an order to stop up or divert any highway where the highway other than a trunk road or motorway is unnecessary or it can be diverted so as to make it nearer or more commodious to the public (section 116(1)). In *Gravesham Borough Council* v *Wilson* [1983], it was said that the word 'commodious' had the dictionary meaning of being larger or roomier. A developer may request the highway authority to make such application (because only the authority may apply), and the authority is entitled, as a condition of granting such a request, to be reimbursed their costs (section 117).

The consent of the district council and parish or community council is required before an application can be made by the highway authority in respect of an unclassified highway (section 116(3)). At the court hearing, the highway authority, frontagers to the highway, users of the highway and, where the highway is classified, the minister and the district council, parish or community council, are entitled to be heard (section 116 (7)). In *Ramblers' Association* v *Kent County Council* [1990], it was said that it would be difficult to conclude that a highway was unnecessary where there is evidence of use of the way unless the public will have an equally convenient route; recreational use should be taken into account. Circular Rights of Way 1/09 advises local authorities not to use section 116 procedures, but alternative procedures, in relation to the stopping up of footpaths and bridleways.

22.5.2. *Stopping up footpaths and bridleways: Highways Act 1980*

A public path extinguishment order may be made by the county council or a district council where a path or way is not needed for public use. If opposed, the order must be confirmed by the Secretary of State (section 118). An order should not usually be made on the ground that some obstruction made the highway impossible to use (*R* v *Secretary of State for the Environment, ex parte Stewart* [1980]). There are provisions for compensating parties with interests harmed by the closure. In *Hertfordshire CC* v *Secretary of State for the Environment, Food and Rural Affairs* [2006], the existence of a public path creation agreement could be taken into account in the decision-making process unless the agreement was dependent on the confirmation of the extinguishment order. In *Young* v *Secretary of State for the Environment, Food and Rural Affairs* [2002], three tests need to be satisfied in deciding whether to make an order: First, the consideration of expediency, second, whether or not the path is substantially less convenient to the public in consequence of the diversion, and third, whether it is expedient having regard to the factors set out in section 119(6).

22.5.3. *Diversion of footpaths and bridleways: Highways Act 1980*

An owner, lessee or occupier of land crossed by a footpath or bridleway may request the county or district council to make a public path diversion order on the ground that it is expedient to do so in the interests of the owner, lessee or of the public. The order must be confirmed by the Secretary of State if opposed (section 119). The applicant may have to agree to reimburse the council making the order any costs they incur, or any compensation they may have to pay to those with interests harmed by the diversion.

A footpath within 300 yards of Chequers was diverted under this power on the ground that there was an assassination risk to the prime minister (*Roberton* v *Secretary of State for the Environment* [1976]).

22.5.4. Stopping up and diversion of highways: Town and Country Planning Act 1990

Where planning permission has been granted for development, the Secretary of State may make an order to stop up or divert a highway to enable the development to be carried out (section 247(1)). The important point here is that the order is necessary to enable the development to proceed in accordance with planning permission.

The order cannot be made retrospectively where the development has already been carried out (*Ashby* v *Secretary of State for the Environment* [1980]). The order may make provision for a new highway as replacement or improvement of an existing one; and for a financial contribution towards the costs of this from any authority or developer.

A highway may be converted into a footpath or bridleway (e.g., a pedestrian policy) by an order made by the local planning authority and confirmed by the Secretary of State (section 249). The order may allow service vehicles at specified times, and compensation may be payable to frontagers affected. In *Saleem* v *Bradford Metropolitan Borough Council* [1984], the Lands Tribunal did not accept that the order closing the highway to vehicular traffic was the direct cause of the claimant's loss.

22.5.5. Footpaths and bridleways affected by development: Town and Country Planning Act 1990

The local planning authority may make an order to stop up or divert a footpath or bridleway to enable development to be carried out in accordance with planning permission granted under the Act (section 257). The order may make provision for alternative routes and for a contribution towards the costs of work in connection therewith. An order may be made retrospectively to the development (see the *Ashby* case in the preceding text).

22.5.6. Extinguishment of right of way

There is, finally, a general power in section 258 of the 1990 Act to extinguish any rights of way over land held by a local planning authority for planning purposes. There is also a power in section 32 of the Acquisition of Land Act 1981 to extinguish footpaths and bridleways in connection with the compulsory purchase of land, and there is power for a local authority to extinguish rights of way under Part IX of the Housing Act 1985 where land is acquired for clearance purposes; the approval of the Secretary of State is required.

22.5.7. *Stopping up or interfering with private rights*

Where land has been acquired or appropriated by the local planning authority, for planning purposes, and development is carried out in accordance with planning permission, private rights, such as easements, may lawfully be interfered with: section 237 of the Town and Country Planning Act 1990. It matters not whether the development is carried out by the local planning authority or a person deriving title under them. There is a right to compensation (section 237(4)), but the claimant must prove that the interference with his private right will cause depreciation in value to his interest (*Ward* v *Wychavon District Council* [1986]).

Part VII

The betterment problem

23 The betterment problem

The Uthwatt Report of 1942 (Final Report of the Expert Committee on Compensation and Betterment, Cmnd 6386) recognised that a system of comprehensive planning and development control would have a significant effect on land values. The report identified the principle of betterment 'which is that, if property is enhanced in value by reason of some action, whether positive or negative, by the state or a local authority, the state or the local authority should be entitled to recover the whole or some part of the increase in value'. Betterment has been recovered in the past by a direct charge on the owner of the property bettered (London County Council (Tower Bridge Southern Approach) Act 1895); by a set-off in respect of the betterment against compensation payable for the acquisition of or injurious affection of other land of the same owner (Land Compensation Act 1961); and by recoupment – that is, the purchase of the land to be bettered at a price that ignores this effect, and then its resale at a price reflecting betterment (see the 1895 Act above).

The report considered that the principle of betterment commanded general acceptance although there were great difficulties in its practical application. It recommended that local authorities should have wide general powers to buy land compulsorily to recoup betterment. Set-off and a direct charge were not favoured; the former because of its casual effect and historically unproductive nature, the latter because of the then known difficulties in collecting betterment by direct charge under the Town and Country Planning Act 1932. The report also recommended that betterment should be collected by a scheme for a periodic levy on increases in annual site values.

These recommendations of the Uthwatt Committee were not wholly accepted by the Labour Government after World War II; the Town and Country Planning Act 1947, in nationalising development rights, and providing a claim for compensation for the loss of those rights (see p. 231), provided that betterment would be collected by a development charge. This charge was 100 per cent of the betterment: the difference between the value of the land with planning permission, and its existing use value.

It was said at the time that the development charge discouraged development, because it denied any financial incentive to release land for that purpose, and, where development took place, it was an additional cost borne by the developer and passed on to the ultimate user of developed land. Strictly speaking, the charge only represented the betterment derived from planning permission, and the higher cost to the ultimate user was not the charge, but the land value consequence of a comprehensive system of planning and development control: if only specified land is to be released for development, that land will increase in value to a greater extent than if there was no planning and development control.

The Conservative Government abolished the development charge from November 1952 by the Town and Country Planning Act 1953, but made no further or alternative provision for the collection of betterment. With another change of government, a further attempt to collect betterment was made with the Land Commission Act 1967: a 40 per cent betterment levy was charged on net development value. The Act also set up the Land Commission, a body that was given wide powers of compulsory purchase of land suitable for development; this was a recognition of one of the recommendations of the Uthwatt Report. The Land Commission Act 1967 was repealed with yet another change of government in 1970 – too early for any real conclusions to be made about the effectiveness of the policy it represented.

The Community Land Act 1975 represented another attempt by the next Labour Government to deal with the betterment problem and 'to enable the community to control the development of land in accordance with its needs and priorities' (White Paper – *Land* – Cmnd 57307 – 1974). The Act contained not only wide powers for local planning authorities to acquire land needed for development, but also enabled the Secretary of State, by order, to impose a duty on authorities to acquire certain development land. Betterment was to be collected for an interim period as a development land tax under the Development Land Tax Act 1976; eventually, acquisition of development land under the 1975 Act would be on a compensation basis that excluded development value. This solution to the problem was fairly close to the original recommendation of the Uthwatt Committee that betterment should be collected by recoupment, i.e., purchase of development and its resale.

The Conservative Government of the 1980s repealed the Community Land Act 1975, but did not repeal the measure concerned with the collection of betterment, the Development Land Tax Act 1976, until 1985. The Labour Governments in power between 1997 and 2010 did introduce any legislation relating to the collection of betterment.

It is possible that a consensus has been reached over the problem of betterment; certainly, the main political parties are no longer in serious disagreement over a solution. However, at local government level, the parties have probably never been further apart, with considerable evidence that the whole planning process is no longer regarded as an objective exercise in the administration of a regulatory device upon development where the initiative lies with the landowner and developer. Planning and development control is regarded by many local councillors as a political process which should be initiated and controlled through local government.

In the case of large development proposals, developers often propose, or are asked to provide, planning gain. Planning gain is an additional benefit unrelated to the proposed development. It is a crude form of betterment recapture. In its favour, it is often voluntarily provided by a developer; however, it is also frequently offered to 'purchase' a planning permission that might not otherwise be granted. There was guidance on planning gain set out in Circular 05/05. This Circular was cancelled and replace by the National Planning Policy Framework 2012. The tests for deciding whether an offer of planning gain satisfies the tests of material considerations and weight were considered by the House of Lords in *Tesco Stores Ltd* v *Secretary of State for the Environment, Transport and the Regions* [1995].

In order to bring some consistency of approach, and to recover the costs of consequential infrastructure provision, the Planning Act 2008 contains provisions for the payment of a Community Infrastructure Levy upon the grant and/or implementation of certain planning permissions: this is more fully described in Chapter 9.

The United Kingdom enjoys the most comprehensive and restrictive planning and development control legislation among the free democratic countries of the world. Undoubtedly, it has brought great benefits, but we have paid a high price for it.

Appendix A

1987 No 764

Town and Country Planning (Use Classes) Order 1987

1 Citation and commencement

This Order may be cited as the Town and Country Planning (Use Classes) Order 1987 and shall come into force on 1st June 1987.

2 Interpretation

In this Order, unless the context otherwise requires:—

"care" means personal care for people in need of such care by reason of old age, disablement, past or present dependence on alcohol or drugs or past or present mental disorder, and in class C2 also includes the personal care of children and medical care and treatment;

"day centre" means premises which are visited during the day for social or recreational purposes or for the purposes of rehabilitation or occupational training, at which care is also provided;

. . .

"industrial process" means a process for or incidental to any of the following purposes:—

    (a)   the making of any article or part of any article (including a ship or vessel, or a film, video or sound recording);

    (b)   the altering, repairing, maintaining, ornamenting, finishing, cleaning, washing, packing, canning, adapting for sale, breaking up or demolition of any article; or

    (c)   the getting, dressing or treatment of minerals;

in the course of any trade or business other than agriculture, and other than a use carried out in or adjacent to a mine or quarry;

"Schedule" means the Schedule to this Order;

"site" means the whole area of land within a single unit of occupation.

3 Use Classes

(1)   Subject to the provisions of this Order, where a building or other land is used for a purpose of any class specified in the Schedule, the use of that building or that other land for any other purpose of the same class shall not be taken to involve development of the land.

(2)   References in paragraph (1) to a building include references to land occupied with the building and used for the same purposes.

(3) A use which is included in and ordinarily incidental to any use in a class specified in the Schedule is not excluded from the use to which it is incidental merely because it is specified in the Schedule as a separate use.

(4) Where land on a single site or on adjacent sites used as parts of a single undertaking is used for purposes consisting of or including purposes falling [within classes B1 and B2] in the Schedule, those classes may be treated as a single class in considering the use of that land for the purposes of this Order, so long as the area used for a purpose falling within class B2 is not substantially increased as a result.

(5) ...

(6) No class specified in the Schedule includes use—

    (a)   as a theatre,

    (b)   as an amusement arcade or centre, or a funfair,

    (c)   as a launderette,

    (d)   for the sale of fuel for motor vehicles,

    (e)   for the sale or display for sale of motor vehicles,

    (f)   for a taxi business or business for the hire of motor vehicles,

    (g)   as a scrapyard, or a yard for the storage or distribution of minerals or the breaking of motor vehicles,

    (h)   for any work registrable under the Alkali, etc. Works Regulation Act 1906,

    (i)   as a hostel,

    (j)   as a waste disposal installation for the incineration, chemical treatment (as defined in Annex I to Directive 2008/98/EC under heading D9) or landfill of hazardous waste as defined (in relation to England) in regulation 6 of the Hazardous Waste (England and Wales) Regulations 2005 or (in relation to Wales) in regulation 6 of the Hazardous Waste (Wales) Regulations 2005,

    (k)   as a retail warehouse club being a retail club where goods are sold, or displayed for sale, only to persons who are members of that club,

    (l)   as a night-club,

    (m)   as a casino.

(7) Where a building or other land is situated in Wales, class B8 (storage or distribution) does not include use of that building or land for the storage of, or as a distribution centre for, radioactive material or radioactive waste.

(8) *For the purpose of paragraph (7), "radioactive material" and "radioactive waste" have the meanings assigned to those terms in the Radioactive Substances Act 1993.*

(9) For the purpose of paragraph (7), "radioactive material" and "radioactive waste" have the same meaning as in the Environmental Permitting (England and Wales) Regulations 2010.

4 Change of use of part of building or land

In the case of a building for a purpose within class C3 (dwellinghouses) in the Schedule, the use as a separate dwellinghouse of any part of the building or of any land occupied with and used for the same purposes as the building is not, by virtue of this Order, to be taken as not amounting to development.

5 Revocations

SCHEDULE

Article 3, 4

Part A

Class A1. Shops

Use for all or any of the following purposes—

    (a)   for the retail sale of goods other than hot food,

    (b)   as a post office,

    (c)   for the sale of tickets or as a travel agency,

    (d)   for the sale of sandwiches or other cold food for consumption off the premises,

    (e)   for hairdressing,

    (f)   for the direction of funerals,

    (g)   for the display of goods for sale,

    (h)   for the hiring out of domestic or personal goods or articles,

    (i)   for the washing or cleaning of clothes or fabrics on the premises,

    (j)   for the reception of goods to be washed, cleaned or repaired.

    (k)   as an internet café; where the primary purpose of the premises is to provide facilities for enabling members of the public to access the internet,

where the sale, display or service is to visiting members of the public.

Class A2. Financial and professional services

Use for the provision of—

    (a)   financial services, or

    (b)   professional services (other than health or medical services), or

    (c)   any other services (including use as a betting office) which it is appropriate to provide in a shopping area,

where the services are provided principally to visiting members of the public.

Class A3. Food and drink

Use for the sale of food and drink for consumption on the premises or of hot food for consumption off the premises.

Class A3. Restaurants and cafes
Use for the sale of food and drink for consumption on the premises.

Class A4. Drinking establishments
Use as a public house, wine-bar or other drinking establishment

Class A5. Hot food takeaways
Use for the sale of hot food for consumption off the premises.

Part B

Class B1. Business
Use for all or any of the following purposes—
 (a) as an office other than a use within class A2 (financial and professional services),
 (b) for research and development of products or processes, or
 (c) for any industrial process,
being a use which can be carried out in any residential area without detriment to the amenity of that area by reason of noise, vibration, smell, fumes, smoke, soot, ash, dust or grit.

Class B2. General industrial
Use for the carrying on of an industrial process other than one falling within Class B1 above . . . .

Class B3. Special Industrial Group A
...

Class B4–Class B7
...

Class B8. Storage or distribution
Use for storage or as a distribution centre.

Part C

Class C1. Hotels
Use as a hotel or as a boarding or guest house where, in each case, no significant element of care is provided.

Class C2. Residential institutions

Use for the provision of residential accommodation and care to people in need of care (other than a use within Class C3 (dwellinghouses)).

Use as a hospital or nursing home.

Use as a residential school, college or training centre.

Class C2A. Secure residential institutions

Use for the provision of secure residential accommodation, including use as a prison, young offenders institution, detention centre, secure training centre, custody centre, short-term holding centre, secure hospital, secure local authority accommodation or use as military barracks.

Class C3. Dwellinghouses

Use as a dwellinghouse (whether or not as a sole or main residence) by—

 (a) a single person or by people to be regarded as forming a single household;

 (b) not more than six residents living together as a single household where care is provided for residents; or

 (c) not more than six residents living together as a single household where no care is provided to residents (other than a use within Class C4).

Interpretation of Class C3

For the purposes of Class C3(a) "single household" shall be construed in accordance with section 258 of the Housing Act 2004.

Class C4. Houses in multiple occupation

Use of a dwellinghouse by not more than six residents as a "house in multiple occupation".

Interpretation of Class C4

For the purposes of Class C4 a "house in multiple occupation" does not include a converted block of flats to which section 257 of the Housing Act 2004 applies but otherwise has the same meaning as in section 254 of the Housing Act 2004.

Part D

Class D1. Non-residential institutions

Any use not including a residential use—

 (a) for the provision of any medical or health services except the use of premises attached to the residence of the consultant or practitioner,

 (b) as a creche, day nursery or day centre,

 (c) for the provision of education,

(d)   for the display of works of art (otherwise than for sale or hire),

(e)   as a museum,

(f)   as a public library or public reading room,

(g)   as a public hall or exhibition hall,

(h)   for, or in connection with, public worship or religious instruction,

(i)   as a law court.

Class D2. Assembly and leisure

Use as—

(a)   a cinema,

(b)   a concert hall,

(c)   a bingo hall or casino,

(d)   a dance hall,

(e)   a swimming bath, skating rink, gymnasium or area for other indoor or outdoor sports or recreations, not involving motorised vehicles or firearms.

Appendix B

The following Order is as it applies in England: 1995 No 418 Town and Country Planning (General Permitted Development) Order 1995

Made - - - 22nd February 1995

The Secretary of State for the Environment, as respects England, and the Secretary of State for Wales, as respects Wales, in exercise of the powers conferred on them by sections 59, 60, 61, 74 and 333(7) of the Town and Country Planning Act 1990, section 54 of the Coal Industry Act 1994 and of all other powers enabling them in that behalf, hereby make the following Order–

1 Citation, commencement and interpretation

(1) This Order may be cited as the Town and Country Planning (General Permitted Development) Order 1995 and shall come into force on 3rd June 1995.

(2) In this Order, unless the context otherwise requires–

"the Act" means the Town and Country Planning Act 1990;

"the 1960 Act" means the Caravan Sites and Control of Development Act 1960;

"aerodrome" means an aerodrome as defined in article 106 of the Air Navigation Order 1989 (interpretation) which is–

    (a) licensed under that Order,

    (b) a Government aerodrome,

    (c) one at which the manufacture, repair or maintenance of aircraft is carried out by a person carrying on business as a manufacturer or repairer of aircraft,

    (d) one used by aircraft engaged in the public transport of passengers or cargo or in aerial work, or

    (e) one identified to the Civil Aviation Authority before 1st March 1986 for inclusion in the UK Aerodrome Index,

    and, for the purposes of this definition, the terms "aerial work", "Government aerodrome" and "public transport" have the meanings given in article 106;

"aqueduct" does not include an underground conduit;

"area of outstanding natural beauty" means an area designated as such by an order made by the [Countryside Agency], as respects England, or the Countryside Council for Wales, as respects Wales, under section 87 of the National Parks and Access to the Countryside Act 1949 (designation of areas of outstanding natural beauty) as confirmed by the Secretary of State;

"building"–

    (a)   includes any structure or erection and, except in Parts 24, [25, 33 and 40], [Class A of Part 31 and Class C of Part 38], of Schedule 2, includes any part of a building, as defined in this article; and

    (b)   does not include plant or machinery and, in Schedule 2, except in Class B of Part 31 and Part 33, does not include any gate, fence, wall or other means of enclosure;

"caravan" has the same meaning as for the purposes of Part I of the 1960 Act (caravan sites);

"caravan site" means land on which a caravan is stationed for the purpose of human habitation and land which is used in conjunction with land on which a caravan is so stationed;

"classified road" means a highway or proposed highway which–

    (a)   is a classified road or a principal road by virtue of section 12(1) of the Highways Act 1980 (general provision as to principal and classified roads); or

    (b)   is classified by the Secretary of State for the purposes of any enactment by virtue of section 12(3) of that Act;

["Crown land" has the meaning given by section 293 of the Act;]

"cubic content" means the cubic content of a structure or building measured externally;

"dwellinghouse" does not include a building containing one or more flats, or a flat contained within such a building;

["electronic communication" has the meaning given in section 15(1) of the Electronic Communications Act 2000;]

"erection", in relation to buildings as defined in this article, includes extension, alteration, or re-erection;

"existing", in relation to any building or any plant or machinery or any use, means (except in the definition of "original") existing immediately before the carrying out, in relation to that building, plant, machinery or use, of development described in this Order;

"flat" means a separate and self-contained set of premises constructed or adapted for use for the purpose of a dwelling and forming part of a building from some other part of which it is divided horizontally;

"floor space" means the total floor space in a building or buildings;

"industrial process" means a process for or incidental to any of the following purposes–

    (a)   the making of any article or part of any article (including a ship or vessel, or a film, video or sound recording);

    (b)   the altering, repairing, maintaining, ornamenting, finishing, cleaning, washing, packing, canning, adapting for sale, breaking up or demolition of any article; or

    (c)   the getting, dressing or treatment of minerals in the course of any trade or business other than agriculture, and other than a process carried out on land used as a mine or adjacent to and occupied together with a mine;

"land drainage" has the same meaning as in section 116 of the Land Drainage Act 1976 (interpretation);

"listed building" has the same meaning as in section 1 of the Planning (Listed Buildings and Conservation Areas) Act 1990 (listing of buildings of special architectural or historic interest);

"by local advertisement" means by publication of the notice in at least one newspaper circulating in the locality in which the area or, as the case may be, the whole or relevant part of the conservation area to which the direction relates is situated;

"machinery" includes any structure or erection in the nature of machinery;

"microwave" means that part of the radio spectrum above 1,000 MHz;

"microwave antenna" means a satellite antenna or a terrestrial microwave antenna;

"mine" means any site on which mining operations are carried out;

"mining operations" means the winning and working of minerals in, on or under land, whether by surface or underground working;

"notifiable pipe-line" means a pipe-line, as defined in section 65 of the Pipe-lines Act 1962 (meaning of pipeline), which contains or is intended to contain a hazardous substance, as defined in regulation 2(1) of the Notification Regulations (interpretation), except–

(a)   a pipe-line the construction of which has been authorised under section 1 of the Pipe-lines Act 1962 (cross-country pipe-lines not to be constructed without the Minister's authority); or

(b)   a pipe-line which contains or is intended to contain no hazardous substance other than–

   (i)   a flammable gas (as specified in item 1 of Part II of Schedule 1 to the Notification Regulations (classes of hazardous substances not specifically named in Part I)) at a pressure of less than 8 bars absolute; or

   (ii)   a liquid or mixture of liquids, as specified in item 4 of Part II of that Schedule;

"Notification Regulations" means the Notification of Installations Handling Hazardous Substances Regulations 1982;

"operational Crown building" means a building which is operational Crown land;

"operational Crown land" means–

(a)   Crown land which is used for operational purposes; and

(b)   Crown land which is held for those purposes,

but does not include–

   (i)   land which, in respect of its nature and situation, is comparable rather with land in general than with land which is used, or held, for operational purposes;

    (ii)   Crown land–
        (aa) belonging to Her Majesty in right of the Crown and forming part of the Crown Estate;
        (bb) in which there is an interest belonging to Her Majesty in right of Her private estates;
        (cc) in which there is an interest belonging to Her Majesty in right of the Duchy of Lancaster; or
        (dd) belonging to the Duchy of Cornwall;

"operational purposes" means the purposes of carrying on the functions of the Crown or of either House of Parliament;

"original" means–

(a)   in relation to a building, other than a building which is Crown land, existing on 1st July 1948, as existing on that date and, in relation to a building, other than a building which is Crown land, built on or after 1st July 1948, as so built;

(b)   in relation to a building which is Crown land on 7th June 2006, as existing on that date and, in relation to a building built on or after 7th June 2006 which is Crown land on the date of its completion, as so built;]

"plant" includes any structure or erection in the nature of plant;

"private way" means a highway not maintainable at the public expense and any other way other than a highway;

"proposed highway" has the same meaning as in section 329 of the Highways Act 1980 (further provision as to interpretation);

"public service vehicle" means a public service vehicle within the meaning of section 1 of the Public Passenger Vehicles Act 1981 (definition of public service vehicles) or a tramcar or trolley vehicle within the meaning of section 192(1) of the Road Traffic Act 1988 (general interpretation);

"satellite antenna" means apparatus designed for transmitting microwave radio energy to satellites or receiving it from them, and includes any mountings or brackets attached to such apparatus;

"scheduled monument" has the same meaning as in section 1(11) of the Ancient Monuments and Archaeological Areas Act 1979 (schedule of monuments);

"by site display" means by the posting of the notice by firm affixture to some object, sited and displayed in such a way as to be easily visible and legible by members of the public;

"site of archaeological interest" means land which is included in the schedule of monuments compiled by the Secretary of State under section 1 of the Ancient Monuments and Archaeological Areas Act 1979 (schedule of monuments), or is within an area of land which is designated as an area of archaeological importance under section 33 of that Act (designation of areas of archaeological importance), or which is within a site registered in any record adopted by resolution by a county council [in England or by a local planning authority in Wales and known in England as the County Sites and Monuments Record and in Wales as the Sites and Monuments Record for the local planning authority area;

"site of special scientific interest" means land to which section 28(1) of the Wildlife and Countryside Act 1981 (areas of special scientific interest) applies;

"solar PV" means solar photovoltaics;

"statutory undertaker" includes, in addition to any person mentioned in section 262(1) of the Act (meaning of statutory undertakers), [a universal service provider (within the meaning of [Part 3 of the Postal Services Act 2011]) in connection with the provision of a universal postal service (within the meaning of [that Part])], the Civil Aviation Authority, the [Environment Agency], any water undertaker, any [gas transporter], and any licence holder within the meaning of section 64(1) of the Electricity Act 1989 (interpretation etc of Part I);

"terrestrial microwave antenna" means apparatus designed for transmitting or receiving terrestrial microwave radio energy between two fixed points;

"trunk road" means a highway or proposed highway which is a trunk road by virtue of section 10(1) or 19 of the Highways Act 1980 (general provisions as to trunk roads, and certain special roads and other highways to become trunk roads) or any other enactment or any instrument made under any enactment;

"the Use Classes Order" means the Town and Country Planning (Use Classes) Order 1987 [; and]

["World Heritage Site" means a property appearing on the World Heritage List kept under article 11(2) of the 1972 UNESCO Convention for the Protection of the World Cultural and Natural Heritage].

(3)   Unless the context otherwise requires, any reference in this Order to the height of a building or of plant or machinery shall be construed as a reference to its height when measured from ground level; and for the purposes of this paragraph "ground level" means the level of the surface of the ground immediately adjacent to the building or plant or machinery in question or, where the level of the surface of the ground on which it is situated or is to be situated is not uniform, the level of the highest part of the surface of the ground adjacent to it.

(4)   ...

(5)   The land referred to elsewhere in this Order as article 1(5) land is the land described in Part 2 of Schedule 1 to this Order (National Parks, areas of outstanding natural beauty and conservation areas etc).

(6)   The land referred to elsewhere in this Order as article 1(6) land is the land described in Part 3 of Schedule 1 to this Order (National Parks and adjoining land and the Broads).

(7)   Paragraphs (8) to (12) apply where an electronic communication is used by a person for the purpose of fulfilling any requirement in this Order or in any

Schedule to this Order to give or send any statement, notice or other document to any other person ("the recipient").

(8)   The requirement shall be taken to be fulfilled where the notice or other document transmitted by means of the electronic communication is–

(a) capable of being accessed by the recipient,

(b) legible in all material respects, and

(c) sufficiently permanent to be used for subsequent reference.

(9)   In paragraph (8), "legible in all material respects" means that the information contained in the notice or document is available to the recipient to no lesser extent than it would be if sent or given by means of a document in printed form.

(10)   Where the electronic communication is received by the recipient outside the recipient's business hours, it shall be taken to have been received on the next working day; and for this purpose "working day" means a day which is not a Saturday, Sunday, Bank Holiday or other public holiday.

(11)   A requirement in this Order or in any Schedule to this Order that any document should be in writing is fulfilled where that document meets the criteria in paragraph (8), and "written" and cognate expressions are to be construed accordingly.

(12)   References in this Order or in any Schedule to this Order to plans, drawings, notices or other documents, or to copies of such documents, include references to such documents or copies of them in electronic form.

(13)   For the purposes of this Order, development earned out by or on behalf of any person in whom control of accommodation in any part of the Palace of Westminster or its precincts is vested shall be treated (so far as it would not otherwise be treated) as development by or on behalf of the Crown.

2 Application

(1)   This Order applies to all land in England and Wales, but where land is the subject of a special development order, whether made before or after the commencement of this Order, this Order shall apply to that land only to such extent and subject to such modifications as may be specified in the special development order.

(2)   Nothing in this Order shall apply to any permission which is deemed to be granted under section 222 of the Act (planning permission not needed for advertisements complying with regulations).

3 Permitted development

(1)   Subject to the provisions of this Order and regulations 60 to 63 of the Conservation (Natural Habitats, &c) Regulations 1994 (general development orders), planning permission is hereby granted for the classes of development described as permitted development in Schedule 2.

(2)   Any permission granted by paragraph (1) is subject to any relevant exception, limitation or condition specified in Schedule 2.

(3)   References in the following provisions of this Order to permission granted by Schedule 2 or by any Part, Class or paragraph of that Schedule are references to the permission granted by this article in relation to development described in that Schedule or that provision of that Schedule.

(4)   Nothing in this Order permits development contrary to any condition imposed by any planning permission granted or deemed to be granted under Part III of the Act otherwise than by this Order.

(5)   "The permission granted by Schedule 2 shall not apply if–

  (a)   in the case of permission granted in connection with an existing building, the building operations involved in the construction of that building are unlawful;

  (b)   in the case of permission granted in connection with an existing use, that use is unlawful.

(6)   "The permission granted by Schedule 2 shall not, except in relation to development permitted by Parts 9, 11, 13 or 30, authorise any development which requires or involves the formation, laying out or material widening of a means of access to an existing highway which is a trunk road or classified road, or creates an obstruction to the view of persons using any highway used by vehicular traffic, so as to be likely to cause danger to such persons.

(7)   Any development falling within Part 11 of Schedule 2 authorised by an Act or order subject to the grant of any consent or approval shall not be treated for the purposes of this Order as authorised unless and until that consent or approval is obtained, except where the Act was passed or the order made after 1st July 1948 and it contains provision to the contrary.

(8)   Schedule 2 does not grant permission for the laying or construction of a notifiable pipe-line, except in the case of the laying or construction of a notifiable pipe-line by a [gas transporter] in accordance with Class F of Part 17 of that Schedule.

(9)   Except as provided in Part 31, Schedule 2 does not permit any development which requires or involves the demolition of a building, but in this paragraph "building" does not include part of a building.

(10)   Subject to paragraph (12), Schedule 1 development or Schedule 2 development within the meaning of *the Town and Country Planning (Environmental Impact Assessment) (England and Wales) Regulations* 1999 [the Town and Country Planning (Environmental Impact Assessment) Regulations 2011] ("the EIA Regulations") is not permitted by this Order unless:

  (a)   the local planning authority has adopted a screening opinion under regulation 5 of those Regulations that the development is not EIA development;

  (b)   the Secretary of State has made a screening direction under regulation 4(7) or 6(4) of those Regulations that the development is not EIA development; or

  (c)   the Secretary of State has given a direction under regulation 4(4) of those Regulations that the development is exempted from the application of those Regulations.

(11) Where:
(a) the local planning authority has adopted a screening opinion pursuant to regulation 5 of the EIA Regulations that development is EIA development and the Secretary of State has in relation to that development neither made a screening direction to the contrary under regulation 4(7) [regulation 4(8)] or 6(4) of those Regulations nor directed under regulation 4(4) of those Regulations that the development is exempted from the application of those Regulations; or
(b) the Secretary of State has directed that development is EIA development,
that development shall be treated, for the purposes of paragraph (10), as development which is not permitted by this Order.

(12) Paragraph (10) does not apply to–
(a) . . .
(b) development which consists of the carrying out by a drainage body within the meaning of [the Environmental Impact Assessment (Land Drainage Improvement Works) Regulations 1999];
(c) . . .
(d) development for which permission is granted by Part 7, Class D of Part 8, Part 11, Class B of Part 12, Class F(a) of Part 17, Class A or Class B of Part 20 or Class B of Part 21 of Schedule 2;
(e) development for which permission is granted by Class C or Class D of Part 20, Class A of Part 21 or Class B of Part 22 of Schedule 2 where the land in, on or under which the development is to be carried out is–
(i) in the case of Class C or Class D of Part 20, on the same authorised site,
(ii) in the case of Class A of Part 21, on the same premises or, as the case may be, the same ancillary mining land,
(iii) in the case of Class B of Part 22, on the same land or, as the case may be, on land adjoining that land,
as that in, on or under which development of any description permitted by the same Class has been carried out before [14th March 1999];
(f) the completion of any development begun before [14th March 1999];
(g) development for which permission is granted by Class B of Part 13.

(13) Where a person uses electronic communications for making any application required to be made under any of Parts 6, 7, 22, 23, 24, 30 or 31 of Schedule 2, that person shall be taken to have agreed–
(a) to the use of electronic communications for all purposes relating to his application which are capable of being effected using such communications;
(b) that his address for the purpose of such communications is the address incorporated into, or otherwise logically associated with, his application; and
(c) that his deemed agreement under this paragraph shall subsist until he gives notice in writing that he wishes to revoke the agreement (and such

revocation shall be final and shall take effect on a date specified by him but not less than seven days after the date on which the notice is given).

4 Directions restricting permitted development

[(1)   If the Secretary of State or the local planning authority is satisfied that it is expedient that development described in any Part, Class or paragraph in Schedule 2, other than Class B of Part 22 or Class B of Part 23, should not be carried out unless permission is granted for it on an application, the Secretary of State or (as the case may be) the local planning authority, may make a direction under this paragraph that the permission granted by article 3 shall not apply to–

(a) all or any development of the Part, Class or paragraph in question in an area specified in the direction; or

(b) any particular development, falling within that Part, Class or paragraph, which is specified in the direction,

and the direction shall specify that it is made under this paragraph.

(2)   A direction under paragraph (1) shall not affect the carrying out of–

(a) development permitted by Part 11 authorised by an Act passed after 1st July 1948 or by an order requiring the approval of both Houses of Parliament approved after that date;

(b) development permitted by Class B of Part 13;

(c) development mentioned in Part 24, unless the direction specifically so provides;

(d) development in an emergency other than development permitted by Part 37;

(e) development permitted by Part 37 or 38.

(3)   A direction made or having effect as if made under this article shall not, unless the direction so provides, affect the carrying out by a statutory undertaker of the following descriptions of development–

(a) the maintenance of bridges, buildings and railway stations;

(b) the alteration and maintenance of railway track, and the provision and maintenance of track equipment, including signal boxes, signalling apparatus and other appliances and works required in connection with the movement of traffic by rail;

(c) the maintenance of docks, harbours, quays, wharves, canals and towing paths;

(d) the provision and maintenance of mechanical apparatus or appliances (including signalling equipment) required for the purposes of shipping or in connection with the embarking, disembarking, loading, discharging or transport of passengers, livestock or goods at a dock, quay, harbour, bank, wharf or basin;

(e) any development required in connection with the improvement, maintenance or repair of watercourses or drainage works;

(f) the maintenance of buildings, runways, taxiways or aprons at an aerodrome; or

(g) the provision, alteration and maintenance of equipment, apparatus and works at an aerodrome, required in connection with the movement of traffic by air (other than buildings, the construction, erection, reconstruction or alteration of which is permitted by Class A of Part 18).

(4)   In this article and in articles 5 and 6 "local planning authority" means the local planning authority whose function it would be to determine an application for planning permission for the development to which the direction relates or is proposed to relate.]

5 Procedure for article 4(1) directions

(1)   Subject to article 6, notice of any direction made under article 4(1) shall, as soon as practicable after the direction has been made, be given by the local planning authority–

(a) by local advertisement;

(b) by site display at no fewer than two locations within the area to which the direction relates, or, if the direction is made under article 4(1)(b), on the site of the particular development to which the direction relates, for a period of not less than six weeks; and

(c) subject to paragraph (2), by serving the notice on the owner and occupier of every part of the land within the area or site to which the direction relates.

(2)   In a case where this paragraph applies, the local planning authority need not serve notice on an owner or occupier in accordance with paragraph (1)(c), if they consider that–

(a) individual service on that owner or occupier is impracticable because it is difficult to identify or locate that person or

(b) the number of owners or occupiers within the area to which the direction relates makes individual service impracticable.

(3)   Paragraph (2) shall not apply where the owner or occupier is a statutory undertaker or the Crown.

(4)   The notice referred to in paragraph (1) shall–

(a) include a description of the development and the area to which the direction relates, or the site to which it relates, as the case may be, and a statement of the effect of the direction;

(b) specify that the direction is made under article 4(1) of this Order;

(c) name a place where a copy of the direction, and a copy of a map defining the area to which it relates, or the site to which it relates, as the case may be, may be seen at all reasonable hours;

(d) specify a period of at least 21 days, stating the date on which that period begins, within which any representations concerning the direction may be made to the local planning authority; and

(e) specify the date on which it is proposed that the direction will come into force, which must be at least 28 days but no longer than two years after the date referred to in sub-paragraph (d).

(5)   Where a notice given by site display is, without any fault or intention of the local planning authority, removed, obscured or defaced before the period referred to in paragraph (4)(d) has elapsed, the authority shall be treated as having complied with the requirements of that paragraph if they have taken reasonable steps for the protection of the notice, including, if need be, its replacement.

(6)   The local planning authority shall send a copy of the direction and the notice under paragraph (1), including a copy of a map defining the area to which it relates, or the site to which it relates, as the case may be, to the Secretary of State on the same day that notice of the direction is first published or displayed in accordance with paragraph (1).

(7)   The direction shall come into force in respect of any part of the land within the area to which it relates on the date specified in accordance with paragraph (4) (e) but shall not come into force unless confirmed by the local planning authority in accordance with paragraphs (9) and (10).

(8)   On making a direction under article 4(1)–

(a)   a county planning authority shall give notice of it to any district planning authority in whose district the area or part of the area to which the direction relates is situated; and

(b)   except in metropolitan districts, a district planning authority shall give notice of it to the county planning authority, if any.

(9)   In deciding whether to confirm a direction made under article 4(1), the local planning authority shall take into account any representations received during the period specified in accordance with paragraph (4)(d).

(10)   The local planning authority shall not confirm a direction until after the expiration of–

(a)   a period of at least 28 days following the latest date on which any notice relating to the direction was served or published; or

(b)   such longer period as may be specified by the Secretary of State following the notification by the local planning authority to the Secretary of State of the direction.

(11)   The local planning authority shall, as soon as practicable after a direction has been confirmed–

(a)   give notice of such confirmation and the date on which the direction will come into force; and

(b)   send a copy of the direction as confirmed to the Secretary of State.

(12)   Notice under paragraph (11)(a) shall be given in the manner described in paragraphs (1) and (4)(a) to (c); and paragraphs (2) and (3) shall apply for this purpose as they apply for the purpose of paragraph (1)(c).

(13)   A local planning authority may, by making a subsequent direction, cancel any direction made by them under article 4(1); and the Secretary of State may, subject to paragraphs (3) and (4) of article 6, make a direction cancelling or modifying any direction under article 4(1) made by a local planning authority at any time before or after its confirmation.

(14)   Paragraphs (1) to (12) shall apply in relation to any direction made under paragraph (13) by a local planning authority unless the direction it is cancelling is a direction to which article 6 applied.

(15)   Paragraphs (2) to (10) of article 6 shall apply in relation to any direction made by a local planning authority under paragraph (13) cancelling a direction to which article 6 applied.

(16)   The Secretary of State shall notify the local planning authority as soon as practicable after making a direction under paragraph (13).

(17)   Paragraphs (1) to (3) and (4)(a) to (c) shall apply to any direction made under paragraph (13) by the Secretary of State.

(18)   A direction made under paragraph (13) by the Secretary of State shall come into force in respect of any part of the land within the area to which it relates–

- (a)   on the date on which the notice is served in accordance with paragraph (1)(c) on the occupier of that part of the land or, if there is no occupier, on the owner; or
- (b)   if paragraph (2) applies, on the date on which the notice is first published or displayed in accordance with paragraph (1).

6 Directions with immediate effect

(1)   This article applies where–

- (a)   a direction relating only to development permitted by any of Parts 1 to 4, or Part 31, of Schedule 2 has been made by the local planning authority under article 4(1) and the authority consider that the development to which the direction relates would be prejudicial to the proper planning of their area or constitute a threat to the amenities of their area; or
- (b)   a direction within the whole or part of any conservation area has been made by the local planning authority under article 4(1) which the authority consider should have immediate effect and the development to which the direction relates is described in sub-paragraphs (a) to (j) of paragraph (3);

(2)   Subject to paragraphs (3), (4) and (5) of this article, paragraphs (1) to (3), (4)(a) to (d), (5), and (8) to (10) of article 5 shall apply, in relation to a direction to which this article applies; and the planning authority shall notify the Secretary of State of the direction on the same day that notice is given under paragraph (1) of article 5.

(3)   The Secretary of State may not make a direction under paragraph (13) of article 5 within the whole or part of any conservation area where the development to which the direction relates is described in–

- (a)   Class A of Part 1 of Schedule 2, consisting of the enlargement, improvement or other alteration of a dwellinghouse, where any part of the enlargement, improvement or alteration would front a relevant location;
- (b)   Class C of Part 1 of that Schedule, where the alteration would be to a roof slope which fronts a relevant location;
- (c)   Class D of Part 1 of that Schedule, where the external door in question fronts a relevant location;
- (d)   Class E of Part 1 of that Schedule, where the building or enclosure, swimming or other pool to be provided would front a relevant location,

or where the part of the building or enclosure maintained, improved or altered would front a relevant location;

(e) Class F of Part 1 of that Schedule, where the hard surface would front a relevant location;

(f) Class H of Part 1 of that Schedule, where the part of the building or other structure on which the antenna is to be installed, altered or replaced fronts a relevant location;

(g) Class A of Part 2 of that Schedule, where the gate, fence, wall or other means of enclosure would be within the curtilage of a dwellinghouse and would front a relevant location;

(h) Class G of Part 1 of that Schedule, consisting of the installation, alteration or replacement of a chimney on a dwellinghouse;

(i) Class C of Part 2 of the Schedule, consisting of the painting of the exterior of any part of–

(i) a dwellinghouse; or

(ii) any building or enclosure within the curtilage of a dwellinghouse, which fronts a relevant location;

(j) Class B of Part 31 of that Schedule, where the gate, fence, wall or other means of enclosure is within the curtilage of a dwellinghouse and fronts a relevant location.

(4) The Secretary of State may not modify a direction to which this article applies or a direction which relates to–

(a) a listed building;

(b) a building which is notified to the authority by the Secretary of State as a building of architectural or historic interest; or

(c) development within the curtilage of a listed building,

and does not relate to land of any other description.

(5) [Paragraph (8)(b)] shall not apply in relation to a direction to which paragraph (3) of this article applies or to a direction which relates to–

(a) a listed building;

(b) a building which is notified to the authority by the Secretary of State as a building of architectural or historic interest; or

(c) development within the curtilage of a listed building,

and does not relate to land of any other description.

(6) The direction shall come into force in respect of any part of the land within the area to which it relates–

(a) on the date on which the notice is served in accordance with paragraph (1)(c) of article 5 on the occupier of that part of the land or, if there is no occupier, on the owner; or

(b) if paragraph (2) of article 5 applies, on the date on which the notice is first published or displayed in accordance with paragraph (1) of article 5.

(7) A direction to which this article applies shall expire at the end of the period of six months beginning with the date on which it comes into force unless confirmed by the local planning authority in accordance with paragraphs (9) and (10) of article 5 before the end of the six month period.

(8)   The local planning authority shall, as soon as practicable after a direction has been confirmed–

(a)   give notice of its confirmation; and

(b)   send a copy of the direction as confirmed to the Secretary of State.

(9)   Notice under paragraph (8)(a) shall be given in the manner described in paragraphs (1) and (4)(a) to (c) of article 5; and paragraphs (2) and (3) of that article shall apply for this purpose as they apply for the purpose of paragraph (1) (c) of article 5.

(10)   In this article "relevant location" means a highway, waterway or open space.]

7 Directions restricting permitted development under Class B of Part 22 or Class B of Part 23

(1)   If, on receipt of a notification from any person that he proposes to carry out development within Class B of Part 22 or Class B of Part 23 of Schedule 2, a mineral planning authority are satisfied as mentioned in paragraph (2) below, they may, within a period of 21 days beginning with the receipt of the notification, direct that the permission granted by article 3 of this Order shall not apply to the development, or to such part of the development as is specified in the direction.

(2)   The mineral planning authority may make a direction under this article if they are satisfied that it is expedient that the development, or any part of it, should not be carried out unless permission for it is granted on an application because–

(a)   the land on which the development is to be carried out is within–

(i)   a National Park,

(ii)   an area of outstanding natural beauty,

(iii)   a site of archaeological interest, and the operation to be carried out is not one described in the Schedule to the Areas of Archaeological Importance (Notification of Operations) (Exemption) Order 1984 (exempt operations),

(iv)   a site of special scientific interest, or

(v)   the Broads;

(b)   the development, either taken by itself or taken in conjunction with other development which is already being carried out in the area or in respect of which notification has been given in pursuance of the provisions of Class B of Part 22 or Class B of Part 23, would cause serious detriment to the amenity of the area in which it is to be carried out or would adversely affect the setting of a building shown as Grade I in the list of buildings of special architectural or historic interest compiled by the Secretary of State under section 1 of the Planning (Listed Buildings and Conservation Areas) Act 1990 (listing of buildings of special architectural or historic interest);

(c)   the development would constitute a serious nuisance to the inhabitants of a nearby residential building, hospital or school; or

(d)   the development would endanger aircraft using a nearby aerodrome.

(3)   A direction made under this article shall contain a statement as to the day on which (if it is not disallowed under paragraph (5) below) it will come into force, which shall be 29 days from the date on which notice of it is sent to the Secretary of State in accordance with paragraph (4) below.

(4)   As soon as is reasonably practicable a copy of a direction under this article shall be sent by the mineral planning authority to the Secretary of State and to the person who gave notice of the proposal to carry out development.

(5)   The Secretary of State may, at any time within a period of 28 days beginning with the date on which the direction is made, disallow the direction; and immediately upon receipt of notice in writing from the Secretary of State that he has disallowed the direction, the mineral planning authority shall give notice in writing to the person who gave notice of the proposal that he is authorised to proceed with the development.

8 Directions

Any power conferred by this Order to give a direction includes power to cancel or vary the direction by a subsequent direction.

9 Revocations

The statutory instruments specified in column 1 of Schedule 3 are hereby revoked to the extent specified in column 3.

SCHEDULE 1

Article 1

Part 1

...

Part 2

Land within–

(a)   a National Park;

(b)   an area of outstanding natural beauty;

(c)   an area designated as a conservation area under section 69 of the Planning (Listed Buildings and Conservation Areas) Act 1990 (designation of conservation areas);

(d)   an area specified by the Secretary of State and the Minister of Agriculture, Fisheries and Food for the purposes of section 41(3) of the Wildlife and Countryside Act 1981 (enhancement and protection of the natural beauty and amenity of the countryside);

(e)   the Broad[; and]

[(f)   a World Heritage Site.]

Part 3
Land within a National Park or within the following areas–
  (a) In England, the Broads or land outside the boundaries of a National
       Park, which is within the parishes listed below–
  in the district of Allerdale–
       Blindcrake, Bothel and Threapland, Bridekirk, Brigham, Broughton,
       Broughton Moor, Camerton, Crosscanonby, Dean, Dearham, Gilcrux,
       Great Clifton, Greysouthen, Little Clifton, Loweswater, Oughterside
       and Allerby, Papcastle, Plumbland, Seaton, Winscales;
  in the borough of Copeland–
       Arlecdon and Frizington, Cleator Moor, Distington, Drigg and
       Carleton, Egremont, Gosforth, Haile, Irton with Santon, Lamplugh,
       Lowca, Lowside Quarter, Millom, Millom Without, Moresby, Parton,
       Ponsonby, St Bees, St Bridget's Beckermet, St John's Beckermet,
       Seascale, Weddicar;
  in the district of Eden–
       Ainstable, Asby, Bandleyside, Bolton, Brough, Brough Sowerby,
       Brougham, Castle Sowerby, Catterlen, Clifton, Cliburn, Crackenthorpe,
       Crosby Garrett, Crosby Ravensworth, Culgaith, Dacre, Dufton,
       Glassonby, Great Salkeld, Great Strickland, Greystoke, Hartley,
       Hesket, Hillbeck, Hunsonby, Hutton, Kaber, Kings Meaburn, Kirkby
       Stephen, Kirby Thore, Kirkoswald, Langwathby, Lazonby, Little
       Strickland, Long Marton, Lowther, Mallerstang, Milburn, Morland,
       Mungrisdale, Murton, Musgrave, Nateby, Newbiggin, Newby, Orton,
       Ousby, Ravanstonedale, Shap, Skelton, Sleagill, Sockbridge and Tirril,
       Soulby, Stainmore, Tebay, Temple Sowerby, Thrimby, Waitby,
       Warcop, Wharton, Winton, Yanwath and Eamont Bridge;
  in the borough of High Peak–
       Chapel-en-le-Frith, Charlesworth, Chinley Buxworth and Brownside,
       Chisworth, Green Fairfield, Hartington Upper Quarter, Hayfield, King
       Sterndale, Tintwistle, Wormhill;
  in the district of South Lakeland–
       Aldingham, Angerton, Arnside, Barbon, Beetham, Blawith and
       Subberthwaite, Broughton West, Burton, Casterton, Docker, Egton-
       with-Newland, Fawcett Forest, Firbank, Grayrigg, Helsington,
       Heversham, Hincaster, Holme, Hutton Roof, Killington, Kirkby
       Ireleth, Kirkby Lonsdale, Lambrigg, Levens, Lower Allithwaite,
       Lower Holker, Lowick, Lupton, Mansergh, Mansriggs, Middleton,
       Milnthorpe, Natland, New Hutton, Old Hutton and Holmescales,
       Osmotherley, Pennington, Preston Patrick, Preston Richard,
       Scalthwaiterigg, Sedgwick, Skelsmergh, Stainton, Strickland Ketel,
       Strickland Roger, Urswick, Whinfell, Whitwell and Selside;
  in the district of West Derbyshire–
       Aldwartc, Birchover, Stanton; and

[(b) In Wales, land outside the boundaries of a National Park which is–
   (i) within the communities listed below–
      in the county borough of Aberconwy and Colwyn–
         Caerhun, Dolgairog;
      in the county of Caernarfonshire and Merionethshire–
         Arthog, Betws Garmon, Bontnewydd, Corris, Llanberis, Llanddeiniolen, Llandwrog, Llanfrothen, Llanllyfni, Llanwnda, Penrhyndeudraeth, Waunfawr; or
   (ii) within the specified parts of the communities listed below–
      in the county borough of Aberconwy and Colwyn–
         those parts of the following communities which were on 31st March 1974 within the former rural district of Nant Conway–
         Conwy, Henryd, Llanddoged and Maenan, and Llanrwst;
         that part of the community of Llangwm which was on 31st March 1974 within the former rural district of Penllyn;
      in the county of Caernarfonshire and Merionethshire:
         those parts of the following communities which were on 31st March 1974 within the former rural district of Gwyrfai–
         Caernarfon, Clynnog, Dolbenmaen, Llandygai, Llanaelhaearn, Llanrug, Pentir, Y Felinheli;
         that part of the community of Talsarnau which was on 31st March 1974 within the former rural district of Deudraeth;
         that part of the community of Barmouth which was on 31st March 1974 within the former rural district of Dolgellau;
         that part of the community of Llandderfel which was on 31st March 1974 within the former rural district of Penllyn;
      in the county of Denbighshire, that part of the community of Llandrillo which was on 31st March 1974 within the former rural district of Penllyn.]

SCHEDULE 2

Article 3

Part 1

Class A

Permitted development

A The enlargement, improvement or other alteration of a dwellinghouse.

Development not permitted

A.1 Development is not permitted by Class A if–
   (a) as a result of the works, the total area of ground covered by buildings within the curtilage of the dwellinghouse (other than the original

dwellinghouse) would exceed 50% of the total area of the curtilage (excluding the ground area of the original dwellinghouse);

(b)   the height of the part of the dwellinghouse enlarged, improved or altered would exceed the height of the highest part of the roof of the existing dwellinghouse;

(c)   the height of the eaves of the part of the dwellinghouse enlarged, improved or altered would exceed the height of the eaves of the existing dwellinghouse;

(d)   the enlarged part of the dwellinghouse would extend beyond a wall which–

    (i)   fronts a highway, and

    (ii)   forms either the principal elevation or a side elevation of the original dwellinghouse;

(e)   the enlarged part of the dwellinghouse would have a single storey and–

    (i)   extend beyond the rear wall of the original dwellinghouse by more than 4 metres in the case of a detached dwellinghouse, or 3 metres in the case of any other dwellinghouse, or

    (ii)   exceed 4 metres in height;

(f)   the enlarged part of the dwellinghouse would have more than one storey and–

    (i)   extend beyond the rear wall of the original dwellinghouse by more than 3 metres, or

    (ii)   be within 7 metres of any boundary of the curtilage of the dwellinghouse opposite the rear wall of the dwellinghouse;

(g)   the enlarged part of the dwellinghouse would be within 2 metres of the boundary of the curtilage of the dwellinghouse, and the height of the eaves of the enlarged part would exceed 3 metres;

(h)   the enlarged part of the dwellinghouse would extend beyond a wall forming a side elevation of the original dwellinghouse, and would–

    (i)   exceed 4 metres in height,

    (ii)   have more than one storey, or

    (iii)   have a width greater than half the width of the original dwellinghouse; or

(i)   it would consist of or include–

    (i)   the construction or provision of a veranda, balcony or raised platform,

    (ii)   the installation, alteration or replacement of a microwave antenna,

    (iii)   the installation, alteration or replacement of a chimney, flue or soil and vent pipe, or

    (iv)   an alteration to any part of the roof of the dwellinghouse.

A.2 In the case of a dwellinghouse on article 1(5) land, development is not permitted by Class A if–

(c)   it would consist of or include the cladding of any part of the exterior of the dwellinghouse with stone, artificial stone, pebble dash, render, timber, plastic or tiles;

(d)   the enlarged part of the dwellinghouse would extend beyond a wall forming a side elevation of the original dwellinghouse; or

(e)   the enlarged part of the dwellinghouse would have more than one storey and extend beyond the rear wall of the original dwellinghouse.

Conditions

A.3 Development is permitted by Class A subject to the following conditions–

(a)   the materials used in any exterior work (other than materials used in the construction of a conservatory) shall be of a similar appearance to those used in the construction of the exterior of the existing dwellinghouse;

(b)   any upper-floor window located in a wall or roof slope forming a side elevation of the dwellinghouse shall be–

(i)   obscure-glazed, and

(ii)   non-opening unless the parts of the window which can be opened are more than 1.7 metres above the floor of the room in which the window is installed; and

(c)   where the enlarged part of the dwellinghouse has more than one storey, the roof pitch of the enlarged part shall, so far as practicable, be the same as the roof pitch of the original dwellinghouse.

Class B
Permitted development

B The enlargement of a dwellinghouse consisting of an addition or alteration to its roof.

Development not permitted

B.1 Development is not permitted by Class B if–

(a)   any part of the dwellinghouse would, as a result of the works, exceed the height of the highest part of the existing roof;

(b)   any part of the dwellinghouse would, as a result of the works, extend beyond the plane of any existing roof slope which forms the principal elevation of the dwellinghouse and fronts a highway;

(c)   the cubic content of the resulting roof space would exceed the cubic content of the original roof space by more than–

(i)   40 cubic metres in the case of a terrace house, or

(ii)   50 cubic metres in any other case;

(d)   it would consist of or include–

(i)   the construction or provision of a veranda, balcony or raised platform, or

(ii)   the installation, alteration or replacement of a chimney, flue or soil and vent pipe; or

(e)   the dwellinghouse is on article 1(5) land.

Conditions

B.2 Development is permitted by Class B subject to the following conditions–

    (a)   the materials used in any exterior work shall be of a similar appearance to those used in the construction of the exterior of the existing dwellinghouse;

    (b)   other than in the case of a hip-to-gable enlargement, the edge of the enlargement closest to the eaves of the original roof shall, so far as practicable, be not less than 20 centimetres from the eaves of the original roof; and

    (c)   any window inserted on a wall or roof slope forming a side elevation of the dwellinghouse shall be–

        (i)   obscure-glazed, and

        (ii)   non-opening unless the parts of the window which can be opened are more than 1.7 metres above the floor of the room in which the window is installed.

Interpretation of Class B

B.3 For the purposes of Class B "resulting roof space" means the roof space as enlarged, taking into account any enlargement to the original roof space, whether permitted by this class or not.

Class C
Permitted development

C Any other alteration to the roof of a dwellinghouse.

Development not permitted

C.1 Development is not permitted by Class C if–

    (a)   the alteration would protrude more than 150 millimetres beyond the plane of the slope of the original roof when measured from the perpendicular with the external surface of the original roof;

    (b)   it would result in the highest part of the alteration being higher than the highest part of the original roof; or

    (c)   it would consist of or include–

        (i)   the installation, alteration or replacement of a chimney, flue or soil and vent pipe, or

        (ii)   the installation, alteration or replacement of solar photovoltaics or solar thermal equipment.

Conditions

C.2 Development is permitted by Class C subject to the condition that any window located on a roof slope forming a side elevation of the dwellinghouse shall be–

    (a)   obscure-glazed; and

(b)   non-opening unless the parts of the window which can be opened are more than 1.7 metres above the floor of the room in which the window is installed.

Class D
Permitted development

D The erection or construction of a porch outside any external door of a dwellinghouse.

Development not permitted

D.1 Development is not permitted by Class D if–
(a)   the ground area (measured externally) of the structure would exceed 3 square metres;
(b)   any part of the structure would be more than 3 metres above ground level; or
(c)   any part of the structure would be within 2 metres of any boundary of the curtilage of the dwellinghouse with a highway.

Class E
Permitted development

E The provision within the curtilage of the dwellinghouse of–
(a)   any building or enclosure, swimming or other pool required for a purpose incidental to the enjoyment of the dwellinghouse as such, or the maintenance, improvement or other alteration of such a building or enclosure; or
(b)   a container used for domestic heating purposes for the storage of oil or liquid petroleum gas.

Development not permitted

E.1 Development is not permitted by Class E if–
(a)   the total area of ground covered by buildings, enclosures and containers within the curtilage (other than the original dwellinghouse) would exceed 50% of the total area of the curtilage (excluding the ground area of the original dwellinghouse);
(b)   any part of the building, enclosure, pool or container would be situated on land forward of a wall forming the principal elevation of the original dwellinghouse;
(c)   the building would have more than one storey;
(d)   the height of the building, enclosure or container would exceed–
(i)   4 metres in the case of a building with a dual-pitched roof,
(ii)   2.5 metres in the case of a building, enclosure or container within 2 metres of the boundary of the curtilage of the dwellinghouse, or
(iii) 3 metres in any other case;

(e)   the height of the eaves of the building would exceed 2.5 metres;
(f)   the building, enclosure, pool or container would be situated within the curtilage of a listed building;
(g)   it would include the construction or provision of a veranda, balcony or raised platform;
(h)   it relates to a dwelling or a microwave antenna; or
(i)   the capacity of the container would exceed 3,500 litres.

E.2 In the case of any land within the curtilage of the dwellinghouse which is within–

(a)   a World Heritage Site,
(b)   a National Park,
(c)   an area of outstanding natural beauty, or
(d)   the Broads,

development is not permitted by Class E if the total area of ground covered by buildings, enclosures, pools and containers situated more than 20 metres from any wall of the dwellinghouse would exceed 10 square metres.

E.3 In the case of any land within the curtilage of the dwellinghouse which is article 1(5) land, development is not permitted by Class E if any part of the building, enclosure, pool or container would be situated on land between a wall forming a side elevation of the dwellinghouse and the boundary of the curtilage of the dwellinghouse.

Interpretation of Class E

E.4 For the purposes of Class E, "purpose incidental to the enjoyment of the dwellinghouse as such" includes the keeping of poultry, bees, pet animals, birds or other livestock for the domestic needs or personal enjoyment of the occupants of the dwellinghouse.

Class F
Permitted development

F Development consisting of–

(a)   the provision within the curtilage of a dwellinghouse of a hard surface for any purpose incidental to the enjoyment of the dwellinghouse as such; or
(b)   the replacement in whole or in part of such a surface.

Conditions

F.1 Development is permitted by Class F subject to the condition that where–

(a)   the hard surface would be situated on land between a wall forming the principal elevation of the dwellinghouse and a highway, and
(b)   the area of ground covered by the hard surface, or the area of hard surface replaced, would exceed 5 square metres,

either the hard surface shall be made of porous materials, or provision shall be made to direct run-off water from the hard surface to a permeable or porous area or surface within the curtilage of the dwellinghouse.

Class G
Permitted development

G The installation, alteration or replacement of a chimney, flue or soil and vent pipe on a dwellinghouse.

Development not permitted

G.1 Development is not permitted by Class G if–
 (a) the height of the chimney, flue or soil and vent pipe would exceed the highest part of the roof by 1 metre or more; or
 (b) in the case of a dwellinghouse on article 1(5) land, the chimney, flue or soil and vent pipe would be installed on a wall or roof slope which–
   (i) fronts a highway, and
   (ii) forms either the principal elevation or a side elevation of the dwellinghouse.

Class H
Permitted development

H The installation, alteration or replacement of a microwave antenna on a dwellinghouse or within the curtilage of a dwellinghouse.

Development not permitted

H.1 Development is not permitted by Class H if–
 (a) it would result in the presence on the dwellinghouse or within its curtilage of–
   (i) more than two antennas;
   (ii) a single antenna exceeding 100 centimetres in length;
   (iii) two antennas which do not meet the relevant size criteria;
   (iv) an antenna installed on a chimney, where the length of the antenna would exceed 60 centimetres;
   (v) an antenna installed on a chimney, where the antenna would protrude above the chimney; or
   (vi) an antenna with a cubic capacity in excess of 35 litres;
 (b) in the case of an antenna to be installed on a roof without a chimney, the highest part of the antenna would be higher than the highest part of the roof;
 (c) in the case of an antenna to be installed on a roof with a chimney, the highest part of the antenna would be higher than the highest part of the chimney, or 60 centimetres measured from the highest part of the ridge tiles of the roof, whichever is the lower; or
 (d) in the case of article 1(5) land, it would consist of the installation of an antenna–
   (i) on a chimney, wall or roof slope which faces onto, and is visible from, a highway;

(ii) in the Broads, on a chimney, wall or roof slope which faces onto, and is visible from, a waterway; or

(iii) on a building which exceeds 15 metres in height.

Conditions

H.2 Development is permitted by Class H subject to the following conditions–

(a) an antenna installed on a building shall, so far as practicable, be sited so as to minimise its effect on the external appearance of the building; and

(b) an antenna no longer needed for reception or transmission purposes shall be removed as soon as reasonably practicable.

Interpretation of Class H

H.3 The relevant size criteria for the purposes of paragraph H.1(a)(iii) are that:

(a) only one of the antennas may exceed 60 centimetres in length; and

(b) any antenna which exceeds 60 centimetres in length must not exceed 100 centimetres in length.

H.4 The length of the antenna is to be measured in any linear direction, and shall exclude any projecting feed element, reinforcing rim, mounting or brackets.

Interpretation of Part 1

I For the purposes of Part 1–

"raised" in relation to a platform means a platform with a height greater than 300 millimetres; and

"terrace house" means a dwellinghouse situated in a row of three or more dwellinghouses used or designed for use as single dwellings, where–

(a) it shares a party wall with, or has a main wall adjoining the main wall of, the dwellinghouse on either side; or

(b) if it is at the end of a row, it shares a party wall with or has a main wall adjoining the main wall of a dwellinghouse which fulfils the requirements of sub-paragraph (a).

Part 2

Class A
Permitted development

A The erection, construction, maintenance, improvement or alteration of a gate, fence, wall or other means of enclosure.

Development not permitted

A.1 Development is not permitted by Class A if–

(a) the height of any gate, fence, wall or means of enclosure erected or constructed adjacent to a highway used by vehicular traffic would, after the carrying out of the development, exceed one metre above ground level;

(b)   the height of any other gate, fence, wall or means of enclosure erected or constructed would exceed two metres above ground level;

(c)   the height of any gate, fence, wall or other means of enclosure maintained, improved or altered would, as a result of the development, exceed its former height or the height referred to in sub-paragraph (a) or (b) as the height appropriate to it if erected or constructed, whichever is the greater; or

(d)   it would involve development within the curtilage of, or to a gate, fence, wall or other means of enclosure surrounding, a listed building.

Class B
Permitted development

B The formation, laying out and construction of a means of access to a highway which is not a trunk road or a classified road, where that access is required in connection with development permitted by any Class in this Schedule (other than by Class A of this Part).

Class C
Permitted development

C The painting of the exterior of any building or work.

Development not permitted

C.1 Development is not permitted by Class C where the painting is for the purpose of advertisement, announcement or direction.

Interpretation of Class C

C.2 In Class C, "painting" includes any application of colour.

[Class D
Permitted development

D The installation, alteration or replacement, within an area lawfully used for off-street parking, of an electrical outlet mounted on a wall for recharging electric vehicles.

Development not permitted

D.1 Development is not permitted by Class D if the outlet and its casing would–

(a)   exceed 0.2 cubic metres;

(b)   face onto and be within two metres of a highway;

(c)   be within a site designated as a scheduled monument; or

(d)   be within the curtilage of a listed building.

Conditions

D.2 Development is permitted by Class D subject to the conditions that when no longer needed as a charging point for electric vehicles–

    (a) the development shall be removed as soon as reasonably practicable; and

    (b) the wall on which the development was mounted or into which the development was set shall, as soon as reasonably practicable, and so far as reasonably practicable, be reinstated to its condition before that development was carried out.

Class E
Permitted development

E The installation, alteration or replacement, within an area lawfully used for off-street parking, of an upstand with an electrical outlet mounted on it for recharging electric vehicles.

Development not permitted

E.1 Development is not permitted by Class E if the upstand and the outlet would–

    (a) exceed 1.6 metres in height from the level of the surface used for the parking of vehicles;

    (b) be within two metres of a highway;

    (c) be within a site designated as a scheduled monument;

    (d) be within the curtilage of a listed building; or

    (e) result in more than one upstand being provided for each parking space.

Conditions

E.2 Development is permitted by Class E subject to the conditions that when the development is no longer needed as a charging point for electric vehicles–

    (a) the development shall be removed as soon as reasonably practicable; and

    (b) the land on which the development was mounted or into which the development was set shall, as soon as reasonably practicable, and so far as reasonably practicable, be reinstated to its condition before that development was carried out.]

Part 3

Class A
Permitted development

A Development consisting of a change of use of a building to a use falling within Class A1 (shops) of the Schedule to the Use Classes Order from a use falling within Class A3 (restaurants and cafes), A4 (drinking establishments) or A5 (hot food takeaways) of the Schedule.

Class AA
Permitted development

AA Development consisting of a change of use of a building to a use railing within Class A3 (restaurants and cafes) of the Schedule to the Use Classes Order from a use falling within Class A4 (drinking establishments) or Class A5 (hot food takeaways) of that Schedule.

Class B
Permitted development

B Development consisting of a change of the use of a building–
    (a)   to a use for any purpose falling within Class B1 (business) of the Schedule to the Use Classes Order from any use falling within Class B2 (general industrial) or B8 (storage and distribution) of that Schedule;
    (b)   to a use for any purpose falling within Class B8 (storage and distribution) of that Schedule from any use falling within Class B1 (business) or B2 (general industrial).

Development not permitted

B.1 Development is not permitted by Class B where the change is to or from a use falling within Class B8 of that Schedule, if the change of use relates to more than 235 square metres of floor space in the building.

Class C
Permitted development

C Development consisting of a change of use to a use falling within Class A2 (financial and professional services) of the Schedule to the Use Classes Order from a use falling within Class A3 *(food and drink)* [(restaurants and cafes), Class A4 (drinking establishments) or Class A5 (hot food takeaways)] of that Schedule.

Class D
Permitted development

D Development consisting of a change of use of any premises with a display window at ground floor level to a use falling within Class A1 (shops) of the Schedule to the Use Classes Order from a use falling within Class A2 (financial and professional services) of that Schedule.

Class E
Permitted development

E Development consisting of a change of the use of a building or other land from a use permitted by planning permission granted on an application, to another use which that permission would have specifically authorised when it was granted.

Development not permitted

E.1 Development is not permitted by Class E if–
- (a) the application for planning permission referred to was made before the 5th December 1988;
- (b) it would be carried out more than 10 years after the grant of planning permission; or
- (c) it would result in the breach of any condition, limitation or specification contained in that planning permission in relation to the use in question.

Class F
Permitted development

F Development consisting of a change of the use of a building–
- (a) to a mixed use for any purpose within Class A1 (shops) of the Schedule to the Use Classes Order and as a single flat, from a use for any purpose within Class A1 of that Schedule;
- (b) to a mixed use for any purpose within Class A2 (financial and professional services) of the Schedule to the Use Classes Order and as a single flat, from a use for any purpose within Class A2 of that Schedule;
- (c) where that building has a display window at ground floor level, to a mixed use for any purpose within Class A1 (shops) of the Schedule to the Use Classes Order and as a single flat, from a use for any purpose within Class A2 (financial and professional services) of that Schedule.

Conditions

F.1 Development permitted by Class F is subject to the following conditions–
- (a) some or all of the parts of the building used for any purposes within Class A1 or Class A2, as the case may be, of the Schedule to the Use Classes Order shall be situated on a floor below the part of the building used as a single flat;
- (b) where the development consists of a change of use of any building with a display window at ground floor level, the ground floor shall not be used in whole or in part as the single flat;
- (c) the single flat shall not be used otherwise than as a dwelling (whether or not as a sole or main residence)–

(i)   by a single person or by people living together as a family, or

(ii) by not more than six residents living together as a single household (including a household where care is provided for residents).

Interpretation of Class F

F.2 For the purposes of Class F–

"care" means personal care for people in need of such care by reason of old age, disablement, past or present dependence on alcohol or drugs or past or present mental disorder.

Class G
Permitted development

G Development consisting of a change of the use of a building–

(a)   to a use for any purpose within Class A1 (shops) of the Schedule to the Use Classes Order from a mixed use for any purpose within Class A1 of that Schedule and as a single flat;

(b)   to a use for any purpose within Class A2 (financial and professional services) of the Schedule to the Use Classes Order from a mixed use for any purpose within Class A2 of that Schedule and as a single flat;

(c)   where that building has a display window at ground floor level, to a use for any purpose within Class A1 (shops) of the Schedule to the Use Classes Order from a mixed use for any purpose within Class A2 (financial and professional services) of that Schedule and as a single flat.

Development not permitted

G.1 Development is not permitted by Class G unless the part of the building used as a single flat was immediately prior to being so used used for any purpose within Class A1 or Class A2 of the Schedule to the Use Classes Order.

Class H
Permitted Development

H. Development consisting of a change of use of a building from use as a casino to a use falling within Class D2 (Assembly and leisure) of the Schedule to the Use Classes Order.

Class I
Permitted development

I Development consisting of a change of the use of a building–

(a)   to a use falling within Class C3 (dwellinghouses) of the Schedule to the Use Classes Order from a use falling within Class C4 (houses in multiple occupation) of that Schedule;

(b)   to a use falling within Class C4 of that Schedule from a use falling within Class C3.

Part 4

Class A

Permitted development

A The provision on land of buildings, moveable structures, works, plant or machinery required temporarily in connection with and for the duration of operations being or to be carried out on, in, under or over that land or on land adjoining that land.

Development not permitted

A.1 Development is not permitted by Class A if–
(a)   the operations referred to are mining operations, or
(b)   planning permission is required for those operations but is not granted or deemed to be granted.

Conditions

A.2 Development is permitted by Class A subject to the conditions that, when the operations have been carried out–
(a)   any building, structure, works, plant or machinery permitted by Class A shall be removed, and
(b)   any adjoining land on which development permitted by Class A has been carried out shall, as soon as reasonably practicable, be reinstated to its condition before that development was carried out.

Class B

Permitted development

B The use of any land for any purpose for not more than 28 days in total in any calendar year, of which not more than 14 days in total may be for the purposes referred to in paragraph B.2, and the provision on the land of any moveable structure for the purposes of the permitted use.

Development not permitted

B.1 Development is not permitted by Class B if–
(a)   the land in question is a building or is within the curtilage of a building,
(b)   the use of the land is for a caravan site,
(c)   the land is, or is within, a site of special scientific interest and the use of the land is for–
    (i)   a purpose referred to in paragraph B.2(b) or other motor sports;
    (ii)   clay pigeon shooting; or
    (iii)   any war game,
    or
(d)   the use of the land is for the display of an advertisement.

Interpretation of Class B

B.2 The purposes mentioned in Class B above are–
   (a)   the holding of a market;
   (b)   motor car and motorcycle racing including trials of speed, and practising for these activities.

B.3 In Class B, "war game" means an enacted, mock or imaginary battle conducted with weapons which are designed not to injure (including smoke bombs, or guns or grenades which fire or spray paint or are otherwise used to mark other participants), but excludes military activities or training exercises organised by or with the authority of the Secretary of State for Defence.

Part 5

Class A
Permitted development

A The use of land, other than a building, as a caravan site in the circumstances referred to in paragraph A.2.

Condition

A.1 Development is permitted by Class A subject to the condition that the use shall be discontinued when the circumstances specified in paragraph A.2 cease to exist, and all caravans on the site shall be removed as soon as reasonably practicable.

Interpretation of Class A

A.2 The circumstances mentioned in Class A are those specified in paragraphs 2 to 10 of Schedule 1 to the 1960 Act (cases where a caravan site licence is not required), but in relation to those mentioned in paragraph 10 do not include use for winter quarters.

Class B
Permitted development

B Development required by the conditions of a site licence for the time being in force under the 1960 Act.

Part 6

Class A
Development on Units of 5 Hectares or More
Permitted development

A The carrying out on agricultural land comprised in an agricultural unit of 5 hectares or more in area of–
   (a)   works for the erection, extension or alteration of a building; or

(b)   any excavation or engineering operations,

which are reasonably necessary for the purposes of agriculture within that unit.

Development not permitted

A.1   Development is not permitted by Class A if–

  (a)   the development would be carried out on a separate parcel of land forming part of the unit which is less than 1 hectare in area;

  (b)   it would consist of, or include, the erection, extension or alteration of a dwelling;

  (c)   it would involve the provision of a building, structure or works not designed for agricultural purposes;

  (d)   the ground area which would be covered by–

    (i)   any works or structure (other than a fence) for accommodating livestock or any plant or machinery arising from engineering operations; or

    (ii)   any building erected or extended or altered by virtue of Class A,

    would exceed 465 square metres, calculated as described in paragraph D2 below;

  (e)   the height of any part of any building, structure or works within 3 kilometres of the perimeter of an aerodrome would exceed 3 metres;

  (f)   the height of any part of any building, structure or works not within 3 kilometres of the perimeter of an aerodrome would exceed 12 metres;

  (g)   any part of the development would be within 25 metres of a metalled part of a trunk road or classified road;

  (h)   it would consist of, or include, the erection or construction of, or the carrying out of any works to, a building, structure or an excavation used or to be used for the accommodation of livestock or for the storage of slurry or sewage sludge where the building, structure or excavation is, or would be, within 400 metres of the curtilage of a protected building; or

  (i)   it would involve excavations or engineering operations on or over article 1(6) land which are connected with fish farming.

Conditions

A.2

(1)   Development is permitted by Class A subject to the following conditions–

  (a)   where development is carried out within 400 metres of the curtilage of a protected building, any building, structure, excavation or works resulting from the development shall not be used for the accommodation of livestock except in the circumstances described in paragraph D3 below or for the storage of slurry or sewage sludge;

  (b)   where the development involves–

    (i)   the extraction of any mineral from the land (including removal from any disused railway embankment); or

(ii)   the removal of any mineral from a mineral-working deposit, the mineral shall not be moved off the unit;

(c)   waste materials shall not be brought on to the land from elsewhere for deposit except for use in works described in Class A(a) or in the provision of a hard surface and any materials so brought shall be incorporated forthwith into the building or works in question.

(2)   Subject to paragraph (3), development consisting of–

(a)   the erection, extension or alteration of a building;

(b)   the formation or alteration of a private way;

(c)   the carrying out of excavations or the deposit of waste material (where the relevant area, as defined in paragraph D4 below, exceeds 0.5 hectare); or

(d)   the placing or assembly of a tank in any waters,

is permitted by Class A subject to the following conditions–

(i)   the developer shall, before beginning the development, apply to the local planning authority for a determination as to whether the prior approval of the authority will be required to the siting, design and external appearance of the building, the siting and means of construction of the private way, the siting of the excavation or deposit or the siting and appearance of the tank, as the case may be;

(ii)   the application shall be accompanied by a written description of the proposed development and of the materials to be used and a plan indicating the site together with any fee required to be paid;

(iii)   the development shall not be begun before the occurrence of one of the following–

(aa)   the receipt by the applicant from the local planning authority of a written notice of their determination that such prior approval is not required;

(bb)   where the local planning authority give the applicant notice within 28 days following the date of receiving his application of their determination that such prior approval is required, the giving of such approval; or

(cc)   the expiry of 28 days following the date on which the application was received by the local planning authority without the local planning authority making any determination as to whether such approval is required or notifying the applicant of their determination;

(iv)

(aa)   where the local planning authority give the applicant notice that such prior approval is required the applicant shall display a site notice by site display on or near the land on which the proposed development is to be carried out, leaving the notice in position for not less than

21 days in the period of 28 days from the date on which the local planning authority gave the notice to the applicant;

(bb) where the site notice is, without any fault or intention of the applicant, removed, obscured or defaced before the period of 21 days referred to in sub-paragraph (aa) has elapsed, he shall be treated as having complied with the requirements of that sub-paragraph if he has taken reasonable steps for protection of the notice and, if need be, its replacement;

(v) the development shall, except to the extent that the local planning authority otherwise agree in writing, be carried out–

(aa) where prior approval is required, in accordance with the details approved;

(bb) where prior approval is not required, in accordance with the details submitted with the application; and

(vi) the development shall be carried out–

(aa) where approval has been given by the local planning authority, within a period of five years from the date on which approval was given;

(bb) in any other case, within a period of five years from the date on which the local planning authority were given the information referred to in sub-paragraph (d)(ii).

(3)   The conditions in paragraph (2) do not apply to the extension or alteration of a building if the building is not on article 1(6) land except in the case of a significant extension or a significant alteration.

(4)   Development consisting of the significant extension or the significant alteration of a building may only be carried out once by virtue of Class A(a).

(5)   Where development consists of works for the erection, significant extension or significant alteration of a building and

(a) the use of the building or extension for the purposes of agriculture within the unit permanently ceases within ten years from the date on which the development was substantially completed; and

(b) planning permission has not been granted on an application, or has not been deemed to be granted under Part III of the Act, for development for purposes other than agriculture, within three years from the date on which the use of the building or extension for the purposes of agriculture within the unit permanently ceased,

then, unless the local planning authority have otherwise agreed in writing, the building or, in the case of development consisting of an extension, the extension, shall be removed from the land and the land shall, so far as is practicable, be restored to its condition before the development took place, or to such condition as may have been agreed in writing between the local planning authority and the developer.

(6)   Where an appeal has been made, under the Act, in relation to an application for development described in paragraph 5(b), within the period described in that paragraph, that period shall be extended until the appeal is finally determined or withdrawn.

(7)   Where development is permitted by Class A(a), the developer shall notify the local planning authority, in writing and within 7 days, of the date on which the development was substantially completed.

Class B
Development on Units of Less than 5 Hectares
Permitted development

B   The carrying out on agricultural land comprised in an agricultural unit of not less than 0.4 but less than 5 hectares in area of development consisting of–

(a)   the extension or alteration of an agricultural building;

(b)   the installation of additional or replacement plant or machinery;

(c)   the provision, rearrangement or replacement of a sewer, main, pipe, cable or other apparatus;

(d)   the provision, rearrangement or replacement of a private way;

(e)   the provision of a hard surface;

(f)   the deposit of waste; or

(g)   the carrying out of any of the following operations in connection with fish farming, namely, repairing ponds and raceways; the installation of grading machinery, aeration equipment or flow meters and any associated channel; the dredging of ponds; and the replacement of tanks and nets,

where the development is reasonably necessary for the purposes of agriculture within the unit.

Development not permitted

B.1   Development is not permitted by Class B if–

(a)   the development would be carried out on a separate parcel of land forming part of the unit which is less than 0.4 hectare in area;

(b)   the external appearance of the premises would be materially affected;

(c)   any part of the development would be within 25 metres of a metalled part of a trunk road or classified road;

(d)   it would consist of, or involve, the carrying out of any works to a building or structure used or to be used for the accommodation of livestock or the storage of slurry or sewage sludge where the building or structure is within 400 metres of the curtilage of a protected building; or

(e)   it would relate to fish farming and would involve the placing or assembly of a tank on land or in any waters or the construction of a pond in which fish may be kept or an increase (otherwise than by the removal of silt) in the size of any tank or pond in which fish may be kept.

B.2 Development is not permitted by Class B(a) if–

(a)   the height of any building would be increased;

(b)   the cubic content of the original building would be increased by more than 10%;

(c)   any part of any new building would be more than 30 metres from the original building;

(d)   the development would involve the extension, alteration or provision of a dwelling;

(e)   any part of the development would be carried out within 5 metres of any boundary of the unit; or

(f)   the ground area of any building extended by virtue of Class B(a) would exceed 465 square metres.

B.3 Development is not permitted by Class B(b) if–

(a)   the height of any additional plant or machinery within 3 kilometres of the perimeter of an aerodrome would exceed 3 metres;

(b)   the height of any additional plant or machinery not within 3 kilometres of the perimeter of an aerodrome would exceed 12 metres;

(c)   the height of any replacement plant or machinery would exceed that of the plant or machinery being replaced; or

(d)   the area to be covered by the development would exceed 465 square metres calculated as described in paragraph D2 below.

B.4 Development is not permitted by Class B(e) if the area to be covered by the development would exceed 465 square metres calculated as described in paragraph D2 below.

Conditions

B.5 Development permitted by Class B and carried out within 400 metres of the curtilage of a protected building is subject to the condition that any building which is extended or altered, or any works resulting from the development, shall not be used for the accommodation of livestock except in the circumstances described in paragraph D3 below or for the storage of slurry or sewage sludge.

B.6 Development consisting of the extension or alteration of a building situated on article 1(6) land or the provision, rearrangement or replacement of a private way on such land is permitted subject to–

(a)   the condition that the developer shall, before beginning the development, apply to the local planning authority for a determination as to whether the prior approval of the authority will be required to the siting, design and external appearance of the building as extended or altered or the siting and means of construction of the private way; and

(b)   the conditions set out in paragraphs A2(2)(ii) to (vi) above.

B.7 Development is permitted by Class B(f) subject to the following conditions–

(a) that waste materials are not brought on to the land from elsewhere for deposit unless they are for use in works described in Class B(a), (d) or (e) and are incorporated forthwith into the building or works in question; and

(b) that the height of the surface of the land will not be materially increased by the deposit.

B.8 Development is permitted by Class B(a) subject to the following conditions–

(a) Where development consists of works for the significant extension or significant alteration of a building and

(i) the use of the building or extension for the purposes of agriculture within the unit permanently ceases within ten years from the date on which the development was substantially completed; and

(ii) planning permission has not been granted on an application, or has not been deemed to be granted under Part III of the Act, for development for purposes other than agriculture, within three years from the date on which the use of the building or extension for the purposes of agriculture within the unit permanently ceased,

then, unless the local planning authority have otherwise agreed in writing, the extension, in the case of development consisting of an extension, shall be removed from the land and the land shall, so far as is practicable, be restored to its condition before the development took place, or to such condition as may have been agreed in writing between the local planning authority and the developer.

(b) Where an appeal has been made, under the Act, in relation to an application for development described in paragraph B8(a)(ii), within the period described in that paragraph, that period shall be extended until the appeal is finally determined or withdrawn.

(c) The developer shall notify the local planning authority in writing and within 7 days, of the date on which the development was substantially completed.

Class C
Mineral Working for Agricultural Purposes
Permitted development

C The winning and working on land held or occupied with land used for the purposes of agriculture of any minerals reasonably necessary for agricultural purposes within the agricultural unit of which it forms part.

Development not permitted

C.1 Development is not permitted by Class C if any excavation would be made within 25 metres of a metalled part of a trunk road or classified road.

Condition

C.2 Development is permitted by Class C subject to the condition that no mineral extracted during the course of the operation shall be moved to any place outside the land from which it was extracted, except to land which is held or occupied with that land and is used for the purposes of agriculture.

Interpretation of Part 6

D.1 For the purposes of Part 6–

"agricultural land" means land which, before development permitted by this Part is carried out, is land in use for agriculture and which is so used for the purposes of a trade or business, and excludes any dwellinghouse or garden;

"agricultural unit" means agricultural land which is occupied as a unit for the purposes of agriculture, including–

(a) any dwelling or other building on that land occupied for the purpose of farming the land by the person who occupies the unit, or

(b) any dwelling on that land occupied by a farmworker;

"building" does not include anything resulting from engineering operations;

"fish farming" means the breeding, rearing or keeping of fish or shellfish (which includes any kind of crustacean and mollusc);

"livestock" includes fish or shellfish which are farmed;

"protected building" means any permanent building which is normally occupied by people or would be so occupied, if it were in use for purposes for which it is apt; but does not include–

(i) a building within the agricultural unit; or

(ii) a dwelling or other building on another agricultural unit which is used for or in connection with agriculture;

"significant extension" and "significant alteration" mean any extension or alteration of the building where the cubic content of the original building would be exceeded by more than 10% or the height of the building as extended or altered would exceed the height of the original building;

"slurry" means animal faeces and urine (whether or not water has been added for handling); and

"tank" includes any cage and any other structure for use in fish farming.

D.2 For the purposes of Part 6–

(a) an area calculated as described in this paragraph comprises the ground area which would be covered by the proposed development, together with the ground area of any building (other than a dwelling), or any structure, works, plant, machinery, ponds or tanks within the same unit which are being provided or have been provided within the preceding two years and any part of which would be within 90 metres of the proposed development;

    (b)   400 metres is to be measured along the ground.

D.3 The circumstances referred to in paragraphs A2(1)(a) and B5 are–

    (a)   that no other suitable building or structure, 400 metres or more from the curtilage of a protected building, is available to accommodate the livestock; and

    (b)

        (i)   that the need to accommodate the livestock arises from–

            (aa) quarantine requirements; or

            (bb) an emergency due to another building or structure in which the livestock could otherwise be accommodated being unavailable because it has been damaged or destroyed by fire, flood or storm; or

        (ii) in the case of animals normally kept out of doors, they require temporary accommodation in a building or other structure–

            (aa) because they are sick or giving birth or newly born; or

            (bb) to provide shelter against extreme weather conditions.

D.4 For the purposes of paragraph A2(2)(c), the relevant area is the area of the proposed excavation or the area on which it is proposed to deposit waste together with the aggregate of the areas of all other excavations within the unit which have not been filled and of all other parts of the unit on or under which waste has been deposited and has not been removed.

D.5 In paragraph A2(2)(iv), "site notice" means a notice containing–

    (a)   the name of the applicant,

    (b)   the address or location of the proposed development,

    (c)   a description of the proposed development and of the materials to be used,

    (d)   a statement that the prior approval of the authority will be required to the siting, design and external appearance of the building, the siting and means of construction of the private way, the siting of the excavation or deposit or the siting and appearance of the tank, as the case may be,

    (e)   the name and address of the local planning authority,

and which is signed and dated by or on behalf of the applicant.

D.6 For the purposes of Class B–

    (a)   the erection of any additional building within the curtilage of another building is to be treated as the extension of that building and the additional building is not to be treated as an original building;

    (b)   where two or more original buildings are within the same curtilage and are used for the same undertaking they are to be treated as a single original building in making any measurement in connection with the extension or alteration of either of them.

D.7 In Class C, "the purposes of agriculture" includes fertilising land used for the purposes of agriculture and the maintenance, improvement or alteration of any buildings, structures or works occupied or used for such purposes on land so used.

Part 7

Class A

Permitted development

A The carrying out on land used for the purposes of forestry, including afforestation, of development reasonably necessary for those purposes consisting of–

  (a) works for the erection, extension or alteration of a building;
  (b) the formation, alteration or maintenance of private ways;
  (c) operations on that land, or on land held or occupied with that land, to obtain the materials required for the formation, alteration or maintenance of such ways;
  (d) other operations (not including engineering or mining operations).

Development not permitted

A.1 Development is not permitted by Class A if–

  (a) it would consist of or include the provision or alteration of a dwelling;
  (b) the height of any building or works within 3 kilometres of the perimeter of an aerodrome would exceed 3 metres in height; or
  (c) any part of the development would be within 25 metres of the metalled portion of a trunk road or classified road.

A.2

(1) Subject to paragraph (3), development consisting of the erection of a building or the extension or alteration of a building or the formation or alteration of a private way is permitted by Class A subject to the following conditions–

  (a) the developer shall, before beginning the development, apply to the local planning authority for a determination as to whether the prior approval of the authority will be required to the siting, design and external appearance of the building or, as the case may be, the siting and means of construction of the private way;
  (b) the application shall be accompanied by a written description of the proposed development, the materials to be used and a plan indicating the site together with any fee required to be paid;
  (c) the development shall not be begun before the occurrence of one of the following–
      (i) the receipt by the applicant from the local planning authority of a written notice of their determination that such prior approval is not required;
      (ii) where the local planning authority give the applicant notice within 28 days following the date of receiving his application of their determination that such prior approval is required, the giving of such approval;
      (iii) the expiry of 28 days following the date on which the application was received by the local planning authority without the local planning authority making any determination as to whether

such approval is required or notifying the applicant of their determination;

(d)

    (i)   where the local planning authority give the applicant notice that such prior approval is required the applicant shall display a site notice by site display on or near the land on which the proposed development is to be carried out, leaving the notice in position for not less than 21 days in the period of 28 days from the date on which the local planning authority gave the notice to the applicant;

    (ii)   where the site notice is, without any fault or intention of the applicant, removed, obscured or defaced before the period of 21 days referred to in sub-paragraph (i) has elapsed, he shall be treated as having complied with the requirements of that sub-paragraph if he has taken reasonable steps for protection of the notice and, if need be, its replacement;

(e)   the development shall, except to the extent that the local planning authority otherwise agree in writing, be carried out–

    (i)   where prior approval is required, in accordance with the details approved;

    (ii)   where prior approval is not required, in accordance with the details submitted with the application;

(f)   the development shall be carried out–

    (i)   where approval has been given by the local planning authority, within a period of five years from the date on which approval was given,

    (ii)   in any other case, within a period of five years from the date on which the local planning authority were given the information referred to in sub-paragraph (b).

(2)   In the case of development consisting of the significant extension or the significant alteration of the building such development may be carried out only once.

(3)   Paragraph (1) does not preclude the extension or alteration of a building if the building is not on article 1(6) land except in the case of a significant extension or a significant alteration.

Interpretation of Class A

A.3 For the purposes of Class A–

"significant extension" and "significant alteration" mean any extension or alteration of the building where the cubic content of the original building would be exceeded by more than 10% or the height of the building as extended or altered would exceed the height of the original building; and

"site notice" means a notice containing–

    (a)   the name of the applicant,

    (b)   the address or location of the proposed development,

(c) a description of the proposed development and of the materials to be used,

(d) a statement that the prior approval of the authority will be required to the siting, design and external appearance of the building or, as the case may be, the siting and means of construction of the private way,

(e) the name and address of the local planning authority,

and which is signed and dated by or on behalf of the applicant.

Part 8

...

Class A
Permitted development

A The erection, extension or alteration of an industrial building or a warehouse.

Development not permitted

A.1 Development is not permitted by Class A if–

(a) the height of any part of the new building erected would exceed–

(i) if within ten metres of a boundary of the curtilage of the premises, five metres;

(ii) in all other cases, the height of the highest building within the curtilage of the premises or 15 metres, whichever is lower;

(b) the height of the building as extended or altered would exceed–

(i) if within ten metres of a boundary of the curtilage of the premises, five metres;

(ii) in all other cases, the height of the building being extended or altered;

(c) any part of the development would be within five metres of any boundary of the curtilage of the premises;

(d) the gross floor space of any new building erected would exceed 100 square metres;

(e) the gross floor space of the original building would be exceeded by more than–

(i) 10% in respect of development on any article 1(5) land or 25% in any other case; or

(ii) 500 square metres in respect of development on any article 1(5) land or 1,000 square metres in any other case;

whichever is the lesser;

(f) the development would lead to a reduction in the space available for the parking or turning of vehicles; or

(g) the development would be within the curtilage of a listed building.

Conditions

A.2 Development is permitted by Class A subject to the following conditions–
-   (a) the development must be within the curtilage of an existing industrial building or warehouse;
-   (b) any building as erected, extended or altered shall only be used–
    -   (i)   in the case of an industrial building, for the carrying out of an industrial process for the purposes of the undertaking, for research and development of products or processes, or the provision of employee facilities ancillary to the undertaking;
    -   (ii)   in the case of a warehouse, for storage or distribution for the purposes of the undertaking or the provision of employee facilities ancillary to the undertaking;
-   (c) no building as erected, extended or altered shall be used to provide employee facilities–
    -   (i)   between 7.00 pm and 6.30 am, for employees other than those present at the premises of the undertaking for the purpose of their employment, or
    -   (ii)   at all, if a notifiable quantity of a hazardous substance is present at the premises of the undertaking;
-   (d) any new building erected shall, in the case of article 1(5) land, be constructed using materials which have a similar external appearance to those used for the existing industrial building or warehouse; and
-   (e) any extension or alteration shall, in the case of article 1(5) land, be constructed using materials which have a similar external appearance to those used for the building being extended or altered.

Interpretation of Class A

A.3 For the purposes of Class A–
-   (a) where two or more original buildings are within the same curtilage and are used for the same undertaking, they are to be treated as a single original building in making any measurement;
-   (b) "original building" does not include any building erected at any time under Class A;
-   (c) "employee facilities" means social, care or recreational facilities provided for employees of the undertaking, including crèche facilities provided for the children of such employees;
-   (d) "industrial building" means a building used for the carrying out of an industrial process and includes a building used for the carrying out of such a process on land used as a dock, harbour or quay for the purposes of an industrial undertaking and land used for research and development of products or processes, but does not include a building on land in or adjacent to and occupied together with a mine; and
-   (e) "warehouse" means a building used for any purpose within Class B8 (storage or distribution) of the Schedule to the Use Classes Order but

does not include a building on land in or adjacent to and occupied together with a mine.

Class B
Permitted development

B Development carried out on industrial land for the purposes of an industrial process consisting of–
    (a)   the installation of additional or replacement plant or machinery,
    (b)   the provision, rearrangement or replacement of a sewer, main, pipe, cable or other apparatus, or
    (c)   the provision, rearrangement or replacement of a private way, private railway, siding or conveyor.

Development not permitted

B.1 Development described in Class B(a) is not permitted if–
    (a)   it would materially affect the external appearance of the premises of the undertaking concerned; or
    (b)   any plant or machinery would exceed a height of 15 metres above ground level or the height of anything replaced, whichever is the greater.

Interpretation of Class B

B.2 In Class B, "industrial land" means land used for the carrying out of an industrial process, including land used for the purposes of an industrial undertaking as a dock, harbour or quay but does not include land in or adjacent to and occupied together with a mine.

Class C
Permitted development

C Development consisting of–
    (a)   the provision of a hard surface within the curtilage of an industrial building or warehouse to be used for the purpose of the undertaking concerned; or
    (b)   the replacement in whole or in part of such a surface.

Development not permitted

C.1 Development is not permitted by Class C if the development would be within the curtilage of a listed building.

Conditions

C.2 Development is permitted by Class C subject to the following conditions–
    (a)   where there is a risk of groundwater contamination the hard surface shall not be made of porous materials;

(b)　in all other cases, either–
- (i)　the hard surface shall be made of porous materials, or
- (ii)　provision shall be made to direct run-off water from the hard surface to a permeable or porous area or surface within the curtilage of the industrial building or warehouse.

Interpretation of Class C

C.3　In Class C–

"industrial building" means a building used for the carrying out of an industrial process and includes a building used for the carrying out of such a process on land used as a dock, harbour or quay for the purposes of an industrial undertaking and land used for research and development of products or processes, but does not include a building on land in or adjacent to and occupied together with a mine; and

"warehouse" means a building used for any purpose within Class B8 (storage or distribution) of the Schedule to the Use Classes Order but does not include a building on land in or adjacent to and occupied together with a mine.

Class D
Permitted development

D　The deposit of waste material resulting from an industrial process on any land comprised in a site which was used for that purpose on 1st July 1948 whether or not the superficial area or the height of the deposit is extended as a result.

Development not permitted

D.1　Development is not permitted by Class D if–
- (a)　the waste material is or includes material resulting from the winning and working of minerals; or
- (b)　the use on 1st July 1948 was for the deposit of material resulting from the winning and working of minerals.

Part 9

Class A
Permitted development

A　The carrying out on land within the boundaries of an unadopted street or private way of works required for the maintenance or improvement of the street or way.

Interpretation of Class A

A.1　For the purposes of Class A–

"unadopted street" means a street not being a highway maintainable at the public expense within the meaning of the Highways Act 1980.

Part 10

Class A

Permitted development

The carrying out of any works for the purposes of inspecting, repairing or renewing any sewer, main, pipe, cable or other apparatus, including breaking open any land for that purpose.

Part 11

Class A

Permitted development

A Development authorised by–
- (a) a local or private Act of Parliament,
- (b) an order approved by both Houses of Parliament, or
- (c) an order under section 14 or 16 of the Harbours Act 1964 (orders for securing harbour efficiency etc, and orders conferring powers for improvement, construction etc of harbours)

which designates specifically the nature of the development authorised and the land upon which it may be carried out.

Condition

A.1 Development is not permitted by Class A if it consists of or includes–
- (a) the erection, construction, alteration or extension of any building, bridge, aqueduct, pier or dam, or
- (b) the formation, laying out or alteration of a means of access to any highway used by vehicular traffic,

unless the prior approval of the appropriate authority to the detailed plans and specifications is first obtained.

Prior approvals

A.2 The prior approval referred to in paragraph A.1 is not to be refused by the appropriate authority nor are conditions to be imposed unless they are satisfied that–
- (a) the development (other than the provision of or works carried out to a dam) ought to be and could reasonably be carried out elsewhere on the land; or
- (b) the design or external appearance of any building, bridge, aqueduct, pier or dam would injure the amenity of the neighbourhood and is reasonably capable of modification to avoid such injury.

Interpretation of Class A

A.3 In Class A, "appropriate authority" means–
- (a) in Greater London or a metropolitan county, the local planning authority,

    (b)   in a National Park [in England], outside a metropolitan county, the county planning authority,

    (c)   in any other case [in England], the district planning authority,

    (d)   in Wales, the local planning authority.

Part 12

Class A
Permitted development

A The erection or construction and the maintenance, improvement or other alteration by a local authority or by an urban development corporation of–

    (a)   any small ancillary building, works or equipment on land belonging to or maintained by them required for the purposes of any function exercised by them on that land otherwise than as statutory undertakers;

    (b)   lamp standards, information kiosks, passenger shelters, public shelters and seats, telephone boxes, fire alarms, public drinking fountains, horse troughs, refuse bins or baskets, barriers for the control of people waiting to enter public service vehicles, electric vehicle charging points and any associated infrastructure, and similar structures or works required in connection with the operation of any public service administered by them.

Interpretation of Class A

A.1 For the purposes of Class A "urban development corporation" has the same meaning as in Part 16 of the Local Government, Planning and Land Act 1980 (urban development).

A.2 The reference in Class A to any small ancillary building, works or equipment is a reference to any ancillary building, works or equipment not exceeding 4 metres in height or 200 cubic metres in capacity.

Class B
Permitted development

B The deposit by a local authority of waste material on any land comprised in a site which was used for that purpose on 1st July 1948 whether or not the superficial area or the height of the deposit is extended as a result.

Development not permitted

B.1 Development is not permitted by Class B if the waste material is or includes material resulting from the winning and working of minerals.

Interpretation of Part 12

C For the purposes of Part 12–

    "local authority" includes a parish or community council.

Part 13

...

Class A
Permitted development

A The carrying out by a highway authority–
  (a) on land within the boundaries of a road, of any works required for the maintenance or improvement of the road, where such works involve development by virtue of section 55(2)(b) of the Act; or
  (b) on land outside but adjoining the boundary of an existing highway of works required for or incidental to the maintenance or improvement of the highway.

Class B
Permitted development

B The carrying out by the Secretary of State of works in exercise of his functions under the Highways Act 1980, or works in connection with, or incidental to, the exercise of those functions.

Part 14

Class A
Permitted development

A Development by a drainage body in, on or under any watercourse or land drainage works and required in connection with the improvement, maintenance or repair of that watercourse or those works.

Interpretation of Class A

A.1 For the purposes of Class A–
   "drainage body" has the same meaning as in section 72(1) of the Land Drainage Act 1991 (interpretation) other than the Environment Agency.

Part 15

...

Class A
Permitted development

A Development by the [Environment Agency], for the purposes of their functions, consisting of–

(a)   development not above ground level required in connection with conserving, redistributing or augmenting water resources,

(b)   development in, on or under any watercourse or land drainage works and required in connection with the improvement, maintenance or repair of that watercourse or those works,

(c)   the provision of a building, plant, machinery or apparatus in, on, over or under land for the purpose of survey or investigation,

(d)   the maintenance, improvement or repair of works for measuring the flow in any watercourse or channel,

(e)   any works authorised by or required in connection with an order made under section 73 of the Water Resources Act 1991 (power to make ordinary and emergency drought orders),

(f)   any other development in, on, over or under their operational land, other than the provision of a building but including the extension or alteration of a building.

Development not permitted

A.1   Development is not permitted by Class A if–

(a)   in the case of any Class A(a) development, it would include the construction of a reservoir,

(b)   in the case of any Class A(f) development, it would consist of or include the extension or alteration of a building so that–

(i)   its design or external appearance would be materially affected,

(ii)   the height of the original building would be exceeded, or the cubic content of the original building would be exceeded by more than 25%, or

(iii) the floor space of the original building would be exceeded by more than 1,000 square metres,

or

(c)   in the case of any Class A(f) development, it would consist of the installation or erection of any plant or machinery exceeding 15 metres in height or the height of anything it replaces, whichever is the greater.

Condition

A.2   Development is permitted by Class A(c) subject to the condition that, on completion of the survey or investigation, or at the expiration of six months from the commencement of the development concerned, whichever is the sooner, all such operations shall cease and all such buildings, plant, machinery and apparatus shall be removed and the land restored as soon as reasonably practicable to its former condition (or to any other condition which may be agreed with the local planning authority).

Part 16

Class A
Permitted development

A Development by or on behalf of a sewerage undertaker consisting of–

(a) development not above ground level required in connection with the provision, improvement, maintenance or repair of a sewer, outfall pipe, sludge main or associated apparatus;

(b) the provision of a building, plant, machinery or apparatus in, on, over or under land for the purpose of survey or investigation;

(c) the maintenance, improvement or repair of works for measuring the flow in any watercourse or channel;

(d) any works authorised by or required in connection with an order made under section 73 of the Water Resources Act 1991 (power to make ordinary and emergency drought orders);

(e) any other development in, on, over or under their operational land, other than the provision of a building but including the extension or alteration of a building.

Development not permitted

A.1 Development is not permitted by Class A(e) if–

(a) it would consist of or include the extension or alteration of a building so that–

(i) its design or external appearance would be materially affected;

(ii) the height of the original building would be exceeded, or the cubic content of the original building would be exceeded, by more than 25%; or

(iii) the floor space of the original building would be exceeded by more than 1,000 square metres;

or

(b) it would consist of the installation or erection of any plant or machinery exceeding 15 metres in height or the height of anything it replaces, whichever is the greater.

Condition

A.2 Development is permitted by Class A(b) subject to the condition that, on completion of the survey or investigation, or at the expiration of 6 months from the commencement of the development concerned, whichever is the sooner, all such operations shall cease and all such buildings, plant, machinery and apparatus shall be removed and the land restored as soon as reasonably practicable to its former condition (or to any other condition which may be agreed with the local planning authority).

Interpretation of Class A

A.3 For the purposes of Class A–

"associated apparatus", in relation to any sewer, main or pipe, means pumps, machinery or apparatus associated with the relevant sewer, main or pipe;

"sludge main" means a pipe or system of pipes (together with any pumps or other machinery or apparatus associated with it) for the conveyance of the residue of water or sewage treated in a water or sewage treatment works as the case may be, including final effluent or the products of the dewatering or incineration of such residue, or partly for any of those purposes and partly for the conveyance of trade effluent or its residue.

Part 17

Class A
Railway or Light Railway Undertakings
Permitted development

A Development by railway undertakers on their operational land, required in connection with the movement of traffic by rail.

Development not permitted

A.1 Development is not permitted by Class A if it consists of or includes–
   (a)   the construction of a railway,
   (b)   the construction or erection of a hotel, railway station or bridge, or
   (c)   the construction or erection otherwise than wholly within a railway station of–
      (i)   an office, residential or educational building, or a building used for an industrial process, or
      (ii)   a car park, shop, restaurant, garage, petrol filling station or other building or structure provided under transport legislation.

Interpretation of Class A

A.2 For the purposes of Class A, references to the construction or erection of any building or structure include references to the reconstruction or alteration of a building or structure where its design or external appearance would be materially affected.

Class B
Dock, Pier, Harbour, Water Transport, Canal or Inland Navigation Undertakings
Permitted development

B Development on operational land by statutory undertakers or their lessees in respect of dock, pier, harbour, water transport, or canal or inland navigation undertakings, required–

(a) for the purposes of shipping, or

(b) in connection with the embarking, disembarking, loading, discharging or transport of passengers, livestock or goods at a dock, pier or harbour, or with the movement of traffic by canal or inland navigation or by any railway forming part of the undertaking.

Development not permitted

B.1 Development is not permitted by Class B if it consists of or includes–

(a) the construction or erection of a hotel, or of a bridge or other building not required in connection with the handling of traffic,

(b) the construction or erection otherwise than wholly within the limits of a dock, pier or harbour of–

(i)   an educational building, or

(ii)   a car park, shop, restaurant, garage, petrol filling station or other building provided under transport legislation.

Interpretation of Class B

B.2 For the purposes of Class B, references to the construction or erection of any building or structure include references to the reconstruction or alteration of a building or structure where its design or external appearance would be materially affected, and the reference to operational land includes land designated by an order made under section 14 or 16 of the Harbours Act 1964 (orders for securing harbour efficiency etc, and orders conferring powers for improvement, construction etc of harbours), and which has come into force, whether or not the order was subject to the provisions of the Statutory Orders (Special Procedure) Act 1945.

Class C
Works to Inland Waterways
Permitted development

C The improvement, maintenance or repair of an inland waterway (other than a commercial waterway or cruising waterway) to which section 104 of the Transport Act 1968 (classification of the Board's waterways) applies, and the repair or maintenance of a culvert, weir, lock, aqueduct, sluice, reservoir, let-off valve or other work used in connection with the control and operation of such a waterway.

Class D
Dredgings
Permitted development

D The use of any land by statutory undertakers in respect of dock, pier, harbour, water transport, canal or inland navigation undertakings for the spreading of any dredged material.

Class E
Water or Hydraulic Power Undertakings
Permitted development

E Development for the purposes of their undertaking by statutory undertakers for the supply of water or hydraulic power consisting of–

    (a)   development not above ground level required in connection with the supply of water or for conserving, redistributing or augmenting water resources, or for the conveyance of water treatment sludge,

    (b)   development in, on or under any watercourse and required in connection with the improvement or maintenance of that watercourse,

    (c)   the provision of a building, plant, machinery or apparatus in, on, over or under land for the purpose of survey or investigation,

    (d)   the maintenance, improvement or repair of works for measuring the flow in any watercourse or channel,

    (e)   the installation in a water distribution system of a booster station, valve house, meter or switch-gear house,

    (f)   any works authorised by or required in connection with an order made under section 73 of the Water Resources Act 1991 (power to make ordinary and emergency drought orders),

    (g)   any other development in, on, over or under operational land other than the provision of a building but including the extension or alteration of a building.

Development not permitted

E.1 Development is not permitted by Class E if–

    (a)   in the case of any Class E(a) development, it would include the construction of a reservoir,

    (b)   in the case of any Class E(e) development involving the installation of a station or house exceeding 29 cubic metres in capacity, that installation is carried out at or above ground level or under a highway used by vehicular traffic,

    (c)   in the case of any Class E(g) development, it would consist of or include the extension or alteration of a building so that–

        (i)   its design or external appearance would be materially affected;

        (ii)   the height of the original building would be exceeded, or the cubic content of the original building would be exceeded by more than 25%, or

    (iii)   the floor space of the original building would be exceeded by more than 1,000 square metres, or

(d)   in the case of any Class E(g) development, it would consist of the installation or erection of any plant or machinery exceeding 15 metres in height or the height of anything it replaces, whichever is the greater.

Condition

E.2 Development is permitted by Class E(c) subject to the condition that, on completion of the survey or investigation, or at the expiration of six months from the commencement of the development, whichever is the sooner, all such operations shall cease and all such buildings, plant, machinery and apparatus shall be removed and the land restored as soon as reasonably practicable to its former condition (or to any other condition which may be agreed with the local planning authority).

Class F
[Gas Transporters]
Permitted development

F Development by a [gas transporter] required for the purposes of its undertaking consisting of–

(a)   the laying underground of mains, pipes or other apparatus;

(b)   the installation in a gas distribution system of apparatus for measuring, recording, controlling or varying the pressure, flow or volume of gas, and structures for housing such apparatus;

(c)   the construction in any storage area or protective area specified in an order made under section 4 of the Gas Act 1965 (storage authorisation orders), of boreholes, and the erection or construction in any such area of any plant or machinery required in connection with the construction of such boreholes;

(d)   the placing and storage on land of pipes and other apparatus to be included in a main or pipe which is being or is about to be laid or constructed in pursuance of planning permission granted or deemed to be granted under Part III of the Act (control over development);

(e)   the erection on operational land of the [gas transporter] of a building solely for the protection of plant or machinery;

(f)   any other development carried out in, on, over or under the operational land of the [gas transporter].

Development not permitted

F.1 Development is not permitted by Class F if–

(a)   in the case of any Class F(b) development involving the installation of a structure for housing apparatus exceeding 29 cubic metres in capacity, that installation would be carried out at or above ground level, or under a highway used by vehicular traffic,

(b)   in the case of any Class F(c) development–

    (i)   the borehole is shown in an order approved by the Secretary of State for Trade and Industry for the purpose of section 4(6) of the Gas Act 1965; or

    (ii)   any plant or machinery would exceed 6 metres in height, or

(c)   in the case of any Class F(e) development, the building would exceed 15 metres in height, or

(d)   in the case of any Class F(f) development–

    (i)   it would consist of or include the erection of a building, or the reconstruction or alteration of a building where its design or external appearance would be materially affected;

    (ii)   it would involve the installation of plant or machinery exceeding 15 metres in height, or capable without the carrying out of additional works of being extended to a height exceeding 15 metres; or

    (iii)   it would consist of or include the replacement of any plant or machinery, by plant or machinery exceeding 15 metres in height or exceeding the height of the plant or machinery replaced, whichever is the greater.

Conditions

F.2   Development is permitted by Class F subject to the following conditions–

(a)   in the case of any Class F(a) development, not less than eight weeks before the beginning of operations to lay a notifiable pipe-line, the [gas transporter] shall give notice in writing to the local planning authority of its intention to carry out that development, identifying the land under which the pipe-line is to be laid,

(b)   in the case of any Class F(d) development, on completion of the laying or construction of the main or pipe, or at the expiry of a period of nine months from the beginning of the development, whichever is the sooner, any pipes or other apparatus still stored on the land shall be removed and the land restored as soon as reasonably practicable to its condition before the development took place (or to any other condition which may be agreed with the local planning authority),

(c)   in the case of any Class F(e) development, approval of the details of the design and external appearance of the building shall be obtained, before the development is begun, from–

    (i)   in Greater London or a metropolitan county, the local planning authority,

    (ii)   in a National Park [in England], outside a metropolitan county, the county planning authority,

    (iii)   in any other case [in England], the district planning authority,

    [(iv)   in Wales, the local planning authority.]

Class G
Electricity Undertakings
Permitted development

G Development by statutory undertakers for the generation, transmission or supply of electricity for the purposes of their undertaking consisting of–

(a) the installation or replacement in, on, over or under land of an electric line and the construction of shafts and tunnels and the installation or replacement of feeder or service pillars or transforming or switching stations or chambers reasonably necessary in connection with an electric line;

(b) the installation or replacement of any [electronic communications line] which connects any part of an electric line to any electrical plant or building, and the installation or replacement of any support for any such line;

(c) the sinking of boreholes to ascertain the nature of the subsoil and the installation of any plant or machinery reasonably necessary in connection with such boreholes;

(d) the extension or alteration of buildings on operational land;

(e) the erection on operational land of the undertaking or a building solely for the protection of plant or machinery;

(f) any other development carried out in, on, over or under the operational land of the undertaking.

Development not permitted

G.1 Development is not permitted by Class G if–

(a) in the case of any Class G(a) development–

(i) it would consist of or include the installation or replacement of an electric line to which section 37(1) of the Electricity Act 1989 (consent required for overhead lines) applies; or

(ii) it would consist of or include the installation or replacement at or above ground level or under a highway used by vehicular traffic, of a chamber for housing apparatus and the chamber would exceed 29 cubic metres in capacity;

(b) in the case of any Class G(b) development–

(i) the development would take place in a National Park, an area of outstanding natural beauty, or a site of special scientific interest;

(ii) the height of any support would exceed 15 metres; or

(iii) the [electronic communications line] would exceed 1,000 metres in length;

(c) in the case of any Class G(d) development–

(i) the height of the original building would be exceeded;

(ii) the cubic content of the original building would be exceeded by more than 25% or, in the case of any building on article 1(5) land, by more than 10%, or

(iii)   the floor space of the original building would be exceeded by more than 1,000 square metres or, in the case of any building on article 1(5) land, by more than 500 square metres;

(d)   in the case of any Class G(e) development, the building would exceed 15 metres in height, or

(e)   in the case of any Class G(f) development, it would consist of or include–

(i)   the erection of a building, or the reconstruction or alteration of a building where its design or external appearance would be materially affected, or

(ii)   the installation or erection by way of addition or replacement of any plant or machinery exceeding 15 metres in height or the height of any plant or machinery replaced, whichever is the greater.

Conditions

G.2   Development is permitted by Class G subject to the following conditions–

(a)   in the case of any Class G(a) development consisting of or including the replacement of an existing electric line, compliance with any conditions contained in a planning permission relating to the height, design or position of the existing electric line which are capable of being applied to the replacement line;

(b)   in the case of any Class G(a) development consisting of or including the installation of a temporary electric line providing a diversion for an existing electric line, on the ending of the diversion or at the end of a period of six months from the completion of the installation (whichever is the sooner) the temporary electric line shall be removed and the land on which any operations have been carried out to install that line shall be restored as soon as reasonably practicable to its condition before the installation took place;

(c)   in the case of any Class G(c) development, on the completion of that development, or at the end of a period of six months from the beginning of that development (whichever is the sooner) any plant or machinery installed shall be removed and the land shall be restored as soon as reasonably practicable to its condition before the development took place;

(d)   in the case of any Class G(e) development, approval of details of the design and external appearance of the buildings shall be obtained, before development is begun, from–

(i)   in Greater London or a metropolitan county, the local planning authority,

(ii)   in a National Park [in England], outside a metropolitan county, the county planning authority,

(iii)   in any other case [in England], the district planning authority,

(iv)   in Wales, the local planning authority.

Interpretation of Class G

G.3 For the purposes of Class G(a), "electric line" has the meaning assigned to that term by section 64(1) of the Electricity Act 1989 (interpretation etc of Part 1).

G.4 For the purposes of Class G(b), "electrical plant" has the meaning assigned to that term by the said section 64(1) and "[electronic communications line]" means a wire or cable (including its casing or coating) which forms part of [an electronic communications apparatus] within the meaning assigned to that term by paragraph 1 of Schedule 2 to the Telecommunications Act 1984 (the [electronic communications code]).

G.5 For the purposes of Class G(d), (e) and (f), the land of the holder of a licence under section 6(2) of the Electricity Act 1989 (licences authorising supply etc) shall be treated as operational land if it would be operational land within section 263 of the Act (meaning of "operational land") if such licence holders were statutory undertakers for the purpose of that section.

Class H
Tramway or Road Transport Undertakings
Permitted development

H Development required for the purposes of the carrying on of any tramway or road transport undertaking consisting of–

(a) the installation of posts, overhead wires, underground cables, feeder pillars or transformer boxes in, on, over or adjacent to a highway for the purpose of supplying current to public service vehicles;

(b) the installation of tramway tracks, and conduits, drains and pipes in connection with such tracks for the working of tramways;

(c) the installation of telephone cables and apparatus, huts, stop posts and signs required in connection with the operation of public service vehicles;

(d) the erection or construction and the maintenance, improvement or other alteration of passenger shelters and barriers for the control of people waiting to enter public service vehicles;

(e) any other development on operational land of the undertaking.

Development not permitted

H.1 Development is not permitted by Class H if it would consist of–

(a) in the case of any Class H(a) development, the installation of a structure exceeding 17 cubic metres in capacity,

(b) in the case of any Class H(e) development–

(i) the erection of a building or the reconstruction or alteration of a building where its design or external appearance would be materially affected, (ii)　the installation or erection by way of addition or replacement of any plant or machinery which would exceed 15 metres in height or the height of any plant or machinery it replaces, whichever is the greater,

(iii)　development, not wholly within a bus or tramway station, in pursuance of powers contained in transport legislation.

Class I
Lighthouse Undertakings
Permitted development

I Development required for the purposes of the functions of a general or local lighthouse authority under the Merchant Shipping Act 1894 and any other statutory provision made with respect to a local lighthouse authority, or in the exercise by a local lighthouse authority of rights, powers or duties acquired by usage prior to the 1894 Act.

Development not permitted

I.1 Development is not permitted by Class I if it consists of or includes the erection of offices, or the reconstruction or alteration of offices where their design or external appearance would be materially affected.

Class J
[Universal Services Providers]
Permitted development

J Development required for the purposes of [a universal service provider (within the meaning of [Part 3 of the Postal Services Act 2011]) in connection with the provision of a universal postal service (within the meaning of [that Part])] consisting of–

(a)　the installation of posting boxes or self-service machines,

(b)　any other development carried out in, on, over or under the operational land of the undertaking.

Development not permitted

J.1 Development is not permitted by Class J if–

(a)　it would consist of or include the erection of a building, or the reconstruction or alteration of a building where its design or external appearance would be materially affected, or

(b)　it would consist of or include the installation or erection by way of addition or replacement of any plant or machinery which would exceed 15 metres in height or the height of any existing plant or machinery, whichever is the greater.

Interpretation of Part 17

K For the purposes of Part 17–
   "transport legislation" means section 14(1)(d) of the Transport Act 1962 (supplemental provisions relating to the Boards' powers) or section 10(1)(x) of the Transport Act 1968 (general powers of Passenger Transport Executive).

Part 18

Class A
Development at an Airport
Permitted development

A The carrying out on operational land by a relevant airport operator or its agent of development (including the erection or alteration of an operational building) in connection with the provision of services and facilities at a relevant airport.

Development not permitted

A.1 Development is not permitted by Class A if it would consist of or include–
   (a) the construction or extension of a runway;
   (b) the construction of a passenger terminal the floor space of which would exceed 500 square metres;
   (c) the extension or alteration of a passenger terminal, where the floor space of the building as existing at 5th December 1988 or, if built after that date, of the building as built, would be exceeded by more than 15%;
   (d) the erection of a building other than an operational building;
   (e) the alteration or reconstruction of a building other than an operational building, where its design or external appearance would be materially affected.

Condition

A.2 Development is permitted by Class A subject to the condition that the relevant airport operator consults the local planning authority before carrying out any development, unless that development falls within the description in paragraph A.4.

Interpretation of Class A

A.3 For the purposes of paragraph A.1, floor space shall be calculated by external measurement and without taking account of the floor space in any pier or satellite.
A.4 Development falls within this paragraph if–
   (a) it is urgently required for the efficient running of the airport, and
   (b) it consists of the carrying out of works, or the erection or construction of a structure or of an ancillary building, or the placing on land of equipment, and the works, structure, building, or equipment do not exceed 4 metres in height or 200 cubic metres in capacity.

Class B
Air Traffic Services Development at an Airport
Permitted development

B The carrying out on operational land within the perimeter of a relevant airport by a relevant airport operator or its agent of development in connection with the provision of air traffic services.

Class C
Air Traffic Services Development near an Airport
Permitted development

C The carrying out on operational land outside but within 8 kilometres of the perimeter of a relevant airport, by a relevant airport operator or its agent, of development in connection with the provision of air traffic services.

Development not permitted

C.1 Development is not permitted by Class C if–
- (a) any building erected would be used for a purpose other than housing equipment used in connection with the provision of air traffic services;
- (b) any building erected would exceed a height of 4 metres;
- (c) it would consist of the installation or erection of any radar or radio mast, antenna or other apparatus which would exceed 15 metres in height, or, where an existing mast, antenna or apparatus is replaced, the height of that mast antenna or apparatus, if greater.

Class D
Development by an Air Traffic Services Licence Holder within an Airport
Permitted development

D The carrying out by an air traffic services licence holder or its agents within the perimeter of an airport of development in connection with the provision of air traffic services.

Class E
Development by an Air Traffic Services Licence Holder
on Operational Land
Permitted development

E The carrying out on operational land of an air traffic services licence holder by that licence holder or its agents of development in connection with the provision of air traffic services.

Development not permitted

E.1 Development is not permitted by Class E if—

    (a) any building erected would be used for a purpose other than housing equipment used in connection with the provision of air traffic services;

    (b) any building erected would exceed a height of 4 metres; or

    (c) it would consist of the installation or erection of any radar or radio mast, antenna or other apparatus which would exceed 15 metres in height, or, where an existing mast, antenna or apparatus is replaced, the height of that mast, antenna or apparatus, if greater.

Class F
Development by an Air Traffic Services Licence Holder in an Emergency
Permitted development

F.1 The use of land by or on behalf of an air traffic services licence holder in an emergency to station moveable apparatus replacing unserviceable apparatus.

Condition

F.1 Development is permitted by Class F subject to the condition that on or before the expiry of a period of six months beginning with the date on which the use began, the use shall cease, and any apparatus shall be removed, and the land shall be restored to its condition before the development took place, or to any other condition as may be agreed in writing between the local planning authority and the developer.

Class G
Development by an Air Traffic Services Licence Holder Involving Moveable Structures
Permitted development

G The use of land by or on behalf of an air traffic services licence holder to provide services and facilities in connection with the provision of air traffic services and the erection or placing of moveable structures on the land for the purposes of that use.

Condition

G.1 Development is permitted by Class G subject to the condition that, on or before the expiry of the period of six months beginning with the date on which the use began, the use shall cease, and any structure shall be removed, and the land shall be restored to its condition before the development took place, or to any other condition as may be agreed in writing between the local planning authority and the developer.

Class H
Development by the Civil Aviation Authority for Surveys etc
Permitted development

H The use of land by or on behalf of the Civil Aviation Authority for the station-ing and operation of apparatus in connection with the carrying out of surveys or investigations.

Condition

H.1 Development is permitted by Class H subject to the condition that on or before the expiry of the period of six months beginning with the date on which the use began, the use shall cease, and any apparatus shall be removed, and the land shall be restored to its condition before the development took place, or to any other condition as may be agreed in writing between the local planning authority and the developer.

Class I
Use of Airport Buildings Managed by Relevant Airport Operators
Permitted development

I The use of buildings within the perimeter of an airport managed by a relevant airport operator for purposes connected with air transport services or other flying activities at that airport.

Interpretation of Part 18

J For the purposes of Part 18–
    "air traffic services" has the same meaning as in section 98 of the Transport Act 2000 (air traffic services);
    "air traffic services licence holder" means a person who holds a licence under Chapter I of Part I of the Transport Act 2000;
    "operational building" means a building, other than a hotel, required in connection with the movement or maintenance of aircraft, or with the embarking, disembarking, loading, discharge or transport of passengers, livestock or goods at a relevant airport;
    "relevant airport" means an airport to which Part V of the Airports Act 1986 (status of certain airports as statutory undertakers etc) applies; and
    "relevant airport operator" means a relevant airport operator within the meaning of section 57 of the Airports Act 1986 (scope of Part V).

Part 19

Class A
Permitted development

A The carrying out of operations for the erection, extension, installation, rear-rangement, replacement, repair or other alteration of any–

(a)   plant or machinery,
(b)   buildings,
(c)   private ways or private railways or sidings, or
(d)   sewers, mains, pipes, cables or other similar apparatus,
on land used as a mine.

Development not permitted

A.1   Development is not permitted by Class A–
(a)   in relation to land at an underground mine–
    (i)   on land which is not an approved site; or
    (ii)   on land to which the description in paragraph D.1(b) applies, unless a plan of that land was deposited with the mineral planning authority before 5th June 1989;
(b)   if the principal purpose of the development would be any purpose other than–
    (i)   purposes in connection with the winning and working of minerals at that mine or of minerals brought to the surface at that mine; or
    (ii)   the treatment, storage or removal from the mine of such minerals or waste materials derived from them;
(c)   if the external appearance of the mine would be materially affected;
(d)   if the height of any building, plant or machinery which is not in an excavation would exceed–
    (i)   15 metres above ground level; or
    (ii)   the height of the building, plant or machinery, if any, which is being rearranged, replaced or repaired or otherwise altered,
whichever is the greater;
(e)   if the height of any building, plant or machinery in an excavation would exceed–
    (i)   15 metres above the excavated ground level; or
    (ii)   15 metres above the lowest point of the unexcavated ground immediately adjacent to the excavation; or
    (iii)   the height of the building, plant or machinery, if any, which is being rearranged, replaced or repaired or otherwise altered,
whichever is the greatest;
(f)   if any building erected (other than a replacement building) would have a floor space exceeding 1,000 square metres; or
(g)   if the cubic content of any replaced, extended or altered building would exceed by more than 25% the cubic content of the building replaced, extended or altered or the floor space would exceed by more than 1,000 square metres the floor space of that building.

Condition

A.2 Development is permitted by Class A subject to the condition that before the end of the period of 24 months from the date when the mining operations have permanently ceased, or any longer period which the mineral planning authority agree in writing–

    (a)   all buildings, plant and machinery permitted by Class A shall be removed from the land unless the mineral planning authority have otherwise agreed in writing; and

    (b)   the land shall be restored, so far as is practicable, to its condition before the development took place, or restored to such condition as may have been agreed in writing between the mineral planning authority and the developer.

Class B
Permitted development

B The carrying out, on land used as a mine or on ancillary mining land, with the prior approval of the mineral planning authority, of operations for the erection, installation, extension, rearrangement, replacement, repair or other alteration of any–

    (a)   plant or machinery,

    (b)   buildings, or

    (c)   structures or erections.

Development not permitted

B.1 Development is not permitted by Class B–

    (a)   in relation to land at an underground mine–

        (i)   on land which is not an approved site; or

        (ii)   on land to which the description in paragraph D.1(b) applies, unless a plan of that land was deposited with the mineral planning authority before 5th June 1989;

        or

    (b)   if the principal purpose of the development would be any purpose other than–

        (i)   purposes in connection with the operation of the mine,

        (ii)   the treatment, preparation for sale, consumption or utilization of minerals won or brought to the surface at that mine, or

        (iii)   the storage or removal from the mine of such minerals, their products or waste materials derived from them.

B.2 The prior approval referred to in Class B shall not be refused or granted subject to conditions unless the authority are satisfied that it is expedient to do so because–

    (a)   the proposed development would injure the amenity of the neighbourhood and modifications can reasonably be made or conditions reasonably imposed in order to avoid or reduce that injury, or

(b)   the proposed development ought to be, and could reasonably be, sited elsewhere.

Condition

B.3 Development is permitted by Class B subject to the condition that before the end of the period of 24 months from the date when the mining operations have permanently ceased, or any longer period which the mineral planning authority agree in writing–

(a)   all buildings, plant, machinery, structures and erections permitted by Class B shall be removed from the land unless the mineral planning authority have otherwise agreed in writing; and

(b)   the land shall be restored, so far as is practicable, to its condition before the development took place or restored to such condition as may have been agreed in writing between the mineral planning authority and the developer.

Class C
Permitted development

C The carrying out with the prior approval of the mineral planning authority of development required for the maintenance or safety of a mine or a disused mine or for the purposes of ensuring the safety of the surface of the land at or adjacent to a mine or a disused mine.

Development not permitted

C.1 Development is not permitted by Class C if it is carried out by the Coal Authority or any licensed operator within the meaning of section 65 of the Coal Industry Act 1994 (interpretation).

Prior approvals

C.2
(1)   The prior approval of the mineral planning authority to development permitted by Class C is not required if–

(a)   the external appearance of the mine or disused mine at or adjacent to which the development is to be carried out would not be materially affected;

(b)   no building, plant, machinery, structure or erection–
(i)   would exceed a height of 15 metres above ground level, or
(ii)   where any building, plant, machinery, structure or erection is rearranged, replaced or repaired, would exceed a height of 15 metres above ground level or the height of what was rearranged, replaced or repaired, whichever is the greater,
and

(c)   the development consists of the extension, alteration or replacement of an existing building, within the limits set out in paragraph (3).

(2)   The approval referred to in Class C shall not be refused or granted subject to conditions unless the authority are satisfied that it is expedient to do so because–

(a)   the proposed development would injure the amenity of the neighbour-hood and modifications could reasonably be made or conditions reasonably imposed in order to avoid or reduce that injury, or

(b)   the proposed development ought to be, and could reasonably be, sited elsewhere.

(3)   The limits referred to in paragraph C.2(1)(c) are–

(a)   that the cubic content of the building as extended, altered or replaced does not exceed that of the existing building by more than 25%, and

(b)   that the floor space of the building as extended, altered or replaced does not exceed that of the existing building by more than 1,000 square metres.

Interpretation of Part 19

D.1   An area of land is an approved site for the purposes of Part 19 if–

(a)   it is identified in a grant of planning permission or any instrument by virtue of which planning permission is deemed to be granted, as land which may be used for development described in this Part; or

(b)   in any other case, it is land immediately adjoining an active access to an underground mine which, on 5th December 1988, was in use for the purposes of that mine, in connection with the purposes described in paragraph A.1(b)(i) or (ii) or paragraph B.1(b)(i) to (ii) above.

D.2   For the purposes of Part 19–

"active access" means a surface access to underground workings which is in normal and regular use for the transportation of minerals, materials, spoil or men;

"ancillary mining land" means land adjacent to and occupied together with a mine at which the winning and working of minerals is carried out in pursuance of planning permission granted or deemed to be granted under Part III of the Act (control over development);

"minerals" does not include any coal other than coal won or worked during the course of operations which are carried on exclusively for the purpose of exploring for coal or confined to the digging or carrying away of coal that it is necessary to dig or carry away in the course of activities carried on for purposes which do not include the getting of coal or any product of coal;

"the prior approval of the mineral planning authority" means prior written approval of that authority of detailed proposals for the siting, design and external appearance of the building, plant or machinery proposed to be erected, installed, extended or altered;

"underground mine" is a mine at which minerals are worked principally by underground methods.

Part 20

Class A
Permitted development

A Development by a licensee of the Coal Authority, in a mine started before 1st July 1948, consisting of–
- (a)   the winning and working underground of coal or coal-related minerals in a designated seam area; or
- (b)   the carrying out of development underground which is required in order to gain access to and work coal or coal-related minerals in a designated seam area.

Conditions

A.1 Development is permitted by Class A subject to the following conditions–
- (a)   subject to sub-paragraph (b)–
  - (i)   except in a case where there is an approved restoration scheme or mining operations have permanently ceased, the developer shall, before 31st December 1995 or before any later date which the mineral planning authority may agree in writing, apply to the mineral planning authority for approval of a restoration scheme;
  - (ii)   where there is an approved restoration scheme, reinstatement, restoration and aftercare shall be carried out in accordance with that scheme;
  - (iii)   if an approved restoration scheme does not specify the periods within which reinstatement, restoration or aftercare should be carried out, it shall be subject to conditions that–
    - (aa)   reinstatement or restoration, if any, shall be carried out before the end of the period of 24 months from either the date when the mining operations have permanently ceased or the date when any application for approval of a restoration scheme under sub-paragraph (a)(i) has been finally determined, whichever is later, and
    - (bb)   aftercare, if any, in respect of any part of a site, shall be carried out throughout the period of five years from either the date when any reinstatement or restoration in respect of that part is completed or the date when any application for approval of a restoration scheme under sub-paragraph (a)(i) has been finally determined, whichever is later;
  - (iv)   where there is no approved restoration scheme–
    - (aa)   all buildings, plant, machinery, structures and erections used at any time for or in connection with any previous coal-mining operations at that mine shall be removed from any land which is an authorised site unless the

mineral planning authority have otherwise agreed in writing, and

(bb)   that land shall, so far as practicable, be restored to its condition before any previous coalmining operations at that mine took place or to such condition as may have been agreed in writing between the mineral planning authority and the developer,

before the end of the period specified in sub-paragraph (v);

(v)   the period referred to in sub-paragraph (iv) is–

(aa)   the period of 24 months from the date when the mining operations have permanently ceased or, if an application for approval of a restoration scheme has been made under sub- paragraph (a)(i) before that date, 24 months from the date when that application has been finally determined, whichever is later, or

(bb)   any longer period which the mineral planning authority have agreed in writing;

(vi)   for the purposes of sub-paragraph (a), an application for approval of a restoration scheme has been finally determined when the following conditions have been met–

(aa)   any proceedings on the application, including any proceeding on or in consequence of an application under section 288 of the Act (proceedings for questioning the validity of certain orders, decisions and directions), have been determined, and

(bb)   any time for appealing under section 78 (right to appeal against planning decisions and failure to take such decisions), or applying or further applying under section 288, of the Act (where there is a right to do so) has expired;

(b)   sub-paragraph (a) shall not apply to land in respect of which there is an extant planning permission which–

(i)   has been granted on an application under Part III of the Act, and

(ii)   has been implemented.

Interpretation of Class A

A.2 For the purposes of Class A–

"a licensee of the Coal Authority" means any person who is for the time being authorised by a licence under Part II of the Coal Industry Act 1994 to carry on coal-mining operations to which section 25 of that Act (coalmining operations to be licensed) applies;

"approved restoration scheme" means a restoration scheme which is approved when an application made under paragraph A.1(a)(i) is finally determined, as approved (with or without conditions), or as subsequently varied with the written approval of the mineral planning authority (with or without conditions);

"coal-related minerals" means minerals other than coal which are, or may be, won and worked by coalmining operations;

"designated seam area" means land identified, in accordance with paragraph (a) of the definition of "seam plan", in a seam plan which was deposited with the mineral planning authority before 30th September 1993;

"previous coal-mining operations" has the same meaning as in section 54(3) of the Coal Industry Act 1994 (obligations to restore land affected by coal-mining operations) and references in Class A to the use of anything in connection with any such operations shall include references to its use for or in connection with activities carried on in association with, or for purposes connected with, the carrying on of those operations;

"restoration scheme" means a scheme which makes provision for the reinstatement, restoration or aftercare (or a combination of these) of any land which is an authorised site and has been used at any time for or in connection with any previous coal-mining operations at that mine; and

"seam plan" means a plan or plans on a scale of not less than 1 to 25,000 showing–

(a) land comprising the maximum extent of the coal seam or seams that could have been worked from shafts or drifts existing at a mine at 13th November 1992, without further development on an authorised site other than development permitted by Class B of Part 20 of Schedule 2 to the Town and Country Planning General Development Order 1988, as originally enacted;

(b) any active access used in connection with the land referred to in paragraph (a) of this definition;

(c) the National Grid lines and reference numbers shown on Ordnance Survey maps;

(d) a typical stratigraphic column showing the approximate depths of the coal seam referred to in paragraph (a) of this definition.

Class B
Permitted development

B Development by a licensee of the British Coal Corporation, in a mine started before 1st July 1948, consisting of–

(a) the winning and working underground of coal or coal-related minerals in a designated seam area; or

(b) the carrying out of development underground which is required in order to gain access to and work coal or coal-related minerals in a designated seam area.

Interpretation of Class B

B.1 For the purposes of Class B–

"designated seam area" has the same meaning as in paragraph A.2 above;

"coal-related minerals" means minerals other than coal which can only be economically worked in association with the working of coal or which can only be economically brought to the surface by the use of a mine of coal; and

"a licensee of the British Coal Corporation" means any person who is for the time being authorised by virtue of section 25(3) of the Coal Industry Act 1994 (coal-mining operations to be licensed) to carry on coal-mining operations to which section 25 of that Act applies.

Class C
Permitted development

C Any development required for the purposes of a mine which is carried out on an authorised site at that mine by a licensed operator, in connection with coal-mining operations.

Development not permitted

C.1 Development is not permitted by Class C if–

(a) the external appearance of the mine would be materially affected;

(b) any building, plant or machinery, structure or erection or any deposit of minerals or waste–

(i) would exceed a height of 15 metres above ground level, or

(ii) where a building, plant or machinery would be rearranged, replaced or repaired, the resulting development would exceed a height of 15 metres above ground level or the height of what was rearranged, replaced or repaired, whichever is the greater;

(c) any building erected (other than a replacement building) would have a floor space exceeding 1,000 square metres;

(d) the cubic content of any replaced, extended or altered building would exceed by more than 25% the cubic content of the building replaced, extended or altered or the floor space would exceed by more than 1,000 square metres, the floor space of that building;

(e) it would be for the purpose of creating a new surface access to underground workings or of improving an existing access (which is not an active access) to underground workings; or

(f) it would be carried out on land to which the description in paragraph F.2(1)(b) applies, and a plan of that land had not been deposited with the mineral planning authority before 5th June 1989.

Conditions

C.2 Development is permitted by Class C subject to the condition that before the end of the period of 24 months from the date when the mining operations have permanently ceased, or any longer period which the mineral planning authority agree in writing–

    (a)   all buildings, plant, machinery, structures and erections and deposits of minerals or waste permitted by Class C shall be removed from the land unless the mineral planning authority have otherwise agreed in writing; and

    (b)   the land shall, so far as is practicable, be restored to its condition before the development took place or to such condition as may have been agreed in writing between the mineral planning authority and the developer.

Class D

Permitted development

D Any development required for the purposes of a mine which is carried out on an authorised site at that mine by a licensed operator in connection with coal-mining operations and with the prior approval of the mineral planning authority.

Development not permitted

D.1 Development is not permitted by Class D if–

    (a)   it would be for the purpose of creating a new surface access or improving an existing access (which is not an active access) to underground workings; or

    (b)   it would be carried out on land to which the description in paragraph F.2(1)(b) applies, and a plan of that land had not been deposited with the mineral planning authority before 5th June 1989.

Condition

D.2 Development is permitted by Class D subject to the condition that before the end of the period of 24 months from the date when the mining operations have permanently ceased, or any longer period which the mineral planning authority agree in writing–

    (a)   all buildings, plant, machinery, structures and erections and deposits of minerals or waste permitted by Class D shall be removed from the land, unless the mineral planning authority have otherwise agreed in writing; and

    (b)   the land shall, so far as is practicable, be restored to its condition before the development took place or to such condition as may have been agreed in writing between the mineral planning authority and the developer.

Interpretation of Class D

D.3 The prior approval referred to in Class D shall not be refused or granted subject to conditions unless the authority are satisfied that it is expedient to do so because–

    (a)   the proposed development would injure the amenity of the neighbourhood and modifications could reasonably be made or conditions reasonably imposed in order to avoid or reduce that injury, or

    (b)   the proposed development ought to be, and could reasonably be, sited elsewhere.

Class E
Permitted development

E The carrying out by the Coal Authority or a licensed operator, with the prior approval of the mineral planning authority, of development required for the maintenance or safety of a mine or a disused mine or for the purposes of ensuring the safety of the surface of the land at or adjacent to a mine or a disused mine.

Prior approvals

E.1

(1)   The prior approval of the mineral planning authority to development permitted by Class E is not required if–

    (a)   the external appearance of the mine or disused mine at or adjacent to which the development is to be carried out would not be materially affected;

    (b)   no building, plant or machinery, structure or erection–

        (i)   would exceed a height of 15 metres above ground level, or

        (ii)   where any building, plant, machinery, structure or erection is rearranged, replaced or repaired, would exceed a height of 15 metres above ground level or the height of what was rearranged, replaced or repaired, whichever is the greater,

    and

    (c)   the development consists of the extension, alteration or replacement of an existing building, within the limits set out in paragraph (3).

(2)   The approval referred to in Class E shall not be refused or granted subject to conditions unless the authority are satisfied that it is expedient to do so because–

    (a)   the proposed development would injure the amenity of the neighbourhood and modifications could reasonably be made or conditions reasonably imposed in order to avoid or reduce that injury, or

    (b)   the proposed development ought to be, and could reasonably be, sited elsewhere.

(3) The limits referred to in paragraph E.1(1)(c) are–

    (a) that the cubic content of the building as extended, altered or replaced does not exceed that of the existing building by more than 25%, and

    (b) that the floor space of the building as extended, altered or replaced does not exceed that of the existing building by more than 1,000 square metres.

Interpretation of Part 20

F.1 For the purposes of Part 20–

"active access" means a surface access to underground workings which is in normal and regular use for the transportation of coal, materials, spoil or men;

"coal-mining operations" has the same meaning as in section 65 of the Coal Industry Act 1994 (interpretation) and references to any development or use in connection with coal-mining operations shall include references to development or use for or in connection with activities carried on in association with, or for purposes connected with, the carrying on of those operations;

"licensed operator" has the same meaning as in section 65 of the Coal Industry Act 1994;

"normal and regular use" means use other than intermittent visits to inspect and maintain the fabric of the mine or any plant or machinery; and

"prior approval of the mineral planning authority" means prior written approval of that authority of detailed proposals for the siting, design and external appearance of the proposed building, plant or machinery, structure or erection as erected, installed, extended or altered.

F.2

(1) Subject to sub-paragraph (2), land is an authorised site for the purposes of Part 20 if–

    (a) it is identified in a grant of planning permission or any instrument by virtue of which planning permission is deemed to be granted as land which may be used for development described in this Part; or

    (b) in any other case, it is land immediately adjoining an active access which, on 5th December 1988, was in use for the purposes of that mine in connection with coal-mining operations.

(2) For the purposes of sub-paragraph (1), land is not to be regarded as in use in connection with coalmining operations if–

    (a) it is used for the permanent deposit of waste derived from the winning and working of minerals; or

    (b) there is on, over or under it a railway, conveyor, aerial ropeway, roadway, overhead power line or pipeline which is not itself surrounded by other land used for those purposes.

Part 21

Class A
Permitted development

A The deposit, on premises used as a mine or on ancillary mining land already used for the purpose, of waste derived from the winning and working of minerals at that mine or from minerals brought to the surface at that mine, or from the treatment or the preparation for sale, consumption or utilization of minerals from the mine.

Development not permitted

A.1 Development is not permitted by Class A if–

    (a)   in the case of waste deposited in an excavation, waste would be deposited at a height above the level of the land adjoining the excavation, unless that is provided for in a waste management scheme or a relevant scheme;

    (b)   in any other case, the superficial area or height of the deposit (measured as at 21st October 1988) would be increased by more than 10%, unless such an increase is provided for in a waste management scheme or in a relevant scheme.

Conditions

A.2 Development is permitted by Class A subject to the following conditions–

    (a)   except in a case where a relevant scheme or a waste management scheme has already been approved by the mineral planning authority, the developer shall, if the mineral planning authority so require, within three months or such longer period as the authority may specify, submit a waste management scheme for that authority's approval;

    (b)   where a waste management scheme or a relevant scheme has been approved, the depositing of waste and all other activities in relation to that deposit shall be carried out in accordance with the scheme as approved.

Interpretation of Class A

A.3 For the purposes of Class A–

    "ancillary mining land" means land adjacent to and occupied together with a mine at which the winning and working of minerals is carried out in pursuance of planning permission granted or deemed to be granted under Part III of the Act (control over development); and

    "waste management scheme" means a scheme required by the mineral planning authority to be submitted for their approval in accordance with the condition in paragraph A.2(a) which makes provision for–

        (a)   the manner in which the depositing of waste (other than waste deposited on a site for use for filling any mineral excavation in the mine or on ancillary mining land in order to comply with the terms of any planning permission granted on an application or deemed to be granted under Part III of the Act) is to be carried out after the date of the approval of that scheme;

(b)   where appropriate, the stripping and storage of the subsoil and topsoil;

(c)   the restoration and aftercare of the site.

Class B
Permitted development

B The deposit on land comprised in a site used for the deposit of waste materials or refuse on 1st July 1948 of waste resulting from coal-mining operations.

Development not permitted

B.1 Development is not permitted by Class B unless it is in accordance with a relevant scheme approved by the mineral planning authority before 5th December 1988.

Interpretation of Class B

B.2 For the purposes of Class B–

"coal-mining operations" has the same meaning as in section 65 of the Coal Industry Act 1994 (interpretation).

Interpretation of Part 21

C For the purposes of Part 21–

"relevant scheme" means a scheme, other than a waste management scheme, requiring approval by the mineral planning authority in accordance with a condition or limitation on any planning permission granted or deemed to be granted under Part III of the Act (control over development), for making provision for the manner in which the deposit of waste is to be carried out and for the carrying out of other activities in relation to that deposit.

Part 22

Class A
Permitted development

A Development on any land during a period not exceeding 28 consecutive days consisting of–

(a)   the drilling of boreholes;

(b)   the carrying out of seismic surveys; or

(c)   the making of other excavations,

for the purpose of mineral exploration, and the provision or assembly on that land or adjoining land of any structure required in connection with any of those operations.

Development not permitted

A.1 Development is not permitted by Class A if–

    (a)   it consists of the drilling of boreholes for petroleum exploration;

    (b)   any operation would be carried out within 50 metres of any part of an occupied residential building or a building occupied as a hospital or school;

    (c)   any operation would be carried out within a National Park, an area of outstanding natural beauty, a site of archaeological interest or a site of special scientific interest;

    (d)   any explosive charge of more than 1 kilogram would be used;

    (e)   any excavation referred to in paragraph A(c) would exceed 10 metres in depth or 12 square metres in surface area;

    (f)   in the case described in paragraph A(c) more than 10 excavations would, as a result, be made within any area of 1 hectare within the land during any period of 24 months; or

    (g)   any structure assembled or provided would exceed 12 metres in height, or, where the structure would be within 3 kilometres of the perimeter of an aerodrome, 3 metres in height.

Conditions

A.2 Development is permitted by Class A subject to the following conditions–

    (a)   no operations shall be carried out between 6.00 pm and 7.00 am;

    (b)   no trees on the land shall be removed, felled, lopped or topped and no other thing shall be done on the land likely to harm or damage any trees, unless the mineral planning authority have so agreed in writing;

    (c)   before any excavation (other than a borehole) is made, any topsoil and any subsoil shall be separately removed from the land to be excavated and stored separately from other excavated material and from each other;

    (d)   within a period of 28 days from the cessation of operations unless the mineral planning authority have agreed otherwise in writing–

        (i)   any structure permitted by Class A and any waste material arising from other development so permitted shall be removed from the land,

        (ii)   any borehole shall be adequately sealed,

        (iii)   any other excavation shall be filled with material from the site,

        (iv)   the surface of the land on which any operations have been carried out shall be levelled and any topsoil replaced as the uppermost layer, and

        (v)   the land shall, so far as is practicable, be restored to its condition before the development took place, including the carrying out of any necessary seeding and replanting.

Class B
Permitted development

B Development on any land consisting of–
- (a) the drilling of boreholes;
- (b) the carrying out of seismic surveys; or
- (c) the making of other excavations,

for the purposes of mineral exploration, and the provision or assembly on that land or on adjoining land of any structure required in connection with any of those operations.

Development not permitted

B.1 Development is not permitted by Class B if–
- (a) it consists of the drilling of boreholes for petroleum exploration;
- (b) the developer has not previously notified the mineral planning authority in writing of his intention to carry out the development (specifying the nature and location of the development);
- (c) the relevant period has not elapsed;
- (d) any explosive charge of more than 2 kilograms would be used;
- (e) any excavation referred to in paragraph B(c) would exceed 10 metres in depth or 12 square metres in surface area; or
- (f) any structure assembled or provided would exceed 12 metres in height.

Conditions

B.2 Development is permitted by Class B subject to the following conditions–
- (a) the development shall be carried out in accordance with the details in the notification referred to in paragraph B.1(b), unless the mineral planning authority have otherwise agreed in writing;
- (b) no trees on the land shall be removed, felled, lopped or topped and no other thing shall be done on the land likely to harm or damage any trees, unless specified in detail in the notification referred to in paragraph B.1(b) or the mineral planning authority have otherwise agreed in writing;
- (c) before any excavation other than a borehole is made, any topsoil and any subsoil shall be separately removed from the land to be excavated and stored separately from other excavated material and from each other;
- (d) within a period of 28 days from operations ceasing, unless the mineral planning authority have agreed otherwise in writing–
  - (i) any structure permitted by Class B and any waste material arising from other development so permitted shall be removed from the land,
  - (ii) any borehole shall be adequately sealed,

(iii)   any other excavation shall be filled with material from the site,

(iv)   the surface of the land shall be levelled and any topsoil replaced as the uppermost layer, and

(v)   the land shall, so far as is practicable, be restored to its condition before the development took place, including the carrying out of any necessary seeding and replanting,

and

(e)   the development shall cease no later than a date six months after the elapse of the relevant period, unless the mineral planning authority have otherwise agreed in writing.

Interpretation of Class B

B.3   For the purposes of Class B–

"relevant period" means the period elapsing–

(a)   where a direction is not issued under article 7, 28 days after the notification referred to in paragraph B.1(b) or, if earlier, on the date on which the mineral planning authority notify the developer in writing that they will not issue such a direction, or

(b)   where a direction is issued under article 7, 28 days from the date on which notice of that decision is sent to the Secretary of State, or, if earlier, the date on which the mineral planning authority notify the developer that the Secretary of State has disallowed the direction.

Interpretation of Part 22

C   For the purposes of Part 22–

"mineral exploration" means ascertaining the presence, extent or quality of any deposit of a mineral with a view to exploiting that mineral; and

"structure" includes a building, plant or machinery.

Part 23

Class A
Permitted development

A   The removal of material of any description from a stockpile.

Class B
Permitted development

B   The removal of material of any description from a mineral-working deposit other than a stockpile.

Development not permitted

B.1 Development is not permitted by Class B if–
    (a)   the developer has not previously notified the mineral planning author-
          ity in writing of his intention to carry out the development and supplied
          them with the appropriate details;
    (b)   the deposit covers a ground area exceeding 2 hectares, unless the
          deposit contains no mineral or other material which was deposited on
          the land more than 5 years before the development; or
    (c)   the deposit derives from the carrying out of any operations permitted
          under Part 6 of this Schedule or any Class in a previous development
          order which it replaces.

Conditions

B.2 Development is permitted by Class B subject to the following conditions–
    (a)   it shall be carried out in accordance with the details given in the notice
          sent to the mineral planning authority referred to in paragraph B.1 (a)
          above, unless that authority have agreed otherwise in writing;
    (b)   if the mineral planning authority so require, the developer shall within
          a period of three months from the date of the requirement (or such
          other longer period as that authority may provide) submit to them for
          approval a scheme providing for the restoration and aftercare of the
          site;
    (c)   where such a scheme is required, the site shall be restored and aftercare
          shall be carried out in accordance with the provisions of the approved
          scheme;
    (d)   development shall not be commenced until the relevant period has
          elapsed.

Interpretation of Class B

B.3 For the purposes of Class B–
    "appropriate details" means the nature of the development, the exact loca-
        tion of the mineral-working deposit from which the material would be
        removed, the proposed means of vehicular access to the site at which
        the development is to be carried out, and the earliest date at which any
        mineral presently contained in the deposit was deposited on the land;
        and
    "relevant period" means the period elapsing–
    (a)   where a direction is not issued under article 7, 28 days after the
          notification referred to in paragraph B.1(a) or, if earlier, on the
          date on which the mineral planning authority notify the developer
          in writing that they will not issue such a direction; or
    (b)   where a direction is issued under article 7, 28 days from the date
          on which notice of that direction is sent to the Secretary of State,

or, if earlier, the date on which the mineral planning authority notify the developer that the Secretary of State has disallowed the direction.

Interpretation of Part 23

C For the purposes of Part 23–

"stockpile" means a mineral-working deposit consisting primarily of minerals which have been deposited for the purposes of their processing or sale.

Part 24

Class A
Permitted Development

A Development by or on behalf of [an electronic communications code operator] for the purpose of the operator's [electronic communications network] in, on, over or under land controlled by that operator [or] [in accordance with the electronic communications code], consisting of–

(a) the installation, alteration or replacement of any [electronic communications apparatus],

(b) the use of land in an emergency for a period not exceeding six months to station and operate moveable [electronic communications apparatus] required for the replacement of unserviceable [electronic communications apparatus], including the provision of moveable structures on the land for the purposes of that use, or

(c) development ancillary to radio equipment housing.

Development not permitted

A.1 Development is not permitted by Class A(a) if–

(a) in the case of the installation of apparatus (other than on a building or other structure) the apparatus, excluding any antenna, would exceed a height of 15 metres above ground level;

(b) in the case of the alteration or replacement of apparatus already installed (other than on a building or other structure), the apparatus, excluding any antenna, would when altered or replaced exceed the height of the existing apparatus or a height of 15 metres above ground level, whichever is the greater;

(c) in the case of the installation, alteration or replacement of apparatus on a building or other structure, the height of the apparatus (taken by itself) would exceed–

(i) 15 metres, where it is installed, or is to be installed, on a building or other structure which is 30 metres or more in height; or

(ii) 10 metres in any other case;

(d) in the case of the installation, alteration or replacement of apparatus on a building or other structure, the highest part of the apparatus when installed, altered or replaced would exceed the height of the highest part of the building or structure by more than–

    (i)   10 metres, in the case of a building or structure which is 30 metres or more in height;

    (ii)   8 metres, in the case of a building or structure which is more than 15 metres but less than 30 metres in height; or

    (iii) 6 metres in any other case;

(e) in the case of the installation, alteration or replacement of apparatus (other than an antenna) on a mast, the height of the mast would, when the apparatus was installed, altered or replaced, exceed any relerant height limit specified in respect of apparatus in paragraphs A.1(a), (b), (c) and (d), and for the purposes of applying the limit specified in sub-paragraph (c), the words "(taken by itself)" shall be omitted;

(f) in the case of the installation, alteration or replacement of any apparatus other than–

    (i)   a mast,

    (ii)   an antenna,

    (iii) a public call box,

    (iv) any apparatus which does not project above the level of the surface of the ground, or

    (v)   radio equipment housing,

the ground or base area of the structure would exceed 1.5 square metres;

(g) in the case of the installation, alteration or replacement of an antenna on a building or structure (other than a mast) which is less than 15 metres in height; on a mast located on such a building or structure; or, where the antenna is to be located below a height of 15 metres above ground level, on a building or structure (other than a mast) which is 15 metres or more in height–

    (i)   the antenna is to be located on a wall or roof slope facing a highway which is within 20 metres of the building or structure on which the antenna is to be located;

    (ii)   in the case of dish antennas, the size of any dish would exceed 0.9 metres or the aggregate size of all of the dishes on the building, structure or mast would exceed 1.5 metres, when measured in any dimension;

    (iii) in the case of antennas other than dish antennas, the development (other than the installation, alteration or replacement of one small antenna) would result in the presence on the building or structure of more than two antenna systems; or

    (iv) the building or structure is a listed building or a scheduled monument;

(h) in the case of the installation, alteration or replacement of an antenna on a building or structure (other than a mast) which is 15 metres or more in height, or on a mast located on such a building or structure, where the antenna is located at a height of 15 metres or above, measured from ground level–

(i)   in the case of dish antennas, the size of any dish would exceed 1.3 metres or the aggregate size of all of the dishes on the building, structure or mast would exceed 3.5 metres, when measured in any dimension;

(ii)   in the case of antennas other than dish antennas, the development (other than the installation, alteration or replacement of a maximum of two small antennas) would result in the presence on the building or structure of more than three antenna systems; or

(iii)   the building or structure is a listed building or a scheduled monument;

(i)   in the case of development (other than the installation, alteration or replacement of one small antenna on a dwellinghouse or within the curtilage of a dwellinghouse) on any article 1(5) land or any land which is, or is within, a site of special scientific interest, it would consist of–

(i)   the installation or alteration of an antenna or of any apparatus which includes or is intended for the support of such an antenna; or

(ii)   the replacement of such an antenna or such apparatus by an antenna or apparatus which differs from that which is being replaced,

unless the development is carried out in an emergency;

(j)   it would consist of the installation, alteration or replacement of system apparatus within the meaning of section 8(6) of the Road Traffic (Driver Licensing and Information Systems) Act 1989 (definitions of driver information systems etc);

(k)   in the case of the installation of a mast, on a building or structure which is less than 15 metres in height, such a mast would be within 20 metres of a highway;

(l)   in the case of the installation, alteration or replacement of radio equipment housing–

(i)   the development is not ancillary to the use of any other [electronic communications apparatus];

(ii)   the development would exceed 90 cubic metres or, if located on the roof of a building, the development would exceed 30 cubic metres; or

(iii)   on any article 1(5) land, or on any land which is, or is within, a site of special scientific interest, the development would exceed 2.5 cubic metres, unless the development is carried out in an emergency;

(m) in the case of the installation, alteration or replacement on a dwelling-house or within the curtilage of a dwellinghouse of any [electronic communications apparatus], that apparatus–

(i) is not a small antenna;

(ii) being a small antenna, would result in the presence on that dwellinghouse or within the curtilage of that dwellinghouse of more than one such antenna; or

(iii) being a small antenna, is to be located on a roof or on a chimney so that the highest part of the antenna would exceed in height the highest part of that roof or chimney respectively;

(n) in the case of the installation, alteration or replacement on article 1(5) land of a small antenna on a dwellinghouse or within the curtilage of a dwellinghouse, the antenna is to be located–

(i) on a chimney;

(ii) on a building which exceeds 15 metres in height;

(iii) on a wall or roof slope which fronts a highway; or

(iv) in the Broads, on a wall or roof slope which fronts a waterway;

(o) in the case of the installation, alteration or replacement of a small antenna on a building which is not a dwellinghouse or within the curtilage of a dwellinghouse–

(i) the building is on article 1(5) land;

(ii) the building is less than 15 metres in height, and the development would result in the presence on that building of more than one such antenna; or

(iii) the building is 15 metres or more in height, and the development would result in the presence on that building of more than two such antennas.

Conditions

A.2

(1) Class A(a) and Class A(c) development is permitted subject to the condition that any antenna or supporting apparatus, radio equipment housing or development ancillary to radio equipment housing constructed, installed, altered or replaced on a building in accordance with that permission shall, so far as is practicable, be sited so as to minimise its effect on the external appearance of the building.

(2) Class A(a) and Class A(c) development is permitted subject to the condition that any apparatus or structure provided in accordance with that permission shall be removed from the land, building or structure on which it is situated–

(a) if such development was carried out in an emergency on any article 1(5) land or on any land which is, or is within, a site of special scientific interest, at the expiry of the relevant period, or

(b) in any other case, as soon as reasonably practicable after it is no longer required for [electronic communications purposes],

and such land, building or structure shall be restored to its condition before the development took place, or to any other condition as may be agreed in writing between the local planning authority and the developer.

(3)   Class A(b) development is permitted subject to the condition that any apparatus or structure provided in accordance with that permission shall at the expiry of the relevant period be removed from the land and the land restored to its condition before the development took place.

(4)   Class A development–

    (a)   on article 1(5) land or land which is, or is within, a site of special scientific interest, or

    (b)   on any other land and consisting of the construction, installation, alteration or replacement of a mast; or of an antenna on a building or structure (other than a mast) where the antenna (including any supporting structure) would exceed the height of the building or structure at the point where it is installed or to be installed by 4 metres or more; or of a public call box; or of radio equipment housing with a volume in excess of 2.5 cubic metres; or of development ancillary to radio equipment housing–

is permitted subject, except in case of emergency, to the conditions set out in A.3.

A.3

(1)   The developer shall give notice of the proposed development to any person (other than the developer) who is an owner of the land to which the development relates, or a tenant, before making the application required by paragraph (3)–

    (a)   by serving a developer's notice on every such person whose name and address is known to him; and

    (b)   where he has taken reasonable steps to ascertain the names and addresses of every such person, but has been unable to do so, by local advertisement.

(2)   Where the proposed development consists of the installation of a mast within 3 kilometres of the perimeter of an aerodrome, the developer shall notify the Civil Aviation Authority, the Secretary of State for Defence or the aerodrome operator, as appropriate, before making the application required by paragraph (3).

(3)   Before beginning the development, the developer shall apply to the local planning authority for a determination as to whether the prior approval of the authority will be required to the siting and appearance of the development.

(4)   The application shall be accompanied–

    (a)   by a written description of the proposed development and a plan indicating its proposed location together with any fee required to be paid;

    (b)   where paragraph (1) applies, by evidence that the requirements of paragraph (1) have been satisfied; and

    (c)   where paragraph (2) applies, by evidence that the Civil Aviation Authority, the Secretary of State for Defence or the aerodrome operator, as the case may be, has been notified of the proposal.

(5)   Subject to paragraphs (7)(c) and (d), upon receipt of the application under paragraph (4) the local planning authority shall–

(a) for development which, in their opinion, falls within a category set out in the table of article 10 of the Procedure Order, consult the authority or person mentioned in relation to that category, except where–

(i)   the local planning authority are the authority so mentioned; or

(ii)   the authority or person so mentioned has advised the local planning authority that they do not wish to be consulted,

and shall give the consultees at least 14 days within which to comment;

(b) in the case of development which does not accord with the provisions of the development plan in force in the area in which the land to which the application relates is situated or which would affect a right of way to which Part III of the Wildlife and Countryside Act 1981 (public rights of way) applies, shall give notice of the proposed development, in the appropriate form set out in Schedule 3 to the Procedure Order–

(i)

(aa) by site display in at least one place on or near the land to which the application relates for not less than 21 days, and

(ii)

(bb) by local advertisement;

(c) in the case of development which does not fall within paragraph (b) but which involves development carried out on a site having an area of 1 hectare or more, shall give notice of the proposed development, in the appropriate form set out in Schedule 3 to the Procedure Order–

(i)

(aa) by site display in at least one place on or near the land to which the application relates for not less than 21 days, or

(bb) by serving notice on any adjoining owner or occupier, and

(ii)   by local advertisement;

(d) in the case of development which does not fall within (b) or (c), shall give notice of the proposed development, in the appropriate form set out in Schedule 3 to the Procedure Order–

(i)   by site display in at least one place on or near the land to which the application relates for not less than 21 days, or

(ii)   by serving the notice on any adjoining owner or occupier.

(6) The local planning authority shall take into account any representations made to them as a result of consultations or notices given under A.3, when determining the application made under paragraph (3).

(7) The development shall not be begun before the occurrence of one of the following–

(a) the receipt by the applicant from the local planning authority of a written notice of their determination that such prior approval is not required;

    (b)   where the local planning authority gives the applicant written notice that such prior approval is required, the giving of that approval to the applicant, in writing, within a period of 56 days beginning with the date on which they received his application;

    (c)   where the local planning authority gives the applicant written notice that such prior approval is required, the expiry of a period of 56 days beginning with the date on which the local planning authority received his application without the local planning authority notifying the applicant, in writing, that such approval is given or refused; or

    (d)   the expiry of a period of 56 days beginning with the date on which the local planning authority received the application without the local planning authority notifying the applicant, in writing, of their determination as to whether such prior approval is required.

(8)   The development shall, except to the extent that the local planning authority otherwise agree in writing, be carried out–

    (a)   where prior approval has been given as mentioned in paragraph (7)(b) in accordance with the details approved;

    (b)   in any other case, in accordance with the details submitted with the application.

(9)   The development shall be begun–

    (a)   where prior approval has been given as mentioned in paragraph (7)(b), not later than the expiration of five years beginning with the date on which the approval was given;

    (b)   in any other case, not later than the expiration of five years beginning with the date on which the local planning authority were given the information referred to in paragraph (4).

(10)   In a case of emergency, development is permitted by Class A subject to the condition that the operator shall give written notice to the local planning authority of such development as soon as possible after the emergency begins.

Interpretation of Class A

A.4 For the purposes of Class A–

    "aerodrome operator" means the person for the time being having the management of an aerodrome or, in relation to a particular aerodrome, the management of that aerodrome;

    "antenna system" means a set of antennas installed on a building or structure and operated by a single electronic communications code operator or in accordance with the electronic communications code.

    "development ancillary to radio equipment housing" means the construction, installation, alteration or replacement of structures, equipment or means of access which are ancillary to and reasonably required for the purposes of the radio equipment housing;

    ...

    "developer's notice" means a notice signed and dated by or on behalf of the developer and containing–

(i)   the name of the developer;

(ii)   the address or location of the proposed development;

(iii)   a description of the proposed development (including its siting and appearance and the height of any mast);

(iv)   a statement that the developer will apply to the local planning authority for a determination as to whether the prior approval of the authority will be required to the siting and appearance of the development;

(v)   the name and address of the local planning authority to whom the application will be made;

(vi)   a statement that the application shall be available for public inspection at the offices of the local planning authority during usual office hours;

(vii)   a statement that any person who wishes to make representations about the siting and appearance of the proposed development may do so in writing to the local planning authority;

(viii) the date by which any such representations should be received by the local planning authority, being a date not less than 14 days from the date of the notice; and

(ix)   the address to which such representations should be made.

"land controlled by the operator" means land occupied by the operator in right of a freehold interest or a leasehold interest under a lease granted for a term of not less than 10 years;

"local advertisement" means by publication of the notice in a newspaper circulating in the locality in which the land to which the application relates is situated;

"mast" means a radio mast or a radio tower;

"owner" means any person who is the estate owner in respect of the fee simple, or who is entitled to a tenancy granted or extended for a term of years certain of which not less than seven years remain unexpired;

"Procedure Order" means the Town and Country Planning (General Development Procedure) Order 1995;

"relevant period" means a period which expires–

(i)   six months from the commencement of the construction, installation, alteration or replacement of any apparatus or structure permitted by Class A(a) or Class A(c) or from the commencement of the use permitted by Class A(b), as the case may be, or

(ii)   when the need for such apparatus, structure or use ceases,
        whichever occurs first;

"site display" means by the posting of the notice by firm affixture to some object, sited and displayed in such a way as to be easily visible and legible by members of the public;

"small antenna" means an antenna which–

(i)　is for use in connection with a telephone system operating on a point to fixed multi-point basis;

(ii)　does not exceed 50 centimetres in any linear measurement; and

(iii)　does not, in two-dimensional profile, have an area exceeding 1,591 square centimetres,

and any calculation for the purposes of (ii) and (iii) shall exclude any feed element, reinforcing rim mountings and brackets;

...

"tenant" means the tenant of an agricultural holding any part of which is comprised in the land to which the application relates.

Part 25

Class A

Permitted development

A The installation, alteration or replacement on any building or other structure of a height of 15 metres or more of a microwave antenna and any structure intended for the support of a microwave antenna.

Development not permitted

A.1 Development is not permitted by Class A if–

(a)　the building is a dwellinghouse or the building or structure is within the curtilage of a dwellinghouse;

(b)　it would consist of development of a kind described in paragraph A of Part 24;

(c)　it would consist of the installation, alteration or replacement of system apparatus within the meaning of section 8(6) of the Road Traffic (Driver Licensing and Information Systems) Act 1989 (definitions of driver information systems etc);

(d)　it would result in the presence on the building or structure of more than four antennas;

(e)　in the case of an antenna installed on a chimney, the length of the antenna would exceed 60cm;

(f)　in all other cases, the length of the antenna would exceed 130cm;

(g)　it would consist of the installation of an antenna with a cubic capacity in excess of 35 litres;

(h)　the highest part of the antenna or its supporting structure would be more than three metres higher than the highest part of the building or structure on which it is installed or is to be installed;

(i)　in the case of article 1(5) land, it would consist of the installation of an antenna–

(i)　on a chimney, wall or roof slope which faces onto, and is visible from, a highway;

(ii)   in the Broads, on a chimney, wall or roof slope which faces onto, and is visible from, a waterway.

Conditions

A.2 Development is permitted by Class A subject to the following conditions–
(a)   the antenna shall, so far as is practicable, be sited so as to minimise its effect on the external appearance of the building or structure on which it is installed;
(b)   an antenna no longer needed for reception or transmission purposes shall be removed from the building or structure as soon as reasonably practicable.

A.3 The length of an antenna is to be measured in any linear direction, and shall exclude any projecting feed element, reinforcing rim, mounting or brackets.

Class B
Permitted development

B The installation, alteration or replacement on any building or other structure of a height of less than 15 metres of a microwave antenna.

Development not permitted

B.1 Development is not permitted by Class B if–
(a)   the building is a dwellinghouse or other structure within the curtilage of a dwellinghouse;
(b)   it would consist of development of a kind described in paragraph A of Part 24;
(c)   it would consist of the installation, alteration or replacement of system apparatus within the meaning of section 8(6) of the Road Traffic (Driver Licensing and Information Systems) Act 1989 (definitions of driver information systems etc);
(d)   it would result in the presence on the building or structure of–
(i)   more than two antennas;
(ii)   a single antenna exceeding 100 centimetres in length;
(iii)   two antennas which do not meet the relevant size criteria;
(iv)   an antenna installed on a chimney, where the length of the antenna would exceed 60cm;
(v)   an antenna installed on a chimney, where the antenna would protrude over the chimney;
(vi)   an antenna with a cubic capacity in excess of 35 litres;
(e)   in the case of an antenna to be installed on a roof without a chimney, the highest part of the antenna would be higher than the highest part of the roof;
(f)   in the case of an antenna to be installed on a roof with a chimney, the highest part of the antenna would be higher than the highest part of

the chimney stack, or 60cm measured from the highest part of the ridge tiles of the roof, whichever is the lowest;

(g) in the case of article 1(5) land, it would consist of the installation of an antenna–

   (i) on a chimney, wall or roof slope which faces onto, and is visible from, a highway;

   (ii) in the Broads, on a chimney, wall or roof slope which faces onto, and is visible from, a waterway.

Condition

B.2 Development is permitted by Class B subject to the following conditions

(a) the antenna shall, so far as practicable, be sited so as to minimise its effect on the external appearance of the building or structure on which it is installed;

(b) an antenna no longer needed for reception or transmission purposes shall be removed from the building or structure as soon as reasonably practicable.

B.3 The relevant size criteria for the purposes of paragraph B.1(d)(iii) are that:

(a) only one of the antennas may exceed 60 centimetres in length; and

(b) any antenna which exceeds 60 centimetres in length must not exceed 100 centimetres in length.

B.4 The length of an antenna is to be measured in any linear direction and shall exclude any projecting feed element, reinforcing rim, mounting or brackets.

Part 26

Class A

Permitted development

A Development by or on behalf of the Historic Buildings and Monuments Commission for England, consisting of–

(a) the maintenance, repair or restoration of any building or monument;

(b) the erection of screens, fences or covers designed or intended to protect or safeguard any building or monument; or

(c) the carrying out of works to stabilise ground conditions by any cliff, watercourse or the coastline;

where such works are required for the purposes of securing the preservation of any building or monument.

Development not permitted

A.1 Development is not permitted by Class A(a) if the works involve the extension of the building or monument.

Condition

A.2 Except for development also falling within Class A(a), Class A(b) development is permitted subject to the condition that any structure erected in accordance with that permission shall be removed at the expiry of a period of six months (or such longer period as the local planning authority may agree in writing) from the date on which work to erect the structure was begun.

Interpretation of Class A

A.3 For the purposes of Class A–
  "building or monument" means any building or monument in the guardianship of the Historic Buildings and Monuments Commission for England or owned, controlled or managed by them.

Part 27

Class A
Permitted development

A The use of land by members of a recreational organisation for the purposes of recreation or instruction, and the erection or placing of tents on the land for the purposes of the use.

Development not permitted

A.1 Development is not permitted by Class A if the land is a building or is within the curtilage of a dwellinghouse.

Interpretation of Class A

A.2 For the purposes of Class A–
  "recreational organisation" means an organisation holding a certificate of exemption under section 269 of the Public Health Act 1936 (power of local authority to control use of moveable dwellings).

Part 28

Class A
Permitted development

A Development on land used as an amusement park consisting of–
  (a) the erection of booths or stalls or the installation of plant or machinery to be used for or in connection with the entertainment of the public within the amusement park; or
  (b) the extension, alteration or replacement of any existing booths or stalls, plant or machinery so used.

Development not permitted

A.1 Development is not permitted by Class A if–
    (a)   the plant or machinery would–
        (i)   if the land or pier is within 3 kilometres of the perimeter of an aerodrome, exceed a height of 25 metres or the height of the highest existing structure (whichever is the lesser), or
        (ii)   in any other case, exceed a height of 25 metres;
    (b)   in the case of an extension to an existing building or structure, that building or structure would as a result exceed 5 metres above ground level or the height of the roof of the existing building or structure, whichever is the greater, or
    (c)   in any other case, the height of the building or structure erected, extended, altered or replaced would exceed 5 metres above ground level.

Interpretation of Class A

A.2 For the purposes of Class A–
    "amusement park" means an enclosed area of open land, or any part of a seaside pier, which is principally used (other than by way of a temporary use) as a funfair or otherwise for the purposes of providing public entertainment by means of mechanical amusements and side-shows; but, where part only of an enclosed area is commonly so used as a funfair or for such public entertainment, only the part so used shall be regarded as an amusement park; and
    "booths or stalls" includes buildings or structures similar to booths or stalls.

Part 29

Class A
Permitted development

A The installation, alteration or replacement of system apparatus by or on behalf of a driver information system operator.

Development not permitted

A.1 Development is not permitted by Class A if–
    (a)   in the case of the installation, alteration or replacement of system apparatus other than on a building or other structure–
        (i)   the ground or base area of the system apparatus would exceed 1.5 square metres; or
        (ii)   the system apparatus would exceed a height of 15 metres above ground level;
    (b)   in the case of the installation, alteration or replacement of system apparatus on a building or other structure–

(i) the highest part of the apparatus when installed, altered, or replaced would exceed in height the highest part of the building or structure by more than 3 metres; or

(ii) the development would result in the presence on the building or structure of more than two microwave antennas.

Conditions

A.2 Development is permitted by Class A subject to the following conditions–

(a) any system apparatus shall, so far as practicable, be sited so as to minimise its effect on the external appearance of any building or other structure on which it is installed;

(b) any system apparatus which is no longer needed for a driver information system shall be removed as soon as reasonably practicable.

Interpretation of Class A

A.3 For the purposes of Class A–

"driver information system operator" means a person granted an operator's licence under section 10 of the Road Traffic (Driver Licensing and Information Systems) Act 1989 (operators' licences); and

"system apparatus" has the meaning assigned to that term by section 8(6) of that Act (definitions of driver information systems etc).

Part 30

Class A
Permitted development

A Development consisting of–

(a) the setting up and the maintenance, improvement or other alteration of facilities for the collection of tolls;

(b) the provision of a hard surface to be used for the parking of vehicles in connection with the use of such facilities.

Development not permitted

A.1 Development is not permitted by Class A if–

(a) it is not located within 100 metres (measured along the ground) of the boundary of a toll road;

(b) the height of any building or structure would exceed–
(i)   7.5 metres excluding any rooftop structure; or
(ii)   10 metres including any rooftop structure;

(c) the aggregate area of the floor space at or above ground level of any building or group of buildings within a toll collection area, excluding the floor space of any toll collection booth, would exceed 1,500 square metres.

Conditions

A.2 In the case of any article 1(5) land, development is permitted by Class A subject to the following conditions–

(a)   the developer shall, before beginning the development, apply to the local planning authority for a determination as to whether the prior approval of the authority will be required to the siting, design and external appearance of the facilities for the collection of tolls;

(b)   the application shall be accompanied by a written description, together with plans and elevations, of the proposed development and any fee required to be paid;

(c)   the development shall not be begun before the occurrence of one of the following–

(i)   the receipt by the applicant from the local planning authority of a written notice of their determination that such prior approval is not required;

(ii)   where the local planning authority give the applicant notice within 28 days following the date of receiving his application of their determination that such prior approval is required, the giving of such approval; or

(iii)   the expiry of 28 days following the date on which the application was received by the local planning authority without the local planning authority making any determination as to whether such approval is required or notifying the applicant of their determination;

(d)   the development shall, except to the extent that the local planning authority otherwise agree in writing, be carried out–

(i)   where prior approval is required, in accordance with the details approved;

(ii)   where prior approval is not required, in accordance with the details submitted with the application;

and

(e)   the development shall be carried out–

(i)   where approval has been given by the local planning authority, within a period of five years from the date on which the approval was given;

(ii)   in any other case, within a period of five years from the date on which the local planning authority were given the information referred to in sub-paragraph (b).

Interpretation of Class A

A.3 For the purposes of Class A–

"facilities for the collection of tolls" means such buildings, structures, or other facilities as are reasonably required for the purpose of or in connection with the collection of tolls in pursuance of a toll order;

"ground level" means the level of the surface of the ground immediately adjacent to the building or group of buildings in question or, where the level of the surface of the ground on which it is situated or is to be situated is not uniform, the level of the highest part of the surface of the ground adjacent to it;

"rooftop structure" means any apparatus or structure which is reasonably required to be located on and attached to the roof, being an apparatus or structure which is–

(a) so located for the provision of heating, ventilation, air conditioning, water, gas or electricity;

(b) lift machinery; or

(c) reasonably required for safety purposes;

"toll" means a toll which may be charged pursuant to a toll order;

"toll collection area" means an area of land where tolls are collected in pursuance of a toll order, and includes any facilities for the collection of tolls;

"toll collection booth" means any building or structure designed or adapted for the purpose of collecting tolls in pursuance of a toll order;

"toll order" has the same meaning as in Part I of the New Roads and Street Works Act 1991 (new roads in England and Wales); and

"toll road" means a road which is the subject of a toll order.

Part 31

Class A

Permitted development

A Any building operation consisting of the demolition of a building.

Development not permitted

A.1 Development is not permitted by Class A where–

(a) the building has been rendered unsafe or otherwise uninhabitable by the action or inaction of any person having an interest in the land on which the building stands; and

(b) it is practicable to secure safety or health by works of repair or works for affording temporary support.

Conditions

A.2 Development is permitted by Class A subject to the following conditions–

(a) where demolition is urgently necessary in the interests of safety or health and the measures immediately necessary in such interests are the demolition of the building the developer shall, as soon as reasonably practicable, give the local planning authority a written justification of the demolition;

(b)   where the demolition does not fall within sub-paragraph (a) and is not excluded demolition–

(i)   the developer shall, before beginning the development, apply to the local planning authority for a determination as to whether the prior approval of the authority will be required to the method of demolition and any proposed restoration of the site;

(ii)   the application shall be accompanied by a written description of the proposed development, a statement that a notice has been posted in accordance with sub-paragraph (iii) and any fee required to be paid;

(iii) subject to sub-paragraph (iv), the applicant shall display a site notice by site display on or near the land on which the building to be demolished is sited and shall leave the notice in place for not less than 21 days in the period of 28 days beginning with the date on which the application was submitted to the local planning authority;

(iv)   where the site notice is, without any fault or intention of the applicant, removed, obscured or defaced before the period of 21 days referred to in sub-paragraph (iii) has elapsed, he shall be treated as having complied with the requirements of that sub-paragraph if he has taken reasonable steps for protection of the notice and, if need be, its replacement;

(v)   the development shall not be begun before the occurrence of one of the following–

(aa)   the receipt by the applicant from the local planning authority of a written notice of their determination that such prior approval is not required;

(bb)   where the local planning authority give the applicant notice within 28 days following the date of receiving his application of their determination that such prior approval is required, the giving of such approval; or

(cc)   the expiry of 28 days following the date on which the application was received by the local planning authority without the local planning authority making any determination as to whether such approval is required or notifying the applicant of their determination;

(vi)   the development shall, except to the extent that the local planning authority otherwise agree in writing, be carried out–

(aa)   where prior approval is required, in accordance with the details approved;

(bb)   where prior approval is not required, in accordance with the details submitted with the application;

and
(vii) the development shall be carried out–
  (aa) where approval has been given by the local planning authority, within a period of five years from the date on which approval was given;
  (bb) in any other case, within a period of five years from the date on which the local planning authority were given the information referred to in sub-paragraph (ii).

Interpretation of Class A

A.3 For the purposes of Class A–
"excluded demolition" means demolition–
  (a) on land which is the subject of a planning permission, for the redevelopment of the land, granted on an application or deemed to be granted under Part III of the Act (control over development),
  (b) required or permitted to be carried out by or under any enactment, or
  (c) required to be carried out by virtue of a relevant obligation;
"relevant obligation" means–
  (a) an obligation arising under an agreement made under section 106 of the Act, as originally enacted (agreements regulating development or use of land);
  (b) a planning obligation entered into under section 106 of the Act, as substituted by section 12 of the Planning and Compensation Act 1991 (planning obligations), or under section 299A of the Act (Crown planning obligations);
  (c) an obligation arising under or under an agreement made under any provision corresponding to section 106 of the Act, as originally enacted or as substituted by the Planning and Compensation Act 1991, or to section 299A of the Act; and
"site notice" means a notice containing–
  (a) the name of the applicant,
  (b) a description, including the address, of the building or buildings which it is proposed be demolished,
  (c) a statement that the applicant has applied to the local planning authority for a determination as to whether the prior approval of the authority will be required to the method of demolition and any proposed restoration of the site,
  (d) the date on which the applicant proposes to carry out the demolition, and
  (e) the name and address of the local planning authority,
and which is signed and dated by or on behalf of the applicant.

Class B
Permitted development

B Any building operation consisting of the demolition of the whole or any part of any gate, fence, wall or other means of enclosure.

Part 32

...

Class A
Permitted development

A The erection, extension or alteration of a school, college, university or hospital building.

Development not permitted

A.1 Development is not permitted by Class A–
   (a)   if the cumulative gross floor space of any buildings erected, extended or altered would exceed–
      (i)   25% of the gross floor space of the original school, college, university or hospital buildings; or
      (ii)   100 square metres,
      whichever is the lesser;
   (b)   if any part of the development would be within five metres of a boundary of the curtilage of the premises;
   (c)   if, as a result of the development, any land used as a playing field at any time in the five years before the development commenced and remaining in this use could no longer be so used;
   (d)   if the height of any new building erected would exceed five metres;
   (e)   if the height of the building as extended or altered would exceed–
      (i)   if within ten metres of a boundary of the curtilage of the premises, five metres; or
      (ii)   in all other cases, the height of the building being extended or altered;
   (f)   if the development would be within the curtilage of a listed building; or
   (g)   unless–
      (i)   in the case of school, college or university buildings, the predominant use of the existing buildings on the premises is for the provision of education;
      (ii)   in the case of hospital buildings, the predominant use of the existing buildings on the premises is for the provision of any medical or health services.

Conditions

A.2 Development is permitted by Class A subject to the following conditions–

(a) the development must be within the curtilage of an existing school, college, university or hospital;

(b) the development shall only be used as part of, or for a purpose incidental to, the use of that school, college, university or hospital;

(c) any new building erected shall, in the case of article 1(5) land, be constructed using materials which have a similar external appearance to those Used for the original school, college, university or hospital buildings; and

(d) any extension or alteration shall, in the case of article 1(5) land, be constructed using materials which have a similar external appearance to those used for the building being extended or altered.

Interpretation

A.3 For the purposes of Class A–

(a) where two or more original buildings are within the same curtilage and are used for the same institution, they are to be treated as a single original building in making any measurement; and

(b) "original school, college, university or hospital building" means any original building which is a school, college, university or hospital building, as the case may be, other than any building erected at any time under Class A.

Class B
Permitted development

B Development consisting of–

(a) the provision of a hard surface within the curtilage of any school, college, university or hospital to be used for the purposes of that school, college, university or hospital; or

(b) the replacement in whole or in part of such a surface.

Development not permitted

B.1 Development is not permitted by Class B if–

(a) the cumulative area of ground covered by a hard surface within the curtilage of the site (other than hard surfaces already existing on 6th April 2010) would exceed 50 square metres;

(b) as a result of the development, any land used as a playing field at any time in the five years before the development commenced and remaining in this use could no longer be so used; or

(c) the development would be within the curtilage of a listed building.

Conditions

B.2 Development is permitted by Class B subject to the following conditions–
    (a)   where there is a risk of groundwater contamination the hard surface shall not be made of porous materials;
    (b)   in all other cases, either–
        (i)   the hard surface shall be made of porous materials, or
        (ii)   provision shall be made to direct run-off water from the hard surface to a permeable or porous area or surface within the curtilage of the institution.

Part 33

Class A

Permitted development

A The installation, alteration or replacement on a building of a closed circuit television camera to be used for security purposes.

Development not permitted

A.1 Development is not permitted by Class A if–
    (a)   the building on which the camera would be installed, altered or replaced is a listed building or a scheduled monument;
    (b)   the dimensions of the camera including its housing exceed 75 centimetres by 25 centimetres by 25 centimetres;
    (c)   any part of the camera would, when installed, altered or replaced, be less than 250 centimetres above ground level;
    (d)   any part of the camera would, when installed, altered or replaced, protrude from the surface of the building by more than one metre when measured from the surface of the building;
    (e)   any part of the camera would, when installed, altered or replaced, be in contact with the surface of the building at a point which is more than one metre from any other point of contact;
    (f)   any part of the camera would be less than 10 metres from any part of another camera installed on a building;
    (g)   the development would result in the presence of more than four cameras on the same side of the building; or
    (h)   the development would result in the presence of more than 16 cameras on the building.

Conditions

A.2 Development is permitted by Class A subject to the following conditions–
    (a)   the camera shall, so far as practicable, be sited so as to minimise its effect on the external appearance of the building on which it is situated;
    (b)   the camera shall be removed as soon as reasonably practicable after it is no longer required for security purposes.

Interpretation of Class A

A.3 For the purposes of Class A–

"camera", except in paragraph A.1(b), includes its housing, pan and tilt mechanism, infra red illuminator, receiver, mountings and brackets; and

"ground level" means the level of the surface of the ground immediately adjacent to the building or, where the level of the surface of the ground is not uniform, the level of the highest part of the surface of the ground adjacent to it.

Part 34

Class A

Permitted development

A The erection or construction and the maintenance, improvement or other alteration by or on behalf of the Crown of–

    (a) any small ancillary building, works or equipment on Crown land required for operational purposes;

    (b) lamp standards, information kiosks, passenger shelters, shelters and seats, telephone boxes, fire alarms, drinking fountains, refuse bins or baskets, barriers for the control of people and vehicles, and similar structures or works required in connection with the operational purposes of the Crown.

Interpretation of Class A

A.1 The reference in Class A to any smafl ancillary building, works or equipment is a reference to any ancillary building, works or equipment not exceeding 4 metres in height or 200 cubic metres in capacity.

Class B

Permitted development

B The extension or alteration by or on behalf of the Crown of an operational Crown building.

Development not permitted

B.1 Development is not permitted by Class B if–

    (a) the building as extended or altered is to be used for purposes other than those of–

        (i) the Crown; or

        (ii) the provision of employee facilities;

    (b) the height of the building as extended or altered would exceed the height of the original building;

    (c) the cubic content of the original building would be exceeded by more than–

(i)   10%, in respect of development on any article 1(5) land; or

(ii)   25%, in any other case;

(d)   the floor space of the original building would be exceeded by more than–

(i)   500 square metres in respect of development on any article 1(5) land; or

(ii)   1,000 square metres in any other case;

(e)   the external appearance of the original building would be materially affected;

(f)   any part of the building as extended or altered would be within 5 metres of any boundary of the curtilage of the original building; or

(g)   the development would lead to a reduction in the space available for the parking or turning of vehicles.

Interpretation of Class B

B.2 For the purposes of Class B–

(a)   the erection of any additional building within the curtilage of another building (whether by virtue of Class B or otherwise) and used in connection with it is to be treated as the extension of that building, and the additional building is not to be treated as an original building;

(b)   where two or more original buildings are within the same curtilage and are used for the same operational purposes, they are to be treated as a single original building in making any measurement;

(c)   "employee facilities" means social, care or recreational facilities provided for employees or servants of the Crown, including crèche facilities provided for the children of such employees or servants.

Class C
Permitted development

C Development carried out by or on behalf of the Crown on operational Crown land for operational purposes consisting of–

(a)   the installation of additional or replacement plant or machinery;

(b)   the provision, rearrangement or replacement of a sewer, main, pipe, cable or other apparatus; or

(c)   the provision, rearrangement or replacement of a private way, private railway, siding or conveyor.

Development not permitted

C.1 Development described in Class C(a) is not permitted if–

(a)   it would materially affect the external appearance of the premises; or

(b)   any plant or machinery would exceed a height of 15 metres above ground level or the height of anything replaced, whichever is the greater.

Interpretation of Class C

C.2 In Class C, "Crown land" does not include land in or adjacent to and occupied together with a mine.

Class D
Permitted development

D The provision by or on behalf of the Crown of a hard surface within the curtilage of an operational Crown building.

Part 35

Class A
Permitted development

A The carrying out on operational Crown land, by or on behalf of the Crown, of development (including the erection or alteration of an operational building) in connection with the provision of services and facilities at an airbase.

Development not permitted

A.1 Development is not permitted by Class A if it would consist of or include–
  (a) the construction or extension of a runway;
  (b) the construction of a passenger terminal the floor space of which would exceed 500 square metres;
  (c) the extension or alteration of a passenger terminal, where the floor space of the building as existing at 7th June 2006 or, if built after that date, of the building as built, would be exceeded by more than 15%;
  (d) the erection of a building other than an operational building;
  (e) the alteration or reconstruction of a building other than an operational building, where its design or external appearance would be materially affected.

Condition

A.2 Development is permitted by Class A subject to the condition that the relevant airbase operator consults the local planning authority before carrying out any development, unless that development falls within the description in paragraph A.4.

Interpretation of Class A

A.3 For the purposes of paragraph A.1, floor space shall be calculated by external measurement and without taking account of the floor space in any pier or satellite.

A.4 Development fells within this paragraph if–
  (a) it is urgently required for the efficient running of the airbase, and
  (b) it consists of the carrying out of works, or the erection or construction of a structure or of an ancillary building, or the placing on land of

equipment, and the works, structure, building, or equipment do not exceed 4 metres in height or 200 cubic, metres in capacity.

A.5 For the purposes of Class A, "operational building" means an operational Crown building, other than a hotel, required in connection with the movement or maintenance of aircraft, or with the embarking, disembarking, loading, discharge or transport of passengers, military or civilian personnel, goods, military equipment, munitions and other items.

Class B
Permitted development

B The carrying out on operational land within the perimeter of an airbase, by or on behalf of the Crown, of development in connection with the provision of air traffic services.

Class C
Permitted development

C The carrying out on operational land outside but within 8 kilometres of the perimeter of an airbase, by or on behalf of the Crown, of development in connection with the provision of air traffic services.

Development not permitted

C.1 Development is not permitted by Class C if–
   (a) any building erected would be used for a purpose other than housing equipment used in connection with the provision of air traffic services;
   (b) any building erected would exceed a height of 4 metres; or
   (c) it would consist of the installation or erection of any radar or radio mast, antenna or other apparatus which would exceed 15 metres in height, or, where an existing mast, antenna or apparatus is replaced, the height of that mast antenna or apparatus, if greater.

Class D
Permitted development

D The carrying out on operational land, by or on behalf of the Crown, of development in connection with the provision of air traffic services.

Development not permitted

D.1 Development is not permitted by Class D if–
   (a) any building erected would be used for a purpose other than housing equipment used in connection with the provision of air traffic services;
   (b) any building erected would exceed a height of 4 metres; or

(c)   it would consist of the installation or erection of any radar or radio mast, antenna or other apparatus which would exceed 15 metres in height, or, where an existing mast, antenna or apparatus is replaced, the height of that mast, antenna or apparatus, if greater.

Class E
Permitted development

E   The use of land by or on behalf of the Crown in an emergency to station moveable apparatus replacing unserviceable apparatus in connection with the provision of air traffic services.

Condition

E.1   Development is permitted by Class E subject to the condition that on or before the expiry of a period of six months beginning with the date on which the use began, the use shall cease, and any apparatus shall be removed, and the land shall be restored to its condition before the development took place, or to such other state as may be agreed in writing between the local planning authority and the developer.

Class F
Permitted development

F   The use of land by or on behalf of the Crown to provide services and facilities in connection with the provision of air traffic services and the erection or placing of moveable structures on the land for the purposes of that use.

Condition

F.1   Development is permitted by Class F subject to the condition that, on or before the expiry of the period of six months beginning with the date on which the use began, the use shall cease, any structure shall be removed, and the land shall be restored to its condition before the development took place, or to such other state as may be agreed in writing between the local planning authority and the developer.

Class G
Permitted development

G   The use of land by or on behalf of the Crown for the stationing and operation of apparatus in connection with the carrying out of surveys or investigations.

Condition

G.1   Development is permitted by Class G subject to the condition that on or before the expiry of the period of six months beginning with the date on which the use began, the use shall cease, any apparatus shall be removed, and the land shall be restored to its condition before the development took place, or to such other state as may be agreed in writing between the local planning authority and the developer.

Class H
Permitted development

H The use of buildings by or on behalf of the Crown within the perimeter of an airbase for purposes connected with air transport services or other flying activities at that airbase.

Interpretation of Part 35

I For the purposes of Part 35–

"airbase" means the aggregate of the land, buildings and works comprised in a Government aerodrome within the meaning of article 155 of the Air Navigation Order 2005; and

"air traffic services" has the same meaning as in section 98 of the Transport Act 2000 (air traffic services).

Part 36

Class A
Permitted development

A Development by or on behalf of the Crown on operational Crown land, required in connection with the movement of traffic by rail.

Development not permitted

A.1 Development is not permitted by Class A if it consists of or includes–
    (a)   the construction of a railway;
    (b)   the construction or erection of a hotel, railway station or bridge; or
    (c)   the construction or erection otherwise than wholly within a railway station of an office, residential or educational building, car park, shop, restaurant, garage, petrol filling station or a building used for an industrial process.

Interpretation of Class A

A.2 For the purposes of Class A, references to the construction or erection of any building or structure include references to the reconstruction or alteration of a building or structure where its design or external appearance would be materially affected.

Class B
Permitted development

B Development by or on behalf of the Crown or its lessees on operational Crown land where the development is required–
  (a) for the purposes of shipping; or
  (b) at a dock, pier, pontoon or harbour in connection with the embarking, disembarking, loading, discharging or transport of military or civilian personnel, military equipment, munitions, or other items.

Development not permitted

B.1 Development is not permitted by Class B if it consists of or includes the construction or erection of a bridge or other building not required in connection with the handling of traffic.

Interpretation of Class B

B.2 For the purposes of Class B, references to the construction or erection of any building or structure include references to the reconstruction or alteration of a building or structure where its design or external appearance would be materially affected.

Class C
Permitted development

C The use of any land by or on behalf of the Crown for the spreading of any dredged material resulting from a dock, pier, harbour, water transport, canal or inland navigation undertaking.

Class D
Permitted development

D Development by or on behalf of the Crown on operational Crown land, or for operational purposes, consisting of–
  (a) the use of the land as a lighthouse, with all requisite works, roads and appurtenances;
  (b) the extension of, alteration, or removal of a lighthouse; or
  (c) the erection, placing, alteration or removal of a buoy or beacon.

Development not permitted

D.1 Development is not permitted by Class D if it consists of or includes the erection of offices, or the reconstruction or alteration of offices where their design or external appearance would be materially affected.

Interpretation of Class D

D.2 For the purposes of Class D–
  "buoys and beacons" includes all other marks and signs of the sea; and

"lighthouse" includes any floating and other light exhibited for the guidance of ships, and also any sirens and any other description of fog signals.

Part 37

...

Class A
Permitted development

A Development by or on behalf of the Crown on Crown land for the purposes of–

   (a)   preventing an emergency;

   (b)   reducing, controlling or mitigating the effects of an emergency; or

   (c)   taking other action in connection with an emergency.

Conditions

A.1　Development is permitted by Class A subject to the following conditions–

   (a)   the developer shall, as soon as practicable after commencing development, notify the local planning authority of that development; and

   (b)   on or before the expiry of the period of six months beginning with the date on which the development began–

       (i)   any use of that land for a purpose of Class A shall cease and any buildings, plant, machinery, structures and erections permitted by Class A shall be removed; and

      (ii)   the land shall be restored to its condition before the development took place, or to such other state as may be agreed in writing between the local planning authority and the developer.

Interpretation of Class A

A.2

(1)　For the purposes of Class A, "emergency" means an event or situation which threatens serious damage to–

   (a)   human welfare in a place in the United Kingdom;

   (b)   the environment of a place in the United Kingdom; or

   (c)   the security of the United Kingdom.

(2)　For the purposes of paragraph (1)(a) an event or situation threatens damage to human welfare only if it involves, causes or may cause

   (a)   loss of human life;

   (b)   human illness or injury;

   (c)   homelessness;

   (d)   damage to property;

   (e)   disruption of a supply of money, food, water, energy or fuel;

   (f)   disruption of a system of communication;

(g) disruption of facilities for transport; or

(h) disruption of services relating to health.

(3) For the purposes of paragraph (1)(b) an event or situation threatens damage to the environment only if it involves, causes or may cause–

(a) contamination of land, water or air with biological, chemical or radio-active matter; or

(b) disruption or destruction of plant life or animal life.

Part 38

Class A

Permitted development

A The erection, construction, maintenance, improvement or alteration of a gate, fence, wall of other means of enclosure by or on behalf of the Crown on Crown land for national security purposes.

Development not permitted

A.1 Development is not permitted by Class A if the height of any gate, fence, wall or other means of enclosure erected or constructed would exceed 4.5 metres above ground level.

Class B

Permitted development

B The installation, alteration or replacement by or on behalf of the Crown on Crown land of a closed circuit television camera and associated lighting for national security purposes.

Development not permitted

B.1 Development is not permitted by Class B if–

(a) the dimensions of the camera including its housing exceed 75 centimetres by 25 centimetres by 25 centimetres;

(b) the uniform level of lighting provided exceeds 10 lux measured at ground level.

Conditions

B.2 Development is permitted by Class B subject to the following conditions–

(a) the camera shall, so far as practicable, be sited so as to minimise its effect on the external appearance of any building to which it is fixed;

(b) the camera shall be removed as soon as reasonably practicable after it is no longer required for national security purposes.

Interpretation of Class B

B.3 For the purposes of Class B–

"camera" except in paragraph B1(a) includes its housing, pan and tilt mechanism, infra red illuminator, receiver, mountings and brackets; and

"ground level" means the level of the surface of the ground immediately adjacent to the building to which the camera is attached or, where the level of the surface of the ground is not uniform, the level of the lowest part of the surface of the ground adjacent to it.

Class C
Permitted development

C Development by or on behalf of the Crown for national security purposes in, on, over or under Crown land, consisting of–

(a) the installation, alteration or replacement of any electronic communications apparatus;

(b) the use of land in an emergency for a period not exceeding six months to station and operate moveable electronic communications apparatus required for the replacement of unserviceable electronic communications apparatus, including the provision of moveable structures on the land for the purposes of that use; or

(c) development ancillary to radio equipment housing.

Development not permitted

C.1 Development is not permitted by Class C(a) if–

(a) in the case of the installation of apparatus (other than on a building) the apparatus, excluding any antenna, would exceed a height of 15 metres above ground level;

(b) in the case of the alteration or replacement of apparatus already installed (other than on a building), the apparatus, excluding any antenna, would, when altered or replaced, exceed the height of the existing apparatus or a height of 15 metres above ground level, whichever is the greater;

(c) in the case of the installation, alteration or replacement of apparatus on a building, the height of the apparatus (taken by itself) would exceed the height of the existing apparatus or–

(i) 15 metres, where it is installed, or is to be installed, on a building which is 30 metres or more in height; or

(ii) 10 metres in any other case,

whichever is the greater;

(d) in the case of the installation, alteration or replacement of apparatus on a building, the highest part of the apparatus when installed, altered or replaced would exceed the height of the highest part of the building by more than the height of the existing apparatus or–

(iii)   10 metres, where it is installed, or is to be installed, on a building which is 30 metres or more in height;

(iv)   8 metres, in the case of a building which is more than 15 metres but less than 30 metres in height; or

(v)   6 metres in any other case.

whichever is the greater;

(e)   in the case of the installation, alteration or replacement of apparatus (other than an antenna) on a mast, the height of the mast and the apparatus supported by it would, when the apparatus was installed, altered or replaced, exceed any relevant height limit specified in respect of apparatus in paragraphs C1(a), (b), (c) and (d), and for the purposes of applying the limit specified in sub-paragraph (c), the words "(taken by itself)" shall be disregarded;

(f)   in the case of the installation, alteration or replacement of any apparatus other than–

(i)   a mast;

(ii)   an antenna;

(iii)   any apparatus which does not project above the level of the surface of the ground; or

(iv)   radio equipment housing,

the ground or base area of the structure would exceed the ground or base area of the existing structure or 1.5 square metres, whichever is the greater;

(g)   in the case of the installation, alteration or replacement of an antenna on a building (other than a mast) which is less than 15 metres in height; on a mast located on such a building; or, where the antenna is to be located below a height of 15 metres above ground level, on a building (other than a mast) which is 15 metres or more in height–

(i)   the antenna is to be located on a wall or roof slope facing a highway which is within 20 metres of the building on which the antenna is to be located, unless it is essential for operational purposes that the antenna is located in that position; or

(ii)   in the case of dish antennas, the size of any dish would exceed the size of the existing dish when measured in any dimension or 1.3 metres when measured in any dimension, whichever is the greater;

(h)   in the case of the installation, alteration or replacement of a dish antenna on a building (other than a mast) which is 15 metres or more in height, or on a mast located on such a building, where the antenna is located at a height of 15 metres or above, measured from ground level the size of any dish would exceed the size of the existing dish when measured in any dimension or 1.3 metres when measured in any dimension, whichever is the greater;

(i)   in the case of the installation of a mast, on a building which is less than 15 metres in height, such a mast would be within 20 metres of a highway, unless it is essential for operational purposes that the mast is installed in that position;

    (j)   in the case of the installation, alteration or replacement of radio equipment housing–

        (i)   the development is not ancillary to the use of any other electronic communications apparatus; or

        (ii)   the development would exceed 90 cubic metres or, if located on the roof of a building, the development would exceed 30 cubic metres.

C.2 Development consisting of the installation of apparatus is not permitted by Class C(a) on article 1(5) land unless–

    (a)   the land on which the apparatus is to be installed is, or forms part of, a site on which there is existing electronic communication apparatus;

    (b)   the existing apparatus was installed on the site on or before the relevant day; and

    (c)   the site was Crown land on the relevant day.

C.3

(1)   Subject to paragraph (2), development is not permitted by Class C(a) if it will result in the installation of more than one item of apparatus ("the original apparatus") on a site in addition to any item of apparatus already on that site on the relevant day.

(2)   In addition to the original apparatus which may be installed on a site by virtue of Class C(a), for every four items of apparatus which existed on that site on the relevant day, one additional item of small apparatus may be installed.

(3)   In paragraph (2), "small apparatus" means–

    (a)   a dish antenna, other than on a building, not exceeding 5 metres in diameter and 7 metres in height;

    (b)   an antenna, other than a dish antenna and other than on a building, not exceeding 7 metres in height;

    (c)   a hard standing or other base for any apparatus described in sub-paragraphs (a) and (b), not exceeding 7 metres in diameter;

    (d)   a dish antenna on a building, not exceeding 1.3 metres in diameter and 3 metres in height;

    (e)   an antenna, other than a dish antenna, on a building, not exceeding 3 metres in height;

    (f)   a mast on a building, not exceeding 3 metres in height;

    (g)   equipment housing not exceeding 3 metres in height and of which the area, when measured at ground level, does not exceed 9 square metres.

Conditions

C.4

(1)   Class C(a) and Class C(c) development is permitted subject to the condition that any antenna or supporting apparatus, radio equipment housing or development ancillary to radio equipment housing constructed, installed, altered or replaced on a building in accordance with that permission shall, so far as is practicable, be sited so as to minimise its effect on the external appearance of the building.

(2) Class C(a) development consisting of the installation of any additional apparatus on article 1(5) land is permitted subject to the condition that the apparatus shall be installed as close as is reasonably practicable to any existing apparatus.

(3) Class C(b) development is permitted subject to the condition that any apparatus or structure provided in accordance with that permission shall, at the expiry of the relevant period be removed from the land and the land restored to its condition before the development took place.

(4) Class C development–

    (a) on article 1(5) land or land which is, or is within, a site of special scientific interest; or

    (b) on any other land and consisting of the construction, installation, alteration or replacement of a mast; or of an antenna on a building or structure (other than a mast) where the antenna (including any supporting structure) would exceed the height of the building or structure at the point where it is installed or to be installed by 4 metres or more; or of radio equipment housing with a volume in excess of 2.5 cubic metres; or of development ancillary to radio equipment housing–

is permitted subject, except in case of emergency, to the conditions set out in C.5.

C.5

(1) The developer shall, before commencing development, give notice of the proposed development to any person (other than the developer) who is an owner or tenant of the land to which the development relates–

    (a) by serving the appropriate notice on every such person whose name and address is known to him; and

    (b) where he has taken reasonable steps to ascertain the names and addresses of every such person, but has been unable to do so, by local advertisement.

(2) Where the proposed development consists of the installation of a mast within 3 kilometres of the perimeter of an aerodrome, the developer shall, before commencing development, notify the Civil Aviation Authority, the Secretary of State for Defence or the aerodrome operator, as appropriate.

Interpretation of Class C

C.6 For the purposes of Class C–

    "aerodrome operator" means the person who is for the time being responsible for the management of the aerodrome;

    "appropriate notice" means a notice signed and dated by or on behalf of the developer and containing–

    (a) the name of the developer;

    (b) the address or location of the proposed development;

    (c) a description of the proposed development (including its siting and appearance and the height of any mast);

    "development ancillary to radio equipment housing" means the construction, installation, alteration or replacement of structures, equipment or means of access which are ancillary to and reasonably required for the purposes of the radio equipment housing;

"local advertisement" means by publication of the notice in a newspaper circulating in the locality in which the land to which the proposed development relates is situated;

"mast" means a radio mast or a radio tower;

"owner" means any person who is the estate owner in respect of the fee simple, or who is entitled to a tenancy granted or extended for a term of years certain of which not less than seven years remain unexpired;

"relevant day" means–
    (a)   7th June 2006; or
    (b)   where apparatus is installed pursuant to planning permission granted on or after 7th June 2006, the date when that apparatus is finally installed pursuant to that permission,
    whichever is later;

"relevant period" means a period which expires–
    (a)   six months from the commencement of the construction, installation, alteration or replacement of any apparatus or structure permitted by Class C(a) or Class C(c) or from the commencement of the use permitted by Class C(b), as the case may be; or
    (b)   when the need for such apparatus, structure or use ceases,
    whichever occurs first; and

"tenant" means the tenant of an agricultural holding any part of which is comprised in the land to which the proposed development relates.

Part 39

Class A

Permitted development

A The erection of a building where that is necessary for the purpose of housing poultry or other captive birds to protect them from avian influenza.

Development not permitted

A.1 Development is not permitted by Class A if–
    (a)   the development would affect a listed building or its setting;
    (b)   the height of the building would exceed 12 metres;
    (c)   where the development is within three kilometres of an aerodrome, the height of the building would exceed three metres;
    (d)   the area of ground which would be covered by the building would exceed 465 square metres;
    (e)   where development permitted by Class A is carried out more than once on land in the occupation of a particular person, the aggregate of

the area of ground covered by any such development would exceed 465 square metres;

(f) where the development consists of the extension of a building, the area of ground covered by the building as extended would exceed the area of ground covered by the existing building by more than 50 per cent.

Conditions

A.2 Development is permitted by Class A subject to the following conditions–

(a) the development shall not be used for any purpose other than to house poultry or other captive birds to protect them from avian influenza;

(b) the developer shall, as soon as practicable, and in any event no later than 14 days, after commencing development, serve the relevant notice on the local planning authority; and

(c) on or before the relevant date–

(i) any building permitted by Class A shall be removed from the land; and

(ii) the land shall be restored to its condition before the development took place, or restored to such other condition as may be agreed in writing between the local planning authority and the developer.

Interpretation of Class A

A.3 For the purposes of Class A–

"approved body" means a body approved in accordance with Article 2(1) (c) of Directive 92/65/EEC laying down animal health requirements governing trade in and imports into the Community of animals, semen, ova and embryos not subject to animal health requirements laid down in specific Community rules referred to in Annex A(1) to Directive 90/425/EEC;

"avian influenza" means an infection of poultry or other captive birds caused by any influenza A virus of the subtypes H5 or H7 or with an intravenous pathogenicity index in six week old chickens greater than 1.2;

"other captive bird" means a bird kept in captivity which is not poultry and includes a bird kept as a pet; for shows, races, exhibitions or competitions; for breeding; for sale; or for use by an approved body;

"poultry" means birds reared or kept in captivity for the production of meat or eggs for consumption, for the production of other products, for restocking supplies of game or for the purposes of any breeding programme for the production of such categories of birds;

"relevant date" means–

(a) 19th February 2008; or

(b) the date on which the use of the building permitted by Class A ceases to be necessary for the purposes of protecting poultry or other captive birds from avian influenza,

whichever is the earlier;

"relevant notice" means a notice signed and dated by or on behalf of the developer and containing–

(a) the name of the developer;

(b) the address or location of the development (including a site plan and grid reference);

(c) the name and address of the owner and occupier of the land on which the development is being carried out (if not the developer);

(d) a description of the development (including the type of poultry or other captive birds to be protected); and

(e) the date on which the development commenced.

Part 40

. . .

Class A
Permitted development

A The installation, alteration or replacement of solar PV or solar thermal equipment on–

(a) a dwellinghouse or a block of flats; or

(b) a building situated within the curtilage of a dwellinghouse or a block of flats.

Development not permitted

A.1 Development is not permitted by Class A if–

(a) the solar PV or solar thermal equipment would protrude more than 200 millimetres beyond the plane of the wall or the roof slope when measured from the perpendicular with the external surface of the wall or roof slope;

(b) it would result in the highest part of the solar PV or solar thermal equipment being higher than the highest part of the roof (excluding any chimney);

(c) in the case of land within a conservation area or which is a World Heritage Site, the solar PV or solar thermal equipment would be installed on a wall which fronts a highway;

(d) the solar PV or solar thermal equipment would be installed on a site designated as a scheduled monument; or

(e) the solar PV or solar thermal equipment would be installed on a building within the curtilage of the dwellinghouse or block of flats if the dwellinghouse or block of flats is a listed building.

Conditions

A.2 Development is permitted by Class A subject to the following conditions–

(a) solar PV or solar thermal equipment shall, so far as practicable, be sited so as to minimise its effect on the external appearance of the building;

(b) solar PV or solar thermal equipment shall, so far as practicable, be sited so as to minimise its effect on the amenity of the area; and

(c) solar PV or solar thermal equipment no longer needed for microgeneration shall be removed as soon as reasonably practicable.

Class B
Permitted development

B The installation, alteration or replacement of stand alone solar within the curtilage of a dwellinghouse or a block of flats.

Development not permitted

B.1 Development is not permitted by Class B if–

(a) in the case of the installation of stand alone solar, the development would result in the presence within the curtilage of more than one stand alone solar;

(b) any part of the stand alone solar–
 (i) would exceed four metres in height;
 (ii) would, in the case of land within a conservation area or which is a World Heritage Site, be installed so that it is nearer to any highway which bounds the curtilage than the part of the dwellinghouse or block of flats which is nearest to that highway;
 (iii) would be installed within five metres of the boundary of the curtilage;
 (iv) would be installed within the curtilage of a listed building; or
 (v) would be installed on a site designated as a scheduled monument; or

(c) the surface area of the solar panels forming part of the stand alone solar would exceed nine square metres or any dimension of its array (including any housing) would exceed three metres.

Conditions

B.2 Development is permitted by Class B subject to the following conditions–

(a) stand alone solar shall, so far as practicable, be sited so as to minimise its effect on the amenity of the area; and

(b) stand alone solar which is no longer needed for microgeneration shall be removed as soon as reasonably practicable.

Class C
Permitted development

C The installation, alteration or replacement of a ground source heat pump within the curtilage of a dwellinghouse or a block of flats.

Class D
Permitted development

D The installation, alteration or replacement of a water source heat pump within the curtilage of a dwellinghouse or a block of flats.

Class E
Permitted development

E The installation, alteration or replacement of a flue, forming part of a biomass heating system, on a dwellinghouse or a block of flats.

Development not permitted

E.1 Development is not permitted by Class E if–
    (a)   the height of the flue would exceed the highest part of the roof by one metre or more; or
    (b)   in the case of land within a conservation area or which is a World Heritage Site, the flue would be installed on a wall or roof slope which fronts a highway.

Class F
Permitted development

F The installation, alteration or replacement of a flue, forming part of a combined heat and power system, on a dwellinghouse or a block of flats.

Development not permitted

F.1 Development is not permitted by Class F if–
    (a)   the height of the flue would exceed the highest part of the roof by one metre or more; or
    (b)   in the case of land within a conservation area or which is a World Heritage Site, the flue would be installed on a wall or roof slope which fronts a highway.

Class G
Permitted development

G The installation, alteration or replacement of an air source heat pump–
    (a)   on a dwellinghouse or a block of flats;or
    (b)   within the curtilage of a dwellinghouse or a block of flats, including on a building within that curtilage.

Development not permitted

G.1 Development is not permitted by Class G unless the air source heat pump complies with the MCS Planning Standards or equivalent standards.

G.2 Development is not permitted by Class G if–

   (a)   in the case of the installation of an air source heat pump, the development would result in the presence of more than one air source heat pump on the same building or within the curtilage of the building or block of flats;

   (b)   in the case of the installation of an air source heat pump, a wind turbine is installed on the same building or within the curtilage of the dwellinghouse or block of flats;

   (c)   in the case of the installation of an air source heat pump, a stand alone turbine is installed within the curtilage of the dwellinghouse or block of flats;

   (d)   the volume of the air source heat pump's outdoor compressor unit (including any housing) would exceed 0.6 cubic metres;

   (e)   any part of the air source heat pump would be installed within one metre of the boundary of the curtilage of the dwellinghouse or block of flats;

   (f)   the air source heat pump would be installed on a pitched roof;

   (g)   the air source heat pump would be installed on a flat roof where it would be within one metre of the external edge of that roof;

   (h)   the air source heat pump would be installed on a site designated as a scheduled monument;

   (i)   the air source heat pump would be installed on a building or on land within the curtilage of the dwellinghouse or the block of flats if the dwellinghouse or the block of flats is a listed building;

   (j)   in the case of land within a conservation area or which is a World Heritage Site the air source heat pump–

       (i)   would be installed on a wall or a roof which fronts a highway; or

       (ii)   would be installed so that it is nearer to any highway which bounds the curtilage than the part of the dwellinghouse or block of flats which is nearest to that highway; or

   (k)   in the case of land, other than land within a conservation area or which is a World Heritage Site, the air source heat pump would be installed on a wall of a dwellinghouse or block of flats if–

       (i)   that wall fronts a highway; and

       (ii)   the air source heat pump would be installed on any part of that wall which is above the level of the ground storey.

Conditions

G.3 Development is permitted by Class G subject to the following conditions–

   (a)   the air source heat pump shall be used solely for heating purposes;

   (b)   the air source heat pump shall, so far as practicable, be sited so as to minimise its effect on the external appearance of the building;

   (c)   the air source heat pump shall, so far as practicable, be sited so as to minimise its effect on the amenity of the area; and (d)　the air source heat pump when no longer needed for microgeneration shall be removed as soon as reasonably practicable.

Class H
Permitted development

H　The installation, alteration or replacement of a wind turbine on–
  (a)　a detached dwellinghouse; or
  (b)　a detached building situated within the curtilage of a dwellinghouse or a block of flats.

Development not permitted

H.1　Development is not permitted by Class H unless the wind turbine complies with the MCS Planning Standards or equivalent standards.

H.2　Development is not permitted by Class H if–
  (a)　in the case of the installation of a wind turbine the development would result in the presence of more than one wind turbine on the same building or within the curtilage;
  (b)　in the case of the installation of a wind turbine, a stand alone wind turbine is installed within the curtilage of the dwellinghouse or the block of flats;
  (c)　in the case of the installation of a wind turbine, an air source heat pump is installed on the same building or within its curtilage;
  (d)　the highest part of the wind turbine (including blades) would either–
    (i)　protrude more than three metres above the highest part of the roof (excluding the chimney); or
    (ii)　exceed more than 15 metres in height,
    whichever is the lesser;
  (e)　the distance between ground level and the lowest part of any blade of the wind turbine would be less than five metres;
  (f)　any part of the wind turbine (including blades) would be positioned so that it would be within five metres of any boundary of the curtilage of the dwellinghouse or the block of flats;
  (g)　the swept area of any blade of the wind turbine would exceed 3.8 square metres;
  (h)　the wind turbine would be installed on safeguarded land;
  (i)　the wind turbine would be installed on a site designated as a scheduled monument;
  (j)　the wind turbine would be installed within the curtilage of a building which is a listed building;
  (k)　in the case of land within a conservation area, the wind turbine would be installed on a wall or roof slope of–
    (i)　the detached dwellinghouse; or
    (ii)　a building within the curtilage of the dwellinghouse or block of flats,

which fronts a highway; or

(l) the wind turbine would be installed on article 1(5) land other than land within a conservation area.

Conditions

H.3 Development is permitted by Class H subject to the following conditions–

(a) the blades of the wind turbine shall be made of non reflective materials;

(b) the wind turbine shall, so far as practicable, be sited so as to minimise its effect on the external appearance of the building;

(c) the wind turbine shall, so far as practicable, be sited so as to minimise its effect on the amenity of the area; and

(d) the wind turbine when no longer needed for microgeneration shall be removed as soon as reasonably practicable.

Class I
Permitted development

I The installation, alteration or replacement of a stand alone wind turbine within the curtilage of a dwellinghouse or a block of flats.

Development not permitted

I.1 Development is not permitted by Class I unless the stand alone wind turbine complies with the MCS Planning Standards or equivalent standards.

I.2 Development is not permitted by Class I if–

(a) in the case of the installation of a stand alone wind turbine, the development would result in the presence of more than one stand alone wind turbine within the curtilage of the dwellinghouse or block of flats;

(b) in the case of the installation of a stand alone wind turbine, a wind turbine is installed on the dwellinghouse or on a building within the curtilage of the dwellinghouse or the block of flats;

(c) in the case of the installation of a stand alone wind turbine, an air source heat pump is installed on the dwellinghouse or block of flats or within the curtilage of the dwellinghouse or block of flats;

(d) the highest part of the stand alone wind turbine would exceed 11.1 metres in height;

(e) the distance between ground level and the lowest part of any blade of the stand alone wind turbine would be less than five metres;

(f) any part of the stand alone wind turbine (including blades) would be located in a position which is less than a distance equivalent to the overall height (including blades) of the stand alone wind turbine plus 10 % of its height when measured from any point along the boundary of the curtilage;

(g) the swept area of any blade of the stand alone wind turbine exceeds 3.8 square metres;

(h) the stand alone wind turbine would be installed on safeguarded land;

(i) the stand alone wind turbine would be installed on a site designated as a scheduled monument;

(j) the stand alone wind turbine would be installed within the curtilage of a building which is a listed building;

(k) in the case of land within a conservation area, the stand alone wind turbine would be installed so that it is nearer to any highway which bounds the curtilage than the part of the dwellinghouse or block of flats which is nearest to that highway; or

(l) the stand alone wind turbine would be installed on article 1(5) land other than land within a conservation area.

Conditions

1.3 Development is permitted by Class I subject to the following conditions–

(a) the blades of the stand alone wind turbine shall be made of non reflective materials;

(b) the stand alone wind turbine shall, so far as practicable, be sited so as to minimise its effect on the amenity of the area; and

(c) the stand alone wind turbine when no longer needed for microgeneration shall be removed as soon as reasonably practicable.

Interpretation of Part 40

J For the purposes of Part 40–

"aerodrome"–

(a) means any area of land or water designed, equipped, set apart, or commonly used for affording facilities for the landing and departure of aircraft; and

(b) includes any area or space, whether on the ground, on the roof of a building or elsewhere, which is designed, equipped or set apart for affording facilities for the landing and departure of aircraft capable of descending or climbing vertically; but

(c) does not include any area the use of which for affording facilities for the landing and departure of aircraft has been abandoned and has not been resumed;

"block of flats" means a building which consists wholly of flats;

"detached dwellinghouse" or "detached building" means a dwellinghouse or building, as the case may be, which does not share a party wall with a neighbouring building;

"microgeneration" has the same meaning as in section 82(6) of the Energy Act 2004;

"MSC Planning Standards" means the product and installation standards for air source heat pumps and wind turbines specified in Microgeneration Certification Scheme MCS 020;

"safeguarded land" means land which–

(a) is necessary to be safeguarded for aviation or defence purposes; and

(b)   has been notified as such, in writing, to the Secretary of State by an aerodrome operator, NATS (EN ROUTE) PLC or the Secretary of State for Defence for the purposes of this Part;

"solar PV" means solar photovoltaics;

"stand alone solar" means solar PV or solar thermal equipment which is not installed on a building;

"stand alone wind turbine" means a wind turbine which is not fixed to a building.

Part 41

Class A
Permitted development

A   The extension or alteration of an office building.

Development not permitted

A.1   Development is not permitted by Class A if–
(a)   the gross floor space of the original building would be exceeded by more than–
(i)   25%; or
(ii)   50 square metres,
whichever is the lesser;
(b)   the height of the building as extended would exceed–
(i)   if within ten metres of a boundary of the curtilage of the premises, five metres; or
(ii)   in all other cases, the height of the building being extended;
(c)   any part of the development, other than an alteration, would be within five metres of any boundary of the curtilage of the premises;
(d)   any alteration would be on article 1(5) land; or
(e)   the development would be within the curtilage of a listed building.

Conditions

A.2   Development is permitted by Class A subject to the following conditions–
(a)   any office building as extended or altered shall only be used as part of, or for a purpose incidental to, the use of that office building;
(b)   any extension shall, in the case of article 1(5) land, be constructed using materials which have a similar external appearance to those used for the building being extended; and
(c)   any alteration shall be at ground floor level only.

Interpretation of Class A

A.3   For the purposes of Class A–
(a)   where two or more original buildings are within the same curtilage and are used for the same undertaking, they are to be treated as a single original building in making any measurement; and

(b) "office building" means a building used for any purpose within Class B1(a) of the Schedule to the Use Classes Order.

Class B
Permitted development

B Development consisting of–
- (a) the provision of a hard surface within the curtilage of an office building to be used for the purpose of the office concerned; or
- (b) the replacement in whole or in part of such a surface.

Development not permitted

B.1 Development is not permitted by Class B if–
- (a) the cumulative area of ground covered by a hard surface within the curtilage (excluding hard surfaces already existing on 6th April 2010) would exceed 50 square metres; or
- (b) the development would be within the curtilage of a listed building.

Conditions

B.2 Development is permitted by Class B subject to the following conditions–
- (a) where there is a risk of groundwater contamination, the hard surface shall not be made of porous materials;
- (b) in all other cases, either–
  - (i) the hard surface shall be made of porous materials, or
  - (ii) provision shall be made to direct run-off water from the hard surface to a permeable or porous area or surface within the curtilage of the office building.

Interpretation of Class B

B.3 For the purposes of Class B "office building" means a building used for any purpose within Class B1(a) of the Schedule to the Use Classes Order.

Part 42

Class A
Permitted development

A The extension or alteration of a shop or financial or professional services establishment.

Development not permitted

A.1 Development is not permitted by Class A if–
- (a) the gross floor space of the original building would be exceeded by more than–
  - (i) 25%; or
  - (ii) 50 square metres;
  whichever is the lesser.

(b) the height of the building as extended would exceed four metres;

(c) any part of the development, other than an alteration, would be within two metres of any boundary of the curtilage of the premises;

(d) the development would be within the curtilage of a listed building;

(e) any alteration would be on article 1(5) land;

(f) the development would consist of or include the construction or provision of a veranda, balcony or raised platform;

(g) any part of the development would extend beyond an existing shop front;

(h) the development would involve the insertion or creation of a new shop front or the alteration or replacement of an existing shop front; or

(i) the development would involve the installation or replacement of a security grill or shutter on a shop front.

Conditions

A.2 Development is permitted by Class A subject to the following conditions–

(a) any alteration shall be at ground floor level only;

(b) any extension shall, in the case of article 1(5) land, be constructed using materials which have a similar external appearance to those used for the building being extended; and

(c) any extension or alteration shall only be used as part of, or for a purpose incidental to, the use of the shop or financial or professional services establishment.

Interpretation of Class A

A.3 For the purposes of Class A–

(a) where two or more original buildings are within the same curtilage and are used for the same undertaking, they are to be treated as a single original building in making any measurement;

(b) "raised platform" means a platform with a height greater than 300 millimetres; and

(c) "shop or financial or professional services establishment" means a building, or part of a building, used for any purpose within Classes A1 or A2 of the Schedule to the Use Classes Order and includes buildings with other uses in other parts as long as the other uses are not within the parts being altered or extended.

Class B
Permitted development

B The erection or construction of a trolley store within the curtilage of a shop.

Development not permitted

B.1 Development is not permitted by Class B if–

(a) the gross floor space of the building or enclosure erected would exceed 20 square metres;

(b)   any part of the building or enclosure erected would be–
    (i)   within 20 metres of any boundary of the curtilage of; or
    (ii)   above or below,
any building used for any purpose within Part C of the Schedule to the Use Classes Order or as a hostel;

(c)   the height of the building or enclosure would exceed 2.5 metres;

(d)   the development would be within the curtilage of a listed building; or

(e)   the development would be between a shop front and a highway where the distance between the shop front and the boundary of the curtilage of the premises is less than five metres.

Condition

B.2 Development is permitted by Class B subject to the condition that the building or enclosure is only used for the storage of shopping trolleys.

Interpretation of Class B

B.3 For the purposes of Class B–
"shop" means a building used for any purpose within Class A1 of the Schedule to the Use Classes Order; and
"trolley store" means a building or enclosure designed to be used for the storage of shopping trolleys.

Class C
Permitted development

C Development consisting of–
(a)   the provision of a hard surface within the curtilage of a shop or catering, financial or professional services establishment; or
(b)   the replacement in whole or in part of such a surface.

Development not permitted

C.1 Development is not permitted by Class C if–
(a)   the cumulative area of ground covered by a hard surface within the curtilage of the premises (other than hard surfaces already existing on 6th April 2010) would exceed 50 square metres; or
(b)   the development would be within the curtilage of a listed building.

Index